Deleuze and Cosmology

Deleuze Connections

'It is not the elements or the sets which define the multiplicity. What defines it is the AND, as something which has its place between the elements or between the sets. AND, AND, AND – stammering.'

Gilles Deleuze and Claire Parnet, *Dialogues*

Titles Available in the Series
Ian Buchanan and Claire Colebrook (eds), *Deleuze and Feminist Theory*
Ian Buchanan and John Marks (eds), *Deleuze and Literature*
Mark Bonta and John Protevi (eds), *Deleuze and Geophilosophy*
Ian Buchanan and Marcel Swiboda (eds), *Deleuze and Music*
Ian Buchanan and Gregg Lambert (eds), *Deleuze and Space*
Martin Fuglsang and Bent Meier Sørensen (eds), *Deleuze and the Social*
Ian Buchanan and Adrian Parr (eds), *Deleuze and the Contemporary World*
Constantin V. Boundas (ed.), *Deleuze and Philosophy*
Ian Buchanan and Nicholas Thoburn (eds), *Deleuze and Politics*
Chrysanthi Nigianni and Merl Storr (eds), *Deleuze and Queer Theory*
Jeffrey A. Bell and Claire Colebrook (eds), *Deleuze and History*
Laura Cull (ed.), *Deleuze and Performance*
Mark Poster and David Savat (eds), *Deleuze and New Technology*
Simone Bignall and Paul Patton (eds), *Deleuze and the Postcolonial*
Stephen Zepke and Simon O'Sullivan (eds), *Deleuze and Contemporary Art*
Laura Guillaume and Joe Hughes (eds), *Deleuze and the Body*
Daniel W. Smith and Nathan Jun (eds), *Deleuze and Ethics*
Frida Beckman (ed.), *Deleuze and Sex*
David Martin-Jones and William Brown (eds), *Deleuze and Film*
Laurent de Sutter and Kyle McGee (eds), *Deleuze and Law*
Arun Saldanha and Jason Michael Adams (eds), *Deleuze and Race*
Rebecca Coleman and Jessica Ringrose (eds), *Deleuze and Research Methodologies*
Inna Semetsky and Diana Masny (eds), *Deleuze and Education*
Hélène Frichot and Stephen Loo (eds), *Deleuze and Architecture*
Betti Marenko and Jamie Brassett (eds), *Deleuze and Design*
Hélène Frichot, Catharina Gabrielsson and Jonathan Metzger (eds), *Deleuze and the City*
Colin Gardner and Patricia MacCormack (eds), *Deleuze and the Animal*
Markus P.J.Bohlmann and Anna Hickey-Moody (eds), *Deleuze and Children*
Chantelle Gray van Heerden and Aragorn Eloff (eds), *Deleuze and Anarchism*
Michael James Bennett and Tano S. Posteraro (eds), *Deleuze and Evolutionary Theory*
Rick Dolphijn and Rosi Braidotti (eds), *Deleuze and Guattari and Facism*
Anindya Sekhar Purakayastha and Saswat Samay Das (eds), *Deleuze and Guattari and Terror*
Robert W. Luzecky and Daniel W. Smith (eds), *Deleuze and Time*
Janae Sholtz and Alain Beaulieu (eds), *Deleuze and Cosmology*

Visit the Deleuze Connections website at:
www.edinburghuniversitypress.com/series/delco

Deleuze and Cosmology

Between Art, Philosophy, and the Sciences

Edited by Janae Sholtz
and Alain Beaulieu

EDINBURGH
University Press

Edinburgh University Press is one of the leading university presses in the UK. Publishing new research in the arts and humanities, EUP connects people and ideas to inspire creative thinking, open new perspectives and shape the world we live in. For more information, visit www.edinburghuniversitypress.com.

Edinburgh University Press Ltd
13 Infirmary Street, Edinburgh EH1 1LT

Typeset in 10.5/13 Adobe Sabon by
Cheshire Typesetting Ltd, Cuddington, Cheshire
Printed and bound in the UK using 100% renewable electricity by
CPI Group (UK) Ltd

A CIP record for this book is available from the British Library

ISBN 978 1 3995 6089 4 (hardback)
ISBN 978 1 3995 6091 7 (webready PDF)
ISBN 978 1 3995 6090 0 (epub)

EU Authorised Representative:
Easy Access System Europe
Mustamäe tee 50, 10621 Tallinn, Estonia
gpsr.requests@easproject.com

MIX
Paper | Supporting responsible forestry
FSC www.fsc.org FSC® C013604

This book is printed using paper from well-managed forests, recycling and other controlled sources

Contents

Acknowledgments vii
List of Illustrations viii

Introduction: Toward a New Cosmic Sensitivity 1
Janae Sholtz and Alain Beaulieu

Part I: Creative Cosmology: Chaos and the Cosmic

1 What is Invention? Deleuze and Guattari and the New 19
Dorothea Olkowski

2 Deleuze, an Unwitting Milesian? 37
Arnaud Villani

3 Cosmic Triptych: A Deleuzo-Guattarian Exploration of the Transistantial, the Fleeting, and the Akoúômorphical 53
Antoine Renzo

4 Following a Plastic Vortex of Spinning Waste 72
Macarena Rioseco

Part II: Infinite Cosmologies: Perspectives, Foldings, and Time

5 Points of View: Deleuze, Leibniz, and the New Cosmology 101
Debjyoti Sarkar

6 The Lily of the Field Sings the Glory of the Heavens: The Cosmos of the Fold 118
Bryan Noonan

7 The Fold and String Theory 136
Ewa Szumilewicz

8 Aion/Chronos and Living Time 153
Russell Duvernoy

Part III: Cosmologies of the Quantum

9 Deleuze and Decoherence: Approaching the Measurement Problem through Intensity 175
Michael J. Ardoline

10 The Chaosmos of the Quantum, with Gilles Deleuze 192
Arkady Plotnitsky

11 Immanence, Process, and Relationalism in Loop Quantum Gravity 214
Martin Calamari

12 Zig-Zagging Through the Deleuzian Virtual: The Dark Universe of Diagrams, Cosmology, and Quantum 236
jan jagodzinski

Part IV: Comparative Cosmologies

13 Caught Between a Sufficient Cosmos and a Chaosmos: Laruelle, Deleuze, and Guattari on Philosophy, Cosmology, and the Real 261
Chantelle Gray

14 Biological Philosophy and Geophilosophy in the Shadow of Cosmology: Reading Canguilhem with Deleuze and Guattari 279
Samuel Talcott

15 Modality as Virtuality versus Modality as Possibility: Gilles Deleuze and David Lewis 297
Ilona Schweitzer

16 Toward a Process Cosmology: Deleuze, Whitehead, and Spacetime 314
Keith Robinson

Notes on Contributors 336
Index 342

Acknowledgments

We warmly thank the contributors who generously accepted the invitation to be part of this collective work. This publication originates from a multi-year research program entitled "Gilles Deleuze and Cosmology." We would like to extend our gratitude to our colleagues who were part of the research team, with whom we had lively discussions, and who have played a key role in the organization of various scientific activities over the past few years: physicists Ubi Wichoski and Gennady Chitov, as well as the postdoctoral fellows Michael J. Ardoline, Kyle Novak, and Martin Boucher. We would like to thank the participants from our May 2023 conference, *Cosmology Between Philosophy and Science: Deleuze, Guattari, and the Rest of Us*, for the lively debates and important development of ideas, which provided invaluable focus for our project and helped enrich the philosophical work accomplished in this area. We should thank the artists for their generous contribution of works and ongoing collaboration through the virtual art exhibit. Finally, we would like to thank our main sponsors, the Social Sciences and Humanities Research Council (SSHRC) of Canada, Alvernia University, as well as Laurentian University for their support.

Illustrations

Part illustrations

Scientists Discovered A Massive Crack, Gabrielle Vitollo xi
Rhizomes Ascending, Adrian Switzer 17
Autopoiesis: Actual Occasion, Mikey Georgeson 99
asstrOnOmical affected autOmata, Joseph Nechvatal 173
Line of Flight, Stefan Paulus 259
Two stills from *Sequence 24.4.8: Contact and Totality*, Brian Fridge 335

Figures

4.1 Still photograph from video of spinning process: stretching 85
4.2 Photomontage of ten instances of striated strategy for spinning the plastic bags into threads: flattening–measuring–folding, flattening–measuring–folding, cutting–measuring–cutting, stretching, interlacing, and balling up 86
4.3 Rounded, flat, and circular formation with one stitch as its center. Crochet knitting plastic formation. 87
4.4 The formation folds and meanders due to exponential duplication of the stitches. Crochet knitting plastic formation. 87
4.5 Sketch of the potential future of the formation. Graphite, red and black ink on paper. 88
4.6 Rounded coral-like volume. 89
4.7 Crochet knitting with plastics. Still from video. 89
4.8 Plastic bags. 91

4.9	Still photograph from video.	91
4.10	Installation plan.	92
10.1	Sudbury Neutrino Detector	196
10.2	Screen capture images taken from XSNOED (the x-windows SNO Event Display) by Chris Kyba, retrieved from https://sno.phy.queensu.ca/events/	197
10.3	The Higgs boson effect registered in Cern, open-source photograph	197
10.4	Heisenberg equations	208
10.5	Heisenberg equations	208

Scientists Discovered A Massive Crack is from a series of large satellite image paintings in which I depict locations being transformed by climate change. These remote locations are painted directly from my laptop screen as I traverse Google Earth. This painting relates to Deleuze's writings on machine perception, specifically from a cosmic perspective. The satellite telescope generates composite images that allow humans to make sense of something abstract, for example, climate change. While the source of the geological visuals is a machine, I reintroduce the scientific image as a painted object within a human context. The labor-intensive, hand-painted image reintroduces viewers to these climate-change impacted areas, so that they might take pause and experience this phenomenon in a new way. The large scale and hard-edge geometry of these works also creates an unusual and surprising body–object relationship between the viewer and the painting so that the viewer experiences both time and space differently. This satellite painting shows the weirdness of the cosmic satellite perspective and pushes against the human anthropomorphic view of the universe.

—Gabrielle Vitollo

Gabrielle Vitollo, *Scientists Discovered A Massive Crack,* oil and acrylic on canvas, 60 x 44.5 in. (152 x 113 cm), 2021 © Gabrielle Vitollo

Introduction

Toward a New Cosmic Sensitivity

Janae Sholtz and Alain Beaulieu

Gilles Deleuze and Félix Guattari are among the very few contemporary philosophers who have incorporated cosmological questioning into their thought, thus renewing an old and partly lost way of seeing philosophy's task. Indeed, Deleuze's philosophy—with or without Guattari—is sporadically, yet forcefully, animated by cosmological concerns. They arguably have made the most conceptually sophisticated and thought-provoking contribution in recent continental philosophy of cosmology, despite the fact that their cosmological thinking is not systematized in a chapter or a monograph. Their contribution in the field is rather dispersed throughout their work, as if these scattered thoughts were akin to something, the "world" for that matter, that ultimately escapes unity of knowledge or representation, while leaving room to imagination. Among these cosmological traces one can recall Deleuze's great admiration for Whitehead's *Process and Reality: An Essay in Cosmology* (Deleuze 1994), his comment on Bergson's "cosmic memory" (Deleuze 1990), Deleuze and Guattari's borrowing of James Joyce's neologism "chaosmos" (Deleuze and Guattari 1987; 1994), Deleuze's appeal for an outside of any exteriority (Deleuze 1988), his notion of the unfamiliar "house" in its Baroque or "cosmic" version (Deleuze 1992), as well as Guattari's science-fiction film project *In Search of UIQ* (Guattari 2016). Not to mention the original cosmic lexicon that Deleuze and Guattari propose in *A Thousand Plateaus* and *What is Philosophy?* which includes the cosmic artisan, cosmic forces, the becoming-cosmic, the plane of cosmicization, and so on. God and Man might be dead, but the world as both a philosophical idea and a field of experimentation remains fully alive in Deleuze and Guattari's work. The present book intends to explore this cosmic vitality.

So far, secondary literature has proposed a fragmentary reading of Deleuze and Guattari's sense of cosmicity in connection, for instance, with Stoicism (Sellars 1999; 2006) phenomenology (Sholtz 2015; Beaulieu 2022), art/artisanry (Sholtz 2021), music (Bogue 1991; Pinhas 1997; Solomos 2015), and theology (Harris 2010). These studies offer only a partial overview of their cosmological thinking. Nevertheless, they have prepared the ground for a deeper exploration of their cosmological sensitivity, and favor the time to ask the fundamental questions that are discussed in this volume, which include "What is the originality of Deleuze and Guattari's standpoint on cosmological issues?," "What contributions can Deleuze and Guattari's geodynamism make to the philosophy of cosmology?," "How do scientific theories in cosmology interfere with Deleuze and Guattari's chosmological thinking?," "What would be the possible Deleuzo-Guattarian readings of recent advances, discoveries, or challenges in contemporary scientific cosmology?," and "How can Deleuze and Guattari's conceptuality help formulate new cosmological hypotheses?"

Philosophy

Traditionally, cosmology has been closely linked with philosophy. From Plato and Aristotle to Nicholas of Cusa, Giordano Bruno and Kant, to name a few, philosophers have suggested different representations of the universe and different ways for humans to relate to it. After Kant's critique of rational cosmology, and later on with the development of astrophysics, the link between philosophy and cosmology became, in large part, broken. Astronomical knowledge, for instance, has been reduced to its utility for ocean navigation or serves an entertainment function in the context of popular gatherings around eclipses, auroras, shooting stars, and so on. Also, influential scientists, like Stephen Hawking (Hawking and Mlodinow 2010), consider philosophy "dead," for it has not kept up with modern developments in science and is now unable to answer major cosmological questions.

Nonetheless, authentic cosmological sensibility is still present in recent continental philosophy, namely for thinkers like Nietzsche, Whitehead, Bergson, Husserl, Heidegger, Axelos, as well as Deleuze and Guattari. Unlike traditional cosmologies, these philosophers do not speculate on the laws of organization, finality, perfection, or the ontological superiority of the heavens. They do not suppose a "world beyond this one," but rather *this* world becomes the object of philosophical investigation. In this context, for instance, the question "What is the Earth?," formerly considered relatively obvious, gains a philosophical status. We can see it, for instance, in Nietzsche's challenging invitation "to remain earthly" (Nietzsche 2009), in Husserl's investigation of the "unmoving Earth" (Husserl 2002), in Heidegger's Earth experienced as a "dwelling" (Heidegger 1971), and in Deleuze and Guattari's geophilosophy (1994).

Nevertheless, the Earth is not understood in the same way by these philosophers. For instance, a distinction between the phenomenological Earth and the Deleuzo-Guattarian Earth ought to be made. On the one hand, the phenomenological Earth as a spiritual, original, and unmoving ground must be conquered for Husserl in order to overcome what phenomenologists describe as a crisis of the "planetary age," characterized by technical control over objectivized nature. On the other hand, there is the Deleuzo-Guattarian Earth conceived as a celestial body experienced through its movements under the effect of de- and re-territorializing forces, whose infinite potential makes the Earth not an originary ground, an organized totality, or a "pale blue dot" lost in infinite space, to borrow a popular romantic image, but rather a "a galaxy among others" (Deleuze and Guattari 1987) filled with rich and unheard potentialities. One might say, to paraphrase Spinoza's famous idea: "Philosophy once wanted to know about the supra earthly, whereas we still do not know what the earthly can do!" Deleuze and Guattari's galactic Earth is not merely a spiritual entity. It surely encompasses an incorporeal and non-organic component, but this immaterial feature does not subdue a physical aspect—quite the contrary.

Deleuze and Guattari's cosmological sensitivity remains compatible with some scientific advances in cosmology. For them, science is not symptomatic of a crisis to be overcome, but is rather recognized as a form of thought equally valid for studying the strangeness of the "Erewhon" type of world we live in. In other words, for Deleuze and Guattari, cosmic thought implies the invention of a new Earth by constructing a philosophical plane of immanence, which itself can interfere with the scientific plane of reference, for which terrestrial empiricity is part of a chaosmic experimentation with outside forces. This appeal to the world's physicality calls for the study of connections between Deleuze and Guattari's philosophy and scientific cosmological theories. This type of study is at the core of many of the chapters included in this book.

Science

Deleuze and Guattari find in science a good companion for their work, and this companionship opens up to an abundant secondary literature dealing, for instance, with mathematics (DeLanda 2002; Calamari

2015), biology (Ansell-Pearson 1999), botany (Houle 2011), chemistry (Prigogine and Stengers 1984), and cybernetics (Braidotti 2013). However, such studies do not approach extensively the complex ramifications of Deleuzo-Guattarian thinking with regard to cosmology as science, with the notable exception of a recently published book by Bill Ross (2024) which might contribute in paving the way to a new outlook on Deleuzo-Guattarian thinking. For the most part, commentaries on Deleuze, Guattari, and scientific cosmology deal with thermodynamics and chaos theory (Plotnitsky 2006; 2012; Massumi 1993; DeLanda 2002; 2010; de Beistegui 2004; Bell 2006; Olkowski 2012; Protevi 2013; Beaulieu 2022) without taking into account the variety of scientific theories in cosmology integrated in Deleuze and Guattari's analyses. Yet Deleuze's readers will be familiar with spacetime, supernovas, the "cosmic soup" that succeeded the Big Bang, black holes, Einstein's theories, Hubble's constant, and perhaps more implicitly with quantum mechanics and particle physics, from which Deleuze and Guattari derive inspiration for their work.

Furthermore, one can advocate for a Deleuzo-Guattarian critical type of reading of scientific theories in cosmology, such as the "theory of everything," which is said to unify fundamental interactions; "cosmic inflation," which deals with the early evolution of the universe; the astrophysics of particles, which offers other tools to explain the origins of the universe; the "standard model in cosmology," which is currently undergoing a major crisis (Smolin 2006; Panek 2011); hypotheses surrounding "cosmic bubbles" and the "multiverse"; as well as "string theory," "dark matter," and "dark energy." Most of these relatively recent scientific advances in cosmology are discussed in the present book in light of Deleuze and Guattari's philosophical ideas. The task of exploring the ways in which Deleuze and Guattari appropriate cosmological science, and investigating the ways in which they could provide conceptual tools for the formulation of new cosmological hypotheses in science, remains in large part ahead of us.

Art

The present collection of essays focuses primarily either on philosophical aspects of Deleuze and Guattari's chaosmological thinking or on interconnections between their philosophy and scientific theories in cosmology. However, many of our contributors highlight the inextricable aesthetic and creative nature of Deleuze and Guattari's thought, explicating the ways that philosophy and science either contain creative potential or could be transformed through engagement with their creative and inherently art-oriented cosmological concepts. One should keep in mind that Deleuze and Guattari have a broad conception of the chaosmos. For them, neither philosophy nor science has the final say on what the "world" might be. Rather, this strange universe remains for its observers and experimenters in the midst of a multifaceted process of constitution. Its unity is doomed to escape us, and it finds, instead, expression in a complex, dynamic, and unending series of assemblages between the three forms of thought discussed in *What is Philosophy?* (1994), namely, philosophy, science, and art. Thus, Deleuze and Guattari's redefinition of cosmological questioning is equally inspired by modern artists such as Klee, Stockhausen, Messiaen, and others. This original outlook is succinctly encapsulated in the following assertion: "If there is a modern age, it is, of course, the age of the cosmic" (Deleuze and Guattari 1987, 342).

We would like to add a few words regarding this connection between art and the cosmos to better clarify Deleuze and Guattari's original perspective on cosmological issues. The bountiful cosmological terminology surrounding their thinking is not only developed in contact with scientific theories, such as quantum mechanics, chaos theory, and fractal theory; it is equally inspired by the work of modern artists mainly from the period following the Second World War. For Deleuze and Guattari, the distinguishing characteristic of music, and art in general, does not have to do with describing a world order (classical art) or conveying subjective feelings (romantic art), but rather with grasping invisible and impersonal forces that are intensifying bodies and matter (modern art) (1987, ch. 11 "Of the Refrain"). So, there are different cosmological logics at play in the three ages. Whereas classical artists attempted to keep chaos outside the world that they are representing, and romantic artists tended to confront themselves with chaotic torments, modern artworks strive to grasp non-terrestrial forces from the Outside that are affecting the materiality of this world. In other words, past artistic ages render visible what can actually be represented, either an order (harmony, melody, proportion, and so on) in the case of classical art, or lived experiences (joy, sadness, agony, and so on) in the case of romantic art. By contrast, modern art tackles invisible forces that escape both representation and livability. Positively, modern art encounters a chaosmic type of cosmicity that affects this world. It relates neither classically with an extra-cosmic order, nor romantically with an infra-cosmic self, but with the world itself as chaosmos. In that sense, modern art is cosmic; it is at last in tune with the world.

This emblematic conception of modern art as "cosmic" is inspired by a variety of artists. These include Klee, who presents his work as a depiction of nature affected by non-terrestrial forces (1975) and who famously wrote in his "Creative Credo" that "Art does not reproduce the visible but makes visible" (1961, 76), as well as Stockhausen, who claims to have "made certain vibrations audible for the first time in human history" and who adds: "Among the celestial spheres there are certainly areas of vibration which have to be perceived with completely different kinds of equipment if they are to be available for structuring as music" (1989, 121). Modern artistic cosmicity finds ramifications in various spheres of creation, disparate in terms of form and style but merging under the endeavor of making invisible forces visible.

Finally, while Deleuze defines all art through its ability to capture forces, and modern art, in particular, through its project of making invisible forces visible, Deleuze and Guattari designate music as having a special relation to the cosmos and the cosmic: "music is not the privilege of human beings: the universe, the cosmos, is made of refrains" (1987, 309). The refrain is essentially a matter of composing speeds and slowness into rhythmic beings that may be repeated or from which variations may arise, and which, in turn, form the very composition of the universe, bodies, territories, and selves; we are all made of refrains. Where other art forms resist representation, or figuration (for instance, in the case of Francis Bacon), to instill rhythm, ideally to make rhythm itself the Figure (Deleuze 2003, 60), rhythm and vibration is the content proper to music itself. Rather than making rhythm visible, music composes out of these speeds and slownesses. In "Of the Refrain," Deleuze and Guattari claim that the refrain is the "bloc of content proper to music" (1987, 299). Music operates through sound, the punctuation and variation of vibrations that are then formed into repeatable rhythms, which may be related to the speeds and slownesses that Deleuze and Guattari place at the center of their description of the plane of immanence and the cosmic. The territorialization and deterritorialization found in music follows the patterning, fluctuating, and infinitely vibrating nature of a cosmos/universe filled with cosmic potential. In other words, music shares a sensuous and direct link to the power of the cosmos—an infinite composition of forces, speeds, and slownesses that inform substantial being.

Deleuze's contemplation of the "superiority of music" (2003, 47) in *Francis Bacon: The Logic of Sensation* is another revelation of music's special connection to the cosmic. Deleuze explains that music addresses something very different than the materiality of bodies: "it gives a disembodied and dematerialized body to the most spiritual of entities" (2003, 47). While painting addresses the reality of bodies by making visible the invisible forces within materiality itself such that materiality might begin to escape its form—"the body escapes from itself"—music, as the modulation of speeds and slownesses, operates through no other sensual means than movement. As Deleuze says, "music begins where painting ends ... it involves our bodies, and bodies in general, in another element ... it disembodies bodies" (2003, 47). In other words, music is the most incorporeal of all the art forms; thus, it allows us to think of the genesis of form beyond substance, perhaps even helping us to think of the illusive power of abstract machines and planes of immanence. As it is deterritorialized more than other plastic arts, it has the capacity to deterritorialize us toward the cosmic to an even greater degree.

Deleuze and Guattari are especially intrigued by modern artists for whom this variation is open and who explore this openness by incorporating chance, indeterminacy, and heterogeneous elements of sound. Referencing the work of John Cage, whose emphasis on silence and indeterminacy breaks traditional models of form and intentionality (see Cage 1973), they write:

> Certain modern musicians oppose the transcendent plan(e) of organization, which is said to have dominated all of Western classical music, to the immanent sound plane, which is always given along with that to which it gives rise, brings the imperceptible to perception, and carries only differential speeds and slownesses in a kind of molecular lapping. (1987, 267)

It is worth noting that the form of music itself as an infinite process of variation is intimately connected to experimentation, and Deleuze and Guattari's theories of sound and music have been particularly relevant to the field of experimental electronic music, dedicated as it is to new and unique forms of sonic experimentation: "moments that are marked by a single powerful trait: the impulse to deterritorialize musical form and substance" (Cox 2003). Experimental electronic music has opened up possibilities in registering imperceptible sonorous events and formalizing rhythmic beings derived from all manner of cosmic entities—taking this relation of the cosmic and music to new levels.

While an in-depth study of the sources of cosmic art in Deleuze and Guattari's work and the impact that their conception of art has had on contemporary artists remains to be done, our hypothesis is that their thesis on cosmic art might well function outside their philosophical system by providing a valuable explanation for modern art to be abstracted as it expresses natural entities being affected by cosmic forces.

Not only do we find it important to recall this artistic aspect of Deleuze and Guattari's outlook on cosmology, but also, in alignment with what we take to be the deep commitment to the intersection of art, science, and philosophy, we have incorporated the work of several contemporary artists inspired by Deleuze and Guattari's chaosmological thinking in the current volume. These artists are part of a virtual exhibition curated by ourselves as part of the inquiry into Deleuze and cosmology that inspired this volume. For a more detailed discussion of interferences between philosophy and art with regard to the cosmic, readers are invited to visit the virtual exhibition *Gilles Deleuze, Félix Guattari, and Cosmic Art* that we have curated. It is available here: https://deleuzeguattariart.wixsite.com/deleuze-guattari-art

In sum, the present book has a twofold objective: to analyze the modalities of the survival of cosmological questioning in the work of Deleuze and Guattari, as well as to assess the interferences between their philosophy and recent scientific theories in cosmology. In other words, the present volume aims to understand better how Deleuze and Guattari's chaosmological sensibility becomes more original than the traditional philosophy of cosmology, and it also aims to strengthen an interdisciplinary dialogue between Deleuze and Guattari's ideas and physicists working in cosmology. From there, readers are invited to investigate further how their cosmic theories can be applied to increasingly urgent fields such as anthropocene studies and environmental philosophy.

Deleuze and Guattari's sensitivity to the cosmological type of questioning deserves a closer reading. And the timing for undertaking this research is propitious, as the secondary literature over the last decade indicates a clear interest in their insight into cosmological thinking. The book extends, and humbly attempts to go beyond, existing studies by conducting the first collective inquiry into the subject.

Chapter Summaries

Our volume fulfills these objectives through sixteen chapters organized in four main sections. Part I, "Creative Cosmology: Chaos and the Cosmic," explores the creative potential of the chaosmotic and the cosmic. In "What is Invention? Deleuze and Guattari and the New," Dorothea Olkowski contemplates the tension between the desire for the creation of the new and a determinist view of the universe. To what extent can true invention even occur? By contemplating the applicability of arguments found in physics, as well as neuroscience and mathematics, to Deleuze and Guattari's understanding of creativity, Olkowski provides insight into the extent to which cosmological indetermination (or rather deterministic chaos) might offer some inspiration for creation in multiple realms of thought and art. In "Deleuze, an Unwitting Milesian?," Arnaud Villani addresses Deleuze's contribution to the field of positive chaos by connecting the main characters of chaos identified by him—the freedom of isotropic flux and the absolute speed that allows the virtual to exert its power—with the Greek *Physikoi*, Heraclitus, and Parmenides. In "Cosmic Triptych: A Deleuzo-Guattarian Exploration of the Transistantial, the Fleeting, and the Akoúômorphical," Antoine Renzo undertakes an exploration of the cosmological depths within the Deleuzo-Guattarian philosophy of nature, particularly through the lens of Deleuzian logic concerning aberrant movements, the Guattarian notions of transistantiality, and the framework of the reverberative functions of music as *akoúômorphism*—the aesthetic and political capacity of a world to unfold toward another. In "Following a Plastic Vortex of Spinning Waste," Macarena Rioseco explores the affective matter of the gyres and trash vortexes found in the oceans by considering her own art practices, inspired by Deleuze and Guattari's dynamical cosmology and the topology of smooth space as a virtual plane of immanence populated by material–forces relations and material–flows distributions.

Part II, "Infinite Cosmologies: Perspectives, Foldings, and Time," includes chapters which investigate how the existential, cosmological issues of topology, space, and time overlap in Deleuze's topological concept of the fold and the temporal concepts of Aion/Chronos with conversations about the nature of space and time in physics. In "Points of View: Deleuze, Leibniz, and the New Cosmology," Debjyoti Sarkar investigates the notion of point of view in relation to cosmology with specific reference to Leibniz and Deleuze, arguing that the notion of point of view comes as a response to the thinking of an infinite universe, which becomes a crisis, specifically a theological one. While Leibniz responds to this crisis by defining the point of view as an apex of the cone and hence finding a stable position of convergence, Deleuze claims that divergences must be affirmed, opening possibilities of thinking that do not privilege stability or require theological justification. In "The Lily of the Field Sings the Glory of the Heavens: The Cosmos of the Fold," Bryan Noonan explores the cosmological implications of Deleuze's contemplative soul and the fold, painting a picture of Deleuze's cosmos of expression that is both poetic and exegetical. Noonan explicates the strange resonance between Deleuze and thinkers like Leibniz, Riemann, Butler, and Whitehead, teasing out the cosmological implications of an "expressive" universe and folding them into a beautiful cosmological vision. In "The Fold and String Theory," Ewa Szumilewicz finds a link between string theory and Deleuze's philosophy of the fold and the notion of time. She proposes Deleuze's multidimensional notion of time and his concept of the fold as tools with which to draw out the philosophical implications of string theory. Szumilewicz surmises that both Deleuze and string theory demand a challenging reconceptualization of interconnectivity rather than a granularity perspective, and that reconceptualization demands a new status of the object and subject. In "Aion/Chronos and Living Time," Russell Duvernoy begins with the provocation of Einstein's worry about the peculiarity of the Now as impetus for considering Deleuze's repurposed Aion/Chronos distinction in *The Logic of Sense* for how it might contribute to a more reflective pragmatics of time. Employing alternative models of time from Indigenous thinkers (Whyte) and the relational revolution in physics (Smolin), Duvernoy shows that time's variable images have lived ethical and sociopolitical effects.

Part III, "Cosmologies of the Quantum," includes contributions from scholars who have succeeded in placing themselves in two worlds—the complex physics and mathematics of the quantum and the creative, chaosmos of Deleuze and Guattari—illustrating the promise that such combinations hold for transforming our understanding of both. In "Deleuze and Decoherence: Approaching the Measurement Problem through Intensity," Michael J. Ardoline investigates whether it makes sense to posit a Deleuzian interpretation of quantum mechanics. Deleuze's metaphysics of modality is based on three modes (intensity, actuality, and virtuality) and two processes (individuation and actualization), which Ardoline identifies as being structurally isomorphic to the structures and processes (such as superposition and decoherence) of quantum mechanics. While Ardoline's chapter focuses primarily on the measurement problem as the fundamental question that most interpretations of quantum mechanics (such as Many Worlds) are proposed to solve, in "The Chaosmos of the Quantum, with Gilles Deleuze," Arkady Plotnitsky develops the history of quantum mechanics as a counterpart to the Deleuzian concept of creation. Plotnitsky engages with Deleuze and Guattari's view of creative thought as a confrontation between thinking and chaos, along with their concept of concept as the main vehicle of thought, and addresses the centrality of chaos through the original theory of "reality without realism" (RWR) as a fundamental starting point in quantum physics and philosophy, the intrinsic inaccessibility of which ensures an ever-evolving creation of thought and world. In "Immanence, Process, and Relationalism in Loop Quantum Gravity," Martin Calamari examines three ontological claims that are at the foundation of loop quantum gravity (LQG)—that LQG must satisfy the general relativity-inherited principle of background independence; the rejection of the idea of a changeless Parmenidean "Block Universe" and concomitant subscription to process ontology; and the commitment to fundamental, ontological relationality as the explanation of a dynamic universe—and how they resonate with Deleuze's ontology of becoming, difference, and chaosmology. Finally, in "Zig-Zagging Through the Deleuzian Virtual: The Dark Universe of Diagrams, Cosmology, and Quantum," jan jagodzinski provides a breathtaking account of speculative quantum possibilities—exploring the implications of dark matter, dark energy, and negentropic energy, alongside some of Deleuze and Guattari's most ambitious and creative concepts. jagodzinski presents Roger Penrose's conformal cyclic cosmology (CCC) as a "minor" position that challenges the standard model of the Big Bang universe, explaining how Deleuze's embrace of "insufficient reason" is consonant with Penrose's intervention in the debate between quantum physics and general relativity.

The final part, "Comparative Cosmologies," brings Deleuze and Guattari's cosmological thinking into alignment with an even greater array of thinkers and discourses. These chapters discover synergies and creative divergences between Deleuze and Guattari and influential figures who have developed their own significant cosmological engagements. In "Caught Between a Sufficient Cosmos and a Chaosmos: Laruelle, Deleuze, and Guattari on Philosophy, Cosmology, and the Real," Chantelle Gray situates her investigation of the cosmological implications of Deleuze and Guattari's thought within a critical engagement with Laruelle's insistence on the Real. Gray's emphasis on Laruelle's insistence that the Real is foreclosed to thought resonates with Plotnitsky's development of RWR, as indicating a chaotic beyond thought that becomes a cause for creative and constant negotiation. Gray argues that the evolution of Deleuze's scientific formulation of difference and repetition to the more cosmological language of the body without organs and the plane of immanence indicates a sharper distinction between cosmos and chaos, which allows her to propose an unlikely alignment with Laruelle's non-philosophy (via the intercessors Humberto Maturana and Francisco Varela). In "Biological Philosophy and Geophilosophy in the Shadow of Cosmology: Reading Canguilhem with Deleuze and Guattari," Samuel Talcott begins with the claim that when knowledge and capabilities alter, the very meaning of the world, history, and thought must be reworked—noting that the transformation from the scientific and philosophical views of a closed world to an infinite universe have launched many a thinker down this path. Reading Canguilhem with Deleuze and Guattari with this philosophical obligation in mind highlights their shared methodological and conceptual links and demonstrates the power of an immanent pursuit of philosophy that embraces and responds to the unknown without seeking some definitive transcendence. Talcott argues that both Deleuze and Canguilhem support an immanent cosmological philosophy, one which finds reason to attend to the Earth in ongoing encounter with the unknown. In "Modality as Virtuality versus Modality as Possibility: Gilles Deleuze and David Lewis," Ilona Schweitzer brings the analytic and continental traditions together with respect to two powerful metaphysical views, those of David Lewis and Gilles Deleuze. She investigates whether the assumption of virtual modality (Deleuze) and the assumption of possible modality (Lewis) have points of contact, and then asks whether the modal concepts of Deleuze and Lewis can be applied to current trends in the metaphysics of quantum physics and, if so, with what restrictions. Schweitzer's aim is to trace the methodological scope of genetic thinking (Deleuze) and analytical thinking (Lewis) in relation to scientific metaphysics. In the final chapter of the volume, "Toward a Process Cosmology: Deleuze, Whitehead, and Spacetime," Keith Robinson argues that, for both Deleuze and Whitehead, an understanding of the universe along traditional lines of simultaneity and one-ness (the Block Universe) or the many (Many Worlds or the multiverse) does not do justice to the fusional multiplicity and variable continuum that both present as the heart of the dynamic and transforming processes of cosmic interplay. Concentrating on Deleuze's concept of chaosmos, Robinson reveals the shared understanding of spatiotemporal rhythms that create what Whitehead calls the "extensive continuum" and Deleuze the "plane of immanence."

References

Ansell-Pearson, Keith. 1999. *Germinal Life: The Difference and Repetition of Deleuze*. New York: Routledge.

Beaulieu, Alain. 2022. "Back to Earth! A Comparative Study Between Husserl's and Deleuze' Cosmologies." In *From Deleuze and Guattari to Posthumanism: Philosophies of Immanence*, edited by Christine Daigle, and Terrance McDonald, 106–25. London: Bloomsbury.

Bell, Jeffrey A. 2006. *Philosophy at the Edge of Chaos: Deleuze and the Philosophy of Difference*. Toronto: University of Toronto Press.

Bogue, Ronald. 1991. "Rhizomusicosmology." *Substance* 20, no. 3: 85–101.

Braidotti, Rosi. 2013. *The Posthuman*. Cambridge: Polity.

Cage, John. 1973. *Silence: Lectures and Writings by John Cage*. Hanover, NH: Wesleyan University Press.

Calamari, Martin. 2015. "Riemann Weyl in Deleuze's Bergsonism and the Constitution of the Contemporary Physico-Mathematical Space." *Deleuze Studies* 9, no. 1: 59–87.

Cox, Christoph. 2003. "Wie wird Musik zu einem organlosen Körper? Gilles Deleuze und experimentale Elektronika." In *Soundcultures: Über digitale und Elektronische Musik*, edited by Marcus S. Kleiner and Achim Szepanski, 162–93. Frankfurt: Suhrkamp. http://faculty.hampshire.edu/ccox/Cox-Soundcultures.pdf. Accessed September 8, 2025.

De Beistegui, Miguel. 2004. *Truth and Genesis: Philosophy as Differential Ontology*, Bloomington: Indiana University Press.

DeLanda, Manuel. 2002. *Intensive Science and Virtual Philosophy*. New York: Continuum.

DeLanda, Manuel. 2010. *Deleuze: History and Science*. New York/Dresden: Atropos.

Deleuze, Gilles. 1988. *Foucault*. Translated by Sean Hand. Minneapolis: University of Minnesota Press.

Deleuze, Gilles. 1990. *Bergsonism*. Translated by Hugh Tomlinson and Barbara Habberjam. New York: Zone Books.

Deleuze, Gilles. 1992. *The Fold: Leibniz and the Baroque*. Translated by Tom Conley. Minneapolis: University of Minnesota Press.

Deleuze, Gilles. 1994. *Difference and Repetition*. Translated by Paul Patton. New York: Columbia University Press.

Deleuze, Gilles. 2003. *Francis Bacon: The Logic of Sensation*. Translated by Daniel W. Smith. Minneapolis: University of Minnesota Press.

Deleuze, Gilles, and Félix Guattari. 1987. *A Thousand Plateaus: Capitalism and Schizophrenia*. Translated by Brian Massumi. Minneapolis: University of Minnesota Press.

Deleuze, Gilles, and Félix Guattari. 1994. *What is Philosophy?*. Translated by Hugh Tomlinson and Graham Burchell. New York: Columbia University Press.

Guattari, Félix. 2016. *A Love of UIQ*. Translated by Graeme Thomson and Silvia Maglioni. Minneapolis: University of Minnesota Press.

Harris, Paul A. 2010. "Deleuze's Cinematic Universe of Light: A Cosmic Plane of Immanence." *Substance* 39, no. 1: 115–24.

Hawking, Stephen, and Leonard Mlodinow. 2010. *The Grand Design*. New York: Bantam Books.

Heidegger, Martin. 1971. *Poetry, Language, Thought*. Translated by Albert Hofstadter. London: Harper and Row.

Houle, Karen. 2011. "Animal, Vegetable, Mineral: Ethics as Extension or Becoming? The Case of Becoming Plant." *Journal for Critical Animal Studies* 9, no. 1–2: 89–116.

Husserl, Edmund. 2001. "Foundational Investigations of the Phenomenological Origin of the Spatiality of Nature: The Originary Ark, the Earth, does not Move." Translated by Fred Kersten. In *Husserl at the Limits of Phenomenology: Including Texts by Edmund Husserl and Maurice Merleau-Ponty*, edited by Leonard Lawlor and Bettina Bergo, 117–31. Evanston, IL: Northwestern University Press.
Klee, Paul. 1961. *Notebooks*, Vol. 1. Translated by Charlotte Weidler and Joyce Wittenborn. London: Percy Lund.
Klee, Paul. 1975 [1945]. *On Modern Art*. Translated by Paul Findlay. London: Faber and Faber.
Massumi, Brian. 1993. *A User's Guide to Capitalism and Schizophrenia: Deviations from Deleuze and Guattari*. Cambridge, MA: MIT Press.
Nietzsche, Friedrich. 2009 [1883–85]. *Thus Spoke Zarathustra*. Translated by Graham Parkes. Oxford: Oxford University Press.
Olkowski, Dorothea. 2012. *Postmodern Philosophy and the Scientific Turn*. Bloomington: Indiana University Press.
Panek, Richard. 2011. *The 4 Percent Universe: Dark Matter, Dark Energy and the Race to Discover the Rest of Reality*. Boston: Houghton Mifflin Harcourt.
Pinhas, Richard. 1997. "Cosmos, rythme, plan." *Nomad's Land* 2, no. 3: 54–60.
Plotnitsky, Arkady. 2006. "Chaosmologies: Quantum Field Theory, Chaos and Thought in Gilles Deleuze and Félix Guattari's *What is Philosophy?*" *Paragraph* 29, no. 2: 40–56.
Plotnitsky, Arkady. 2012. "From Resonance to Interference: The Architecture of Concepts and the Relationships Among Philosophy, Art and Science in Deleuze and Guattari." *Parallax* 18, no. 1: 19–32.
Prigogine, Ilya, and Isabelle Stengers. 1984. *Order out of Chaos: Man's New Dialogue with Nature*. New York: Bantam Books.
Protevi, John. 2013. *Life War Earth: Deleuze and the Sciences*. Minneapolis: University of Minnesota Press.
Ross, Bill. 2024. *Order and the Virtual: The Philosophy and Science of Deleuzian Cosmology*. Edinburgh: Edinburgh University Press.
Sellars, John. 1999. "The Point of View of the Cosmos: Deleuze, Romanticism, Stoicism." *Pli* 8: 1–24.
Sellars, John. 2006. "An Ethics of the Event: Deleuze's Stoicism." *Angelaki* 11, no. 3: 157–71.
Sholtz, Janae. 2015. *The Invention of a People: Heidegger and Deleuze on Art and the Political*. Edinburgh: Edinburgh University Press.
Sholtz, Janae. 2021. "*A Thousand Plateaus* and Cosmic Artisanry: On Becoming Destroyer of Worlds." *Deleuze and Guattari Studies* 15, no. 2: 197–225.
Smolin, Lee. 2006. *Trouble in Physics*. New York: Houghton Mifflin.
Solomos, Makis. 2015. "De la ritournelle cosmos à la puissance du son. Cinq essais pour écouter Mille plateaux." In *Gilles Deleuze: La pensée musique*, edited by Pascale Criton and Jean-Marie Jouvel, 151–60. Lyon: Centre de documentation de la musique contemporaine.
Stockhausen, Karlheinz. 1989. *Towards a Cosmic Music*. Shaftesbury: Element.

PART I

CREATIVE COSMOLOGY: CHAOS AND THE COSMIC

I am interested in qualitative state-changes as Deleuze and Guattari discern them in mass quantities of events, actions, experiences, etc.—a change some philosophies describe as "epiphenomenalism"; with Deleuze and Guattari this might better be termed "epimaterialism," or "surrealism," to recall the early twentieth-century art movement. Though horizontal in their structure and force, rhizomes en masse undergo qualitative change—a qualitative state-change pictured in *Rhizomes Ascending* as a building-up and emergence from a page of text. The pink-red color of the deterritorial spaces between rhizomes is sensational—expressing in color the intensity of multiple rhizomatic interactions.

—Adrian Switzer

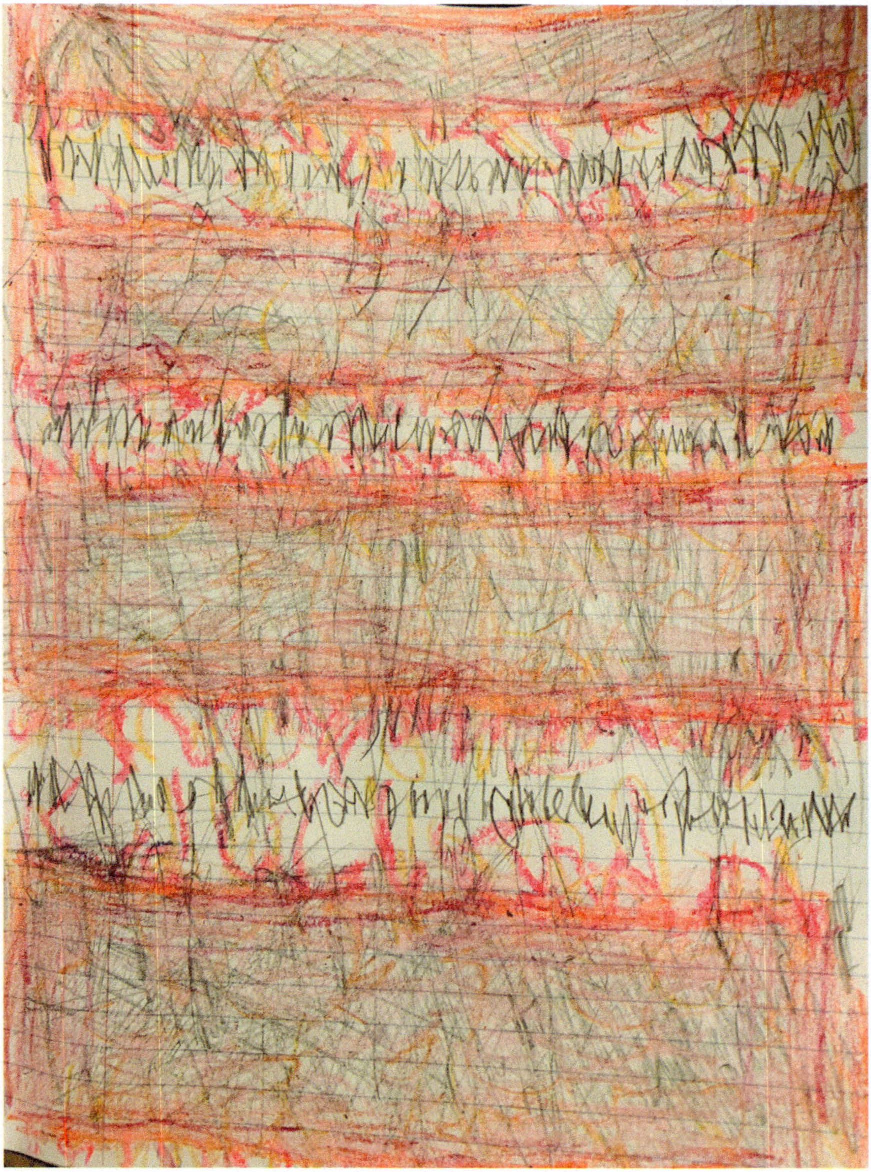

Adrian Switzer, *Rhizomes Ascending*, mixed media (pencil, pen, colored pencil, colored marker), 2022 © Adrian Switzer

Chapter 1

What is Invention? Deleuze and Guattari and the New

Dorothea Olkowski

Creation?

What is invention? What is the creation of something new? Is it, as the *Cambridge English Dictionary* claims, "a product or a way of doing something which has never been made or never existed before"? Or is it, following another of the dictionary's definitions, "something newly designed or created, or the activity of designing or creating new things," or yet, "a story or excuse that is not true, or the act of creating a story or excuse that is not true"?[1] These common dictionary definitions never question the larger context of creation and address it only as a purely personal act. This evades the crucial philosophical and scientific question, which is, does the creation of something new depend on forces beyond the individual, forces beyond the culture or community?

Isaac Newton's (1643–1727) physics described the laws governing the movement of particles in the universe. It set in place a deterministic system praised for being applicable to both earthly and heavenly bodies.

> Given any arrangement of particles that make up the universe, and any choices for their initial motions, Newton's laws can be used to predict the future. But Newton's theory ends up describing an infinite number of different worlds, each one of which arises as a different solution to the theory, which is arrived at by starting with the particles in different position. (Olkowski 2007, 3; see Smolin 2017, 43)

Nevertheless, if everything is determined on the cosmological scale, then, strictly speaking, nothing is new. And yet nearly everyone wants to create or to be considered creative, leading to the necessity of asking if creation is real or to what extent it is real. Clearly, contemporary philosophers and physicists are still struggling with this question. Given Deleuze and Guattari's definition of philosophy as the creation of concepts, we might at least briefly examine some of the arguments of contemporary philosophers and physicists for their applicability to those of Deleuze and Guattari.

Determinism and Physics

Newtonian determinism was reinforced by Pierre Simon Laplace (1749–1827), who theorized what Henri Poincaré (1854–1912) called "Laplace's Demon," a demonic intellect capable of mathematically computing every past and future event in the universe (Laplace 1902). Poincaré, along with Jacques Hadamard (1865–1963), was among the first to refute some of Laplace's claims, perhaps anticipating the proof by Kurt Gödel (1906–78) that mathematics cannot provide solutions for all the problems that it raises (Agassi 2022, 252). Sometime later, the philosopher of science Karl Popper (1902–94) sought to refute Laplacian determinism, not as mathematical but as scientific, because he thought that it had been used to justify moral and political irresponsibility. He argued that if prediction is not only mathematical but includes scientists, who are human predictors, then this changes the game, as human predictors cannot predict their own future predictions.

Popper also takes up Poincaré's argument that, if it is not possible to predict the course of science, then its influence on human lives is unforeseeable, and so it is impossible to predict human futures.[2] Popper adds that if scientific determinism is scientific, then it is also, in fact, predictable and testable, and so, in principle, refutable (Agassi 2022, 253). However (and regrettably for Popper), nothing prevents there being some extension of a scientifically deterministic system that is in some way testable and also comprehensive. As we can see, the question of determinism remains unresolved, and so, consequently, does the question of the creation of undetermined phenomena of any sort (Agassi 2022, 256).

But there is more to this story. As an extension of his critique of Laplace and his extensive investigations of celestial mechanics, Poincaré is credited with opening "a new era" of analysis (calculus). He introduced geometrical, topological, and probabilistic methodologies into his investigation of (shall we say, post-Newtonian?) planetary motion. In his memoir, Poincaré investigated "the great astronomical problem of the stability [or instability] of the solar system" that came to be known as the three-body problem (Chenciner 2012, n.p.).[3] Simply put, when more than two bodies are in one another's orbits they form a chaotic system with unstable movements. In this situation even tiny perturbations in their orbits might be amplified endlessly so that precise predictions about their future are impossible.

This instability opens the door to indeterminism on many levels. What happens when three or more celestial bodies orbit in close proximity to one another? Do the distances between them increase or decrease? What effects do they have on one another? These are the questions that Poincaré pursued. The motion of two celestial bodies in proximity was a problem studied early on by Newton, who calculated the effects of the force of the mutual attraction of gravity using differential equations, which accurately predict the future location of an object based on its current location (Montgomery 2019, 66). The same cannot be accomplished for three bodies, leaving open such apparently simple matters as the effect of the sun on the motion of the moon and Earth. Leonard Euler (1707–83) developed a restricted version of the three-body problem by examining the motion of the moon in relation to the orbits of the Earth and sun (Juhan 2006). Additional studies and limited solutions were carried out by Joseph-Louis Lagrange (1736–1813), along with many others who were fascinated by this problem.

Recent research confirms that chaotic three-body systems disintegrate, eventually expelling one member of the trio while the two remaining bodies form a stable binary system. Systems with hierarchy, where one body dominates, such as that of the sun, Earth, and moon, coexist for longer periods (Ginat and Perets 2021). Although these structures show instability, long-term they conform to deterministic outcomes and predictions, thereby validating determinism on a cosmological scale. If it is the case that the laws of nature are applicable on all scales, the cosmological, the earthly, and also the human (including social, economic, and even neurophysiological realms), then none of these outcomes appears especially fruitful for generating a theory of creation and the creative process which relies on a cosmological theory of indetermination.

This leaves in doubt the degree to which Deleuze and Guattari can make use of this version of cosmological indetermination to formulate or justify the possibility of the creation of something new such as concepts or works of art. What might possibly contribute somewhat to the existence of creative processes is a recent model showing that at least some of the time (5 percent in fact) in computer simulations, chaotic three-body systems are not time-reversible. They cannot simply be run back in time to their initial state, and they can be influenced by something as small as a Planck length (1.6×10^{-35} meters), thus breaking time symmetry and so being judged to be fundamentally unpredictable

(Starr 2020).[4] This is a rather small opening for unpredictability and the genuinely new, but nonetheless it is useful as a thought experiment and might provide some insight into the issue of creation.

Prior to computerized computation, the geometry that Poincaré and others used to do their calculations was that of dynamical systems. The mathematical dynamical system was an idea readily embraced by Deleuze, especially in *Difference and Repetition* (1994). This is the idea of a dynamical system, which in relation to the cosmological three-body problem is known as deterministic chaos. If a dynamical model is called for to describe some observable behavior (such as that of planetary bodies in orbit around one another), the model will consist of a *manifold* or state space and a vector field. A manifold is a playing field or space on which the motion of the system takes place, and a vector field is a set of rules that specify where to go on this space from some starting point. It is a model for the habitual tendencies of the situation as it *evolves* from one state to another. This charting of evolution is what makes the model dynamic. But insofar as the system yields qualitative predictions of long-term behavior, its evolution is deterministic. This makes it useful for predicting the future (Olkowski 2007, 208–12). However, unanticipated and unpredicted variations are still possible and so the dynamical system functions probabilistically—outcomes are likely but not certain. This appears to leave us with a paradoxical situation, a future that is probable and so relatively undetermined, and at the same time a long-term future that is likely determinable and so predictable.

I have written about and criticized this model extensively in *The Universal* (2007) where I cite the work of the physicist Fotini Markopolou, who proposes an alternative, quantum, time-based model of evolution. Recent work derived from the final research of Stephen Hawking might well support this view (Hertog 2023). Nevertheless, I will discuss the Deleuze–Guattari model here briefly to clarify how and why it is of consequence for the question of creation (Olkowski 2007, 210).

The limitation for predicting the evolution of any dynamical structure is this:

> Sufficiently complex dynamical systems require as much informational input as they can provide informational output, in other words, strictly speaking, nothing can be predicted if the input information must be equivalent to the output information. Under these conditions, if the rules of combination do not alter, in some sense, determinism remains, but predictability is lost. (Olkowski 2007, 210)

It is important to know that even for chaotic dynamical systems, the rules of motion are given and carried out along a differentiated trajectory. This motion is the image and realization of Deleuze's concept of differentiation, where every motion is differentiated in relation to the previous ones. It undermines the identity of anything moving through the field because anything in this structure is continuously different from "itself," and so it is unable to be unified. From the point of view of creation, this "limitation" is a gift of contingency and chance. This is because, physically, what might still be contingent are the particular particles themselves, that is, which particles enter into any given trajectory and in what order—in Deleuze and Guattari's terms, which affects, percepts, concepts, prospects, and functives? These terms correspond to their view of the particles (the molecular or molar elements) operative in the making of art, philosophy, science, and logic respectively (see Deleuze and Guattari 1994). In a system sensitive to initial conditions with unstable boundaries such as that proposed by Deleuze and Guattari, some of this cannot be predicted, and every configuration of particles produces a deterministically chaotic world. In other words, predictions about outcomes are no more than probable.

If we think about Deleuze's repudiation of representation in favor of difference, we can see some effects of this. The critique of representation appears in many places in Deleuze's work, but initially, and most thoroughly, it shows up in his critique of Aristotle, for whom *being* is hierarchically primary and distributively common to all categories. For Aristotle, being refers primarily to substance, and all other categories only have being in reference to or by analogy with substance (Olkowski 1999; Aristotle 1970, IV, 2, 1003b). More than analogy is at work here, however. What Deleuze calls *organic representation* is guaranteed by four elements imposed by reflection. Reflection operates by subjecting all differences arising with sensible intuitions to the *identity* of the concept, the *opposition* of predicates, the *analogy to substance* carried out by judgment, and the *resemblance* to perception in order to constitute them as representations (Deleuze 1994, 34–5).

Given this reflective order, the difference of differential dynamical systems cannot even appear as an intelligible structure. If it does appear, it does so as a catastrophe, a break in resemblance, as it implies the impossibility of claiming identity, opposition, analogy, or resemblance where reflection demands that they should occur. Yet much creative work begins with representation. Francis Bacon's paintings generally originated with photographic images, which Bacon then erased and deformed to produce the images characteristic of his work. Deleuze sees the continuation of this organization in all elements of philosophical, artistic, linguistic, and scientific practices (Aristotle 1970, IV, 11, 219a10). It is a theme in both *The Logic of Sense* and *A Thousand Plateaus*. The nomad exists in relation to the State, stuttering exists in relation to signification, Aion in relation to Chronos, and so forth. It may have been Deleuze and Guattari's lifelong work to find structures that enable creation, but instability always requires some degree of stability in some form.

What is a Dynamical System?

What makes the mathematical dynamical system so attractive to Deleuze? Why does he choose it as a leading image of the organization of thought and an emblem of creativity? Dynamical systems have been used to model everything from the motion of planets to the engineering of buildings to the structure of neurons in the brain. Mathematically, dynamical systems are organized by an attractor which acts like a whirlpool that draws the movement of the system in its direction.[5] A *fixed-point* attractor has a single definite destination, but where its trajectory ends depends on where it started, with very different pathways for even minutely different starting points. It's like getting on a slide. Once on it you stay till the end, although in this example there are a lot of different slides to choose from. Once a system is oriented by a second type of attractor called a *limit cycle*, it remains there in its circuit, repeating itself in an endless repetition, like a pendulum that swings eternally (Casti 1994, 28).[6]

Finally, there are the so-called *strange* attractors. They are like strings of yarn that wind over and over one another in a messed-up circuit, without ever making immediate contact with another strand of yarn. These interest Deleuze because their periodic orbits are unstable—their pathways act as repellors rather than attractors. If the system's motion is even minutely perturbed, minutely off its pathway, the strange attractor pushes it even further away from the path or orbit, like a pendulum pushed off its upper attachment (Casti 1994, 29–30). As stated above, the rules governing the system do not alter; they are and remain deterministic. Yet the motions resulting from strange attractors are unpredictable, contingent, random. Another term for strange attractors is the more familiar word *chaos*.[7] We can see that the three-body problem is a version of this, where the three bodies are in unstable orbits in relation to one another until one of them is pushed away. There are limits to what we can and should expect from nature by means of dynamical systems.

A totally chaotic, structurally unstable world would not support living beings, as evolutionary adaptation would have no chance to take hold (Casti 1994, 45). Some amount of stability is necessary for life.

Still, we can ask if and how instability takes advantage of nature's contingencies for the sake of creation. Discontinuity and difference are evident in breaking waves and financial markets, where too many inputs confuse and disrupt the output. The quintessential mathematical manifestation of instability is the catastrophe, the shift from a merely unstable situation to one whose inputs are so disruptive as to shift the outcome onto an entirely new region with a different attractor. The mathematician René Thom is famous for introducing this concept, although he was also reviled by others who denied it as a reality (Casti 1994, 46–7). In the extreme, a catastrophe might amount to what Deleuze calls a black hole, something too disorganized and tumultuous to be of use. More limited versions of catastrophes, versions that are not completely destructive, seem to take place at least occasionally. Anyone can visualize the simplest of these catastrophes by making a sharp S curve out of a flat piece of paper and envisioning something falling from the top upper right-hand edge of the curve to a surface below it, bypassing all the other surfaces that might have broken its fall (Casti 1994, 63–4). This "cusp" catastrophe has been applied to everything from competing political ideologies to the collapse of civilizations. It envisions the move from an unstable political or economic situation to one whose parameters are yet to be defined and so might turn out to be quite problematic.

Given this risk, deterministic chaos, as described above, is perhaps a less dangerous way to accomplish the creation of something new. So, we see Deleuze by and large taking this route when he states that every domain of interest is oriented by an Idea, "an n-dimensional, continuous, defined multiplicity," where dimensions refer to variables upon which a phenomenon depends, and continuity is "a set of relations between changes in these variables" (Deleuze 1994, 182). The important element here is continuity, but what gives this structure an element of contingency is the undefined number and type of dimensions and the elements that populate them. The general orientation of the field is organized by an attractor, generally a fixed attractor, which for the philosopher is an Idea. Ideas may be social, psychological, economic, biological, chemical, physical, and so on. We can see how this is the start of a variable structure that permits the unanticipated arrival of the new without the threat of complete collapse—the black hole. Of course, many situations defined by changes in variables might amount to nothing, so we want to look more closely at what works to the advantage of the creation of something new.

Neuroscience

Like philosophers and psychologists, neuroscientists are curious about creativity. One recent group, consisting of seven researchers, has defined creativity as a cognitive process utilizing divergent thinking, the extrapolation of many possible answers to an initial stimulus or target dataset, and they link this to the cortical thickness of the brain, where greater thickness is correlated with greater "creative achievement" (Jung et al. 2010, 398). The authors further define creativity as a cognitive process producing "something both *novel* and *useful* within a given social context" by means of fluid intelligence, flashes of insight, and flow, that is, full immersion and a feeling of full involvement (Jung et al. 2010, 398, 399). Additionally, they mention the value of accepting the best solution, and perseverance in the face of social acceptance or resistance.

Their review of existing literature concedes that the heterogeneity of outcomes from these studies makes it difficult to draw clear conclusions regarding the role of the frontal, posterior, and lateral hemispheric cortices of the brain. Neurobiological research using brain-imaging techniques such as functional magnetic resonance imaging (fMRI) was utilized to study a "highly creative group," a determination based on "creative story generation" and unique responses to Rorschach inkblots (Jung et al. 2010, 400). It is unclear on what basis small groups of individuals (only men in some cases) were determined to be highly creative and why storytelling and inkblot interpretation qualify as creative. What all of these studies seem to have in common is the effort to localize creativity in specific regions of the brain without, as the authors state, linking this to intelligence and personality. However, these studies fail to understand that any specific cognitive process is only one of many components of creativity, which can never be localized in one part of the brain (Jung et al. 2010, 400).

This led the same group of neuroscientists to a larger-scale study of sixty-five individuals—nearly half of them female—recruited from classrooms at the University of New Mexico, who defined their own creative domains, ranging from the visual and performing arts to science and occupational disciplines. All subjects were administered a series of tests that purportedly measured creative productivity, design fluency, and IQ, including verbal and performance IQ. The results seem to have been mixed. Some parts of the brain showed greater cortical thickening and other parts less. The value and use of these measurements is never made clear. And disturbingly, the authors admit that the subjects might have only been simulating creativity in the lab and exaggerating their self-reported creativity in the world outside. Spontaneous emotional creativity remains a mystery (Jung et al. 2010, 406).

A slightly more informational study carried out by a smaller group consisting of psychology and psychiatry researchers focused on creativity and neuromodulatory systems such as those producing dopamine (Khalil et al. 2019). Once again, the focus was on cognitive processes, but in addition, the study took into consideration moods, motivation, rewards, and regulatory focus. Not surprisingly, the researchers seem to affirm that "creative thinking does not critically depend on a particular single mental process or specific brain region, and it is not mainly associated with right brains, defocused attention, low arousal, or alpha synchronization, as it also has often been hypothesized" (Khalil et al. 2019). They also cite studies that point to the importance of dopamine and neurotransmitters in relation to creativity as well as brain illnesses such as depression, bipolar disorder, psychosis, epilepsy, dementia and autism spectrum disorders, although the precise relation between "madness and genius" remains unclear.

Physics, mathematics, and neuroscience might point to the conditions of possibility for creation and creativity, but they do not on their own guarantee or produce it. Psychology and psychiatry reveal some important neurological and emotional clues, but a great deal of mystery remains. What they all share is a lack of a coherent philosophical theory. It's not enough to carry out empirical studies without a comprehensive theoretical foundation. Even if those studies eventually reveal that the theory is mistaken or incomplete, it is necessary to have a theory as a general guide. Deleuze struggled in many places to connect dynamical systems theory and deterministic chaos to the concept of creation. Yet representation continued to play a role in his descriptions of creative acts and ideas. In the two books on cinema, Deleuze introduces a theory of the brain and brain activity into the question of creation on the human level. This comes to him via Henri Bergson's fascination with non-normative states of mind. What might be the case in the cinema books is that rather than cosmology and dynamical systems standing as the model for the brain and creativity, it is the brain and our neurophysiology that provide a model for cosmology.

Creation and Images

In *Mind-Energy*, Henri Bergson suggests that we are mistaken when we presume that the dream state is a substratum of the normal state of mind. In fact, he argues, the reality of the waking state is obtained by a limitation of the dream state, by concentration and tensing the otherwise diffuse psychical dreamlife. "[T]here does consciousness disport itself, perceiving just to perceive, remembering just to remember, with no care for life, that is, for the action to be accomplished" (Bergson 1920, 155). Cease to will and you pass into the dream state. But the downside is that this state imitates that of insanity. Other states will do this as well. These include a lasting intoxication that alters the chemicals affecting brain tissue such as that left by "certain infectious diseases"; lesions between the frontal and parietal lobes that can produce loss of movement; and damage to the visual cortex and the visual word-forming area behind the ear, which account for image and word recognition respectively (Bergson 1920, 41–2). In each case, the mechanism of the brain is thrown out of gear, out of line with reality, and so out of line with action, and new evidence confirms, as Bergson argued, that even words are recognized as images (Alford 2015).

On the other hand, there are also abnormal states that appear to add something to normal life. Delirium, hallucinations, and obsession introduce new ways of thinking and feeling. In this case, "no psychical state disappears but all are affected" (Bergson 1920, 151–2, 153). But also in this case, the abnormal state is prevented from emerging sooner due to the inhibitory effect of *attention to life*, which eliminates and corrects various brain functions in search of equilibrium. When and wherever this is relaxed, symptoms arise, symptoms that have been waiting for the opportunity to present themselves (Bergson 1920, 154). Thus, the task of psychology, according to Bergson, is to explain why there are not more disordered minds or even why *all* minds are not disordered. Possibly this is even more the case if we consider the opening of *Matter and Memory*, where we are asked to find ourselves in the presence of images, prior to any conceptualization or representation. Is this a little piece of chaos linked to creation and is it possible to have just a little piece of chaos to explore, a brain that says, "I feel" and "I is an other," as well as "I conceive," and "I think"?

Deleuze's *Cinema 2: The Time-Image* takes up the Bergsonian idea of images surrounded by a world—actual images, recollection-images, dream-images, and world-images organized from the most contracted to the most expansive (Deleuze 1989, 68). The entry into images bring us—not to a subject of perception who acts—but more likely to a brain, to what Deleuze refers to as *attentive recognition.* Attentive recognition abandons perception, selects only certain provisional features that may be displaced or replaced, and so abstracts from the action-oriented perception of a thing (Deleuze 1989, 44–5).[8] In the films that interest Deleuze, such as Marcel Carne's *Daybreak*, attentive recognition is produced as a pure optical image. The character played by Jean Gabin knows that he is going to be arrested by the police for having committed a murder. But he no longer knows how to react or is no longer able to react, because optical or sound images call up recollection-images that stifle him. They are the source of what are often called flashbacks—although they are more than this, and insofar as they reflect his destiny they justify his demise, as they carry the tragic hero away to the depths of time (Deleuze 1989, 48). The actual and virtual images reflect one another so that it ceases to be possible to say which is first.

This works well enough in film, but does it function similarly when we are talking about the brain? That is, can the brain sidestep its own subject who perceives and acts? Cosmologically and in engineering, the instability of fields of particles or interacting bodies as well as the catastrophic fall to a completely new manifold enacted by a strange attractor does happen, sometimes with devastating results. To what extent does it show itself in human behavior? And if it does, is it something to be cautious about?

Bergson and Deleuze argue that memory expands along with recollections, dreams, and thoughts, so that an image passes through an "infinite number of planes or circuits," yielding a "mental vision" which may be obliterated by another circuit that creates yet another mental image, and so constitutes layers and layers of "one and the same physical reality" (see Bergson 1988; Bergson 1963; Deleuze 1989, 68). This is not the reality of the perception–action circuit but something completely other, possibly something completely new. Minimally, there is a whole new sense of subjectivity, an extreme redefining of subjectivity—no longer an agent of perception but only a temporal and spiritual feeling (Deleuze 1989, 47). It occurs when a multiplicity of brain circuits transit zones of recollection and return to a deeper state of the present situation, exceeding all recollections. This is not subjective imagining, which is always juxtaposed to the objective real, and it is not a function of association of ideas. As Bergson argues, it is the past brought forward, and it happens only if we place "ourselves" within it by following and adopting the movement by which it expands and emerges from darkness to light (Bergson 1988, 172–3; Deleuze 1989, 54) As Deleuze clarifies, the recollection of the person met last week or last year in a particular place occurs only when I place myself within the recollection.

Deleuze recognizes the films of Orson Welles as typical of those where such recollection occurs. In *Citizen Kane*, Kane "catches up in depth with the friend he will break with," by moving in the past, plunging into the past, so that "it is in the past that he himself moves" (Deleuze 1989, 106). Yet for the most part, reaching into the pure past of pure recollection is a failure. It is a failure of the sensory-motor mechanism and a failure of present perception to link up with a motor-image or even a recollection-image, which is not the deeper state of pure recollection akin to the dream-image, the past in general. As is the case with many failures, Deleuze finds this one to be creative—free to link to other kinds of images—*déjà vu*, the past in general, fantasies, theater, or dream-images—disturbances of memory and failures of recognition (Deleuze 1989, 55).

Bergson's fascination with disturbances of the brain and with its failures provides the key to this success: amnesia, hypnosis, hallucination, madness, visions, nightmares, and dreams all might play a role (Deleuze 1989, 54–5). All these disturbances break with the action-image and break with coherent perception. They form a sort of "automatic subjectivity," in contrast to the action-oriented subjectivity of perception (Deleuze 1989, 55). In cinema, characters can be shown to be involved in an accident, sleeping, dreaming, lost, threatened, extremely stressed, barely alive, in a state so severe as to be severed from ordinary memory, from habit, and even from attentive recognition. Nothing corresponds to these states except the pure past, the past in general moving at dizzying speed and responding to the character's dream state, a state in which the character is powerless to act but exists only within the circuit of pure recollection akin to the unconscious.

Who among us wants to risk this? Is this not, as Bergson points out, mentally and emotionally disruptive? Yet Deleuze seems to imply that if we cling to recollection-images, where we can be assured of not being subject to social and intellectual criticism or worse, then possibly we are not fully engaging our creative capacity. Possibly there is no need to proceed in this risky way, as it is not the same as hallucinating or taking a drug that would make one numb to the external world of perception and action. Perhaps it is more a matter of disregarding the call of the world to act and to be right and so to risk a little stupidity. Perhaps it amounts to disregarding—temporarily or from time to time—what one knows, to envision or sense something new. But this also implies that the former acute perceiver might themselves become something new and that at least some of their perceptions and possible actions will thereby be altered—not always in ways that will be appreciated and celebrated or even reluctantly accepted, and not always in ways that are creative. And there is always the risk of self-delusion. What if this type of suspension is akin to that recently occurring on chatbots, in which certain methods of information gathering and certain methods of persistent questioning produce hallucinations and lies? Even when artificial intelligence functions effectively by taking advantage of a vast, nearly infinite set of memories, in the end it produces little more than generalized secondary or tertiary manifestations of pirated originals.

Stupidity?

With respect to the human brain's relation to perception, as Bergson points out, "[t]he same psychical life, therefore, must be supposed to be repeated an endless number of times on the different stories of memory, and the same act of the mind may be performed at varying heights" (Bergson 1988, 128–9). It may happen that the mind frees itself from current actions and so may do what it likes in relation to present or absent objects. Sometimes the body ceases to adopt the appropriate attitude and cannot call up the memory that would allow it to act correctly in a given situation. Or the memories themselves do not meet the call of perception (Bergson 1988, 132). Bergson links these failures to various pathologies, but Deleuze takes them to be the effect of weak or disjoined connections between optical or sound sensation or perception and a total dream-image, the largest circuit of the brain. This allows the virtual image to actualize in a different image, which itself can be actualized in yet another, and so on.

Given the key role that the ideas of Charles Peirce play in the cinema books, it is interesting that although this pattern seems to correspond to what Peirce called Thirdness, Deleuze lets the reader do the work of making this connection. Given the frequency with which this occurs in Deleuze's work, it has the element of being a deliberate technique, a way of leaving an opening for the reader and another possible avenue of creativity. What can we make of pursuing Thirdness? Thirdness is a matter of thought and meaning or interpretation; it is a mental operation such as the laws of dynamics which predict how bodies will move after forces have been defined (Peirce 1931–35, 1.342, 1.346). For Peirce, this makes thought an active factor in the real world, one that might modify laws. "The interpretant, the 'third' element of semiosis, in turn acts as a sign/representamen (a first), to determine another interpretant (another third), through the *Same* object (a second)—and so on" (Casti 1994, 29–30).[9] Thirdness has the character of *generality* which Deleuze refers to as "the exchange or substitution of particulars ... each of which can be replaced by any other particular idea which *resembles* it in relation to a given word" (Deleuze 1994, 1).

We noted above that Deleuze makes use of the mathematical model to formulate Ideas that are "incarnated in real relations ... and carried over into problems in the form of solutions" (Deleuze 1994, 183). But his chosen mathematical instrument—differential calculus—makes it difficult to locate in it the "Platonic evidence of a dialectic *superior* to mathematics" (1994, 179). Deleuze constructs this dialectic via Peirce by distinguishing the Idea of justice (Thirdness) as the ground, the generality, the differentials of thought; then the *quality* of justice (Firstness, Deleuze's intensity); and finally, the object of the claim possessed by that which grounds, that which is incarnated in the just claimants (Secondness or representation), each of which is organized by the Idea of justice (1994, 62).

Deleuze then correlates his use of mathematical phase space filled with trajectories with Peirce's use of continuity. For Peirce, continuity is generality, understood as conformity to an Idea. But the continuum is a supermultitudinous collection exceeding all existents, and so great that all units lose identity and become indistinct. Deleuze calls this an ideal continuity, and solutions to problems are a rupture in this continuity (1994, 46). It is all that is possible—the virtuality or continuity through which all states of affairs, insofar as they come into existence, will do so (Deleuze and Guattari 1994, 153).

For Peirce as for Deleuze, this structure is not merely linguistic—it is cosmological, as the elements of the semiotic account for the impact of language, the "real," and the cosmological character of Thirdness, which includes *every possible quality*. Deleuze does not use the term Thirdness but instead uses the term "repetition" in active and passive syntheses; yet it is as an open version of Thirdness—an unending evolution—that repetition forms a rich domain of signs, and those signs give rise to Ideas, which in turn are subject to an infinite regress, a becoming that can continue to infinity (Peirce 1931–35, 1.339). In *Cinema 2*, Deleuze offers examples of films where this becoming shows itself, such as when city lights become a pile of lighted cigarettes, which become the columns of a Greek temple, then a silo and a chessboard turn into a view of the Place de la Concorde in Paris (Deleuze 1989, 57–67). Thus, it is characterized as a film technique, or possibly a technique that could be extended to other visual and performing arts—at least to some degree.

At the same time, Deleuze offers an alternative creative method. For the brain, the image does not only expand into larger and larger circuits; it also contracts into the smallest circuits, internal limits, and mental flashbacks, which serve as the double or virtual mirror reflection of the actual image. It thereby creates its own object, the real and the imaginary, the present and the past, *one image with two sides*, each of which is totally reversible in relation to the other. Elsewhere Deleuze refers to this type of structure as that of sense, which does not exist outside of the proposition that expresses it in the act of denoting a thing or state of affairs (Deleuze 1990, 22). Sense is a system turned in one direction to appearance, and in the other to the pure activity of spiritual life—in one direction, that which can be known, that which receives the action of the Idea, and in the other, that which outruns it in its ceaseless spiritual activity. In the cinema books, Deleuze refers us to the process of crystallization and to the crystal image. This is not a function of the mind, but an objective function of some images. A perfect crystal image arises when two mirrors face off—a *mise en abîme*—wherein the virtual image becomes real and the actual becomes virtual. Deleuze refers to the crystal image as the opsign. In film it affects a character who is subjected so strongly to something visual that they do not act, and so perception, action, and thought are broken apart, disrupting habitual repetitions in the same way as certain mosaics that form regular but non-repeating patterns. These formations applied to the brain are subtle and subtly creative.[10]

Deleuze might have derived this concept of the crystal image from a 1969 film, *The Structure of Crystals*, by the Polish filmmaker Krystzof Zanuzzi, in which two scientists, who are friends, reconnect after a separation of several years. One is a noted crystallographer and the other has left the university to work as a rather humble weatherman and to live in the country with his wife, who is a schoolteacher (Zelman 2014). In a crystal, atoms are ordered in repeating patterns, but depending on the chemical composition, they have different symmetries. The life of the country scientist Jan, and his wife Anna, appears at first to be ordered and repetitive, lacking any element of innovation. Yet it is clear that both of them enjoy this molecular way of living, as Jan has taken up this life after a serious climbing accident that left him impaired for six months. On the molar scale, Jan and Anna's life might appear to be repetitive and boring, yet a closer look reveals that they share scientific and intellectual curiosity, intimacy, and a generally happy life.

Interestingly, this scenario seems to be close to the recent scientific discovery of so-called time crystals, quantum systems in which two connected systems are "stuck" in an endless loop, and yet their symmetry is broken. The same sequence does not produce the same result. Time crystals succeed because they are in the lowest energy state and continue in their loop until they interact with the environment (Sutter 2022). If this describes the life of Jan and Anna, it seems that their creative life is not extensively high-speed and erratic, or perhaps not at all. Nor is it the effect of a deranged state of mind. After all, the supermultitudinous continuum is so great that, for Peirce, it is all that is possible, not only in the brain but in the cosmos. Like an artist, poet, scientist, or philosopher, Jan and Anna discern behind the molar scale a molecular immaterial universe. They live in a manner where their actions are not so much chosen as they are a response to life lived as pure sensibility, as the unthought that Deleuze claims forces us to think (Deleuze and Guattari 1994, 75). For Deleuze, this life is a work of art. It lacks all the tricks of the filmmaker, but it is intensely creative.

In a previous work, I argued for the slow-down against what I understood to be Deleuze's call for infinite speeds, the condition of escape from habitual perceptual norms (Olkowski 2007). Yet here it is. Passivity, diffusion, and the slow-down may keep one from being absorbed into the high-speed molar state, thereby awakening a perspective, a point of view shared by nothing and no one. As I observed there,

> [i]f ... you did not instantaneously perceive, conceive and act on what interests you, your conventional responses, your responsibilities, or your important works—or, on a less coarse level ... if you, meaning what is provisionally "you," were not simply enveloped by the myriad forces competing to compose you ... you may have entertained an interval in which to contemplate and to pose a question from out of your own duration. (2007, 33)

This may be the situation of Jan and Anna. A crystallization, a life that is a work of art.

Notes

1. *Cambridge Advanced Learner's Dictionary & Thesaurus*, https://dictionary.cambridge.org/us/dictionary/english/invention (accessed September 8, 2025).
2. Agassi cites examples such as the prediction of Karl Marx, who used the model of the steam engine to predict the increased wealth of the capitalists, when in fact it contributed to the growth and prosperity of the middle classes (Agassi 2022, 253, n. 6).
3. See Poincaré (1890). For a discussion of Poincaré's role in the discovery of chaos theory as well as the confusion around this, see Parker (1998).
4. Originally published as Boekholt et al. (2020).

5. An attractor is defined in mathematics as a set of states, invariant under the dynamics, which neighboring states in a given basin or arena approach asymptotically in the course of dynamic evolution; see http://mathworld.wolfram.com/Attractor.html (accessed September 8, 2025).
6. The first attractor is called a fixed point, the second a limit cycle; it describes both clocks and human heartbeats.
7. It has been suggested that the representation of chaos as fractal geometry fails to account for it as a temporal phenomenon, that is, as repetition over time and undergoing qualitative changes (see Celletti 2008).
8. Deleuze also describes this in his book on Francis Bacon.
9. Vellodi (2014) states that for Peirce an end will be reached, but I leave the system open.
10. I have examined this in detail in Olkowski (2019).

References

Agassi, Joseph. 2022. "Versions of Determinism." *Metascience* 2: 250–60.

Alford, Joseph. 2015. "Our Brain Sees Known Words As Pictures." *IFL Science*. https://www.iflscience.com/our-brain-sees-known-words-pictures-27695. Accessed September 8, 2025.

Aristotle. 1970. *Metaphysics*. Translated by William D. Ross. In *The Basic Works of Aristotle*, edited by Richard McKeon. New York: Random House.

Bergson, Henri. 1920. *Mind-Energy: Lectures and Essays*. Translated by Herbert W. Carr. London: Greenwood Press.

Bergson, Henri. 1963. *Œuvres*. Paris: Presses Universitaires de France.

Bergson, Henri. 1988. *Matter and Memory*. Translated by Nancy M. Paul and Scott W. Palmer. New York: Zone Books.

Boekholt, Tjarda, Simon P. Zwart, and Mauri Valtonen. 2020. "Gargantuan Chaotic Gravitational Three-Body Systems and their Irreversibility to the Planck Length." *Monthly Notices of the Royal Astronomical Society* 493, no. 3: 3932–7. https://doi.org/10.1093/mnras/staa452.

Casti, John. 1994. *Complexification: Explaining a Paradoxical World through the Science of Surprise*. New York: Harper Collins.

Celletti, Marta. 2008. "The Strange Attractor We Call Time: An Exploration of the Concept of Time Through the Thought of Gilles Deleuze and Complexity Theory." *Limina. A Journal of Historical and Cultural Studies* 14: 19–26.

Chenciner, Alain. 2012. "Poincaré and the Three-Body Problem." *Poincaré, 1912–2012, Séminaire Poincaré XVI*: 45–133, http://www.bourbaphy.fr/chenciner.pdf. Accessed September 8, 2025.

Deleuze, Gilles. 1989. *Cinema 2: The Time-Image*. Translated by Hugh Tomlinson and Robert Galeta. Minneapolis: University of Minnesota Press.

Deleuze, Gilles. 1990. *The Logic of Sense*. Translated by Mark Lester and Charles Stivale. New York: Columbia University Press.

Deleuze, Gilles. 1994. *Difference and Repetition*. Translated by Paul Patton. New York: Columbia University Press.

Deleuze, Gilles, and Félix Guattari. 1994. *What is Philosophy?*. Translated by Hugh Tomlinson and Graham Burchell. New York: Columbia University Press.

Ginat, Yonadav B., and Hagai B. Perets. 2021. "Analytical, Statistical Approximate Solution of Dissipative and Nondissipative Binary-Single Stellar Encounters." *Physical Review X*: 11. https://doi.org/10.1103/PhysRevX.11.031020.

Hertog, Thomas. 2023. *On the Origin of Time, Steven Hawking's Final Theory*. New York: Bantam Books.

Juhan, Frank. 2006. "The Three-Body Problem, Special Lecture." https://www.phys.lsu.edu/faculty/gonzalez/Teaching/Phys7221/ThreeBodyProblem.pdf. Accessed September 8, 2025.

Jung, Rex E., Judith M. Segall, Jeremy H. Bockholt, Ranee A. Flores, Shirley M. Smith, Robert S. Chavez, and Richard J. Haier. 2010. "Neuroanatomy of Creativity." *Human Brain Mapping* 31: 398–409.

Khalil, Radwa, Ben Godde, and Ahmed A. Karim. 2019. "The Link Between Creativity, Cognition, and Creative Drives and Underlying Neural Mechanisms." *Frontiers. Neural Circuits* 13. https://doi.org/10.3389/fncir.2019.00018.

Marquis de Laplace, Pierre-Simon. 1902. *A Philosophical Essay on Probabilities*. Translated by Frederick W. Truscott and Frederick L. Emory. London: Chapman and Hall.

Montgomery, Richard. 2019. "The Three-Body Problem." *Scientific American Magazine* 321, no. 2: 66–73.

Olkowski, Dorothea. 1999. *Gilles Deleuze and the Ruin of Representation*. Berkeley: University of California Press.

Olkowski, Dorothea. 2007. *The Universal (In the Realm of the Sensible)*. New York: Columbia University Press.

Olkowski, Dorothea. 2019. "Try Madness, Creation and the Crystalline Brain." In *Deleuze and Guattari's Philosophy of Freedom: Freedom's Refrains*, edited by Dorothea Olkowski and Eftichis Pirovolakis, 27–43. Abingdon: Routledge.

Olkowski, Dorothea. 2021. *Deleuze, Bergson, Merleau-Ponty: The Logic and Pragmatics of Creation, Affective Life, and Perception*. Bloomington: Indiana University Press.

Parker, Matthew W. 1998. "Did Poincaré Really Discover Chaos?." *Studies in the History of. Philosophy, Modern Physics* 29, no. 4: 575–88.

Peirce, Charles S. 1931–35. *The Collected Papers of Charles Sanders Peirce*. Vols. I–II. Edited by Charles Hartshorne and Paul Weiss. Cambridge, MA: Harvard University Press.

Poincaré, Henri. 1890. "Sur le problème des trois corps et les équations de la dynamique." *Acta Mathematica* 13: 1–270.

Smolin, Lee. 2017. *Three Roads to Quantum Gravity*. New York: Basic Books.

Starr, Michelle. 2020. "Just Three Orbiting Black Holes Can Break Time-Reversal Symmetry, Physicists Find." *Physics*. https://www.sciencealert.com/three-black-holes-orbiting-each-other-can-t-always-go-backwards-in-time. Accessed September 8, 2025.

Sutter, Paul M. 2022. "Physicists Link Two Time Crystals in Seemingly Impossible Experiment." *Live Science*. https://www.livescience.com/time-crystals-linked. Accessed September 8, 2025.

Vellodi, Kamini. 2014. "Diagrammatic Thought: Two Forms of Contructivism in C.S. Peirce and Gilles Deleuze." *Parrhesia* 19: 79–95.

Zelman, Julia. 2014. "Friendship as Fragile as Ice, Krzysztof Zanussi's *The Structure of Crystals* (*Struktura Kryształu*, 1969)." *East European Film Bulletin* 42. https://eefb.org/retrospectives/krzysztof-zanussis-the-structure-of-crystals-struktura-krysztalu-1969/. Accessed September 8, 2025.

Chapter 2

Deleuze, an Unwitting Milesian?

Arnaud Villani

The great mistake made about the early Greek philosophers was to regard them as faint-hearted physicists and fledgling scientists. To put an end to this misunderstanding, reiterated for the sole purpose of asserting the superiority of modern science[1]—just as the great mathematician Condorcet could say, with his enormous blunder, that the art of Egyptian busts and Greek *korai* were necessarily outdated given the "progress of the human spirit"[2] in Europe—it must be said not only that some of their intuitions can be confirmed today, but also that they present, 2,600 years earlier, certain aspects that could anticipate our science. I will be working on the conjunction between the thought of chaos in the Milesians, Thales, Anaximander, Anaximenes, and the Sophists, and the complex concepts of chaos and chaosmos in Deleuze.

Deleuze, an Absolutely Not Self-aware Milesian?

What strikes me, as I pointed out during the Deleuze Decade at Cerisy in July 2015 (Villani 2019a), is Deleuze's closeness to Milesian chaos, and yet the disappointment he seems to express about their philosophy as a whole. Deleuze was not a Hellenist, and neither his conceptual characters nor his paradigmatic supports belong to this field. He shared this mistrust of the Greeks with Michel Foucault and contrary to Nietzsche. This view also belongs to a certain Parisianism which believes that Greek and Roman etymology have nothing to teach us, even though it is, as I will show with examples later, the archaeological trace of ancient and almost lost thought.

Now, the publication of Nietzsche's philological works, delayed by the pervasiveness and power of his thought, reveals him to have been a brilliant philologist (as much as he had been a daring composer before he was 20), a Schopenhauerian who really did philology. Deleuze was not a philologist, and all he drew from his brief studies of Greek was an erasable memory. Is that enough for him to use the metaphor of the prehistoric cave from which one emerges haggard and smoky? First, in those days, caves were not places to live but places of ritual and worship. What's more, this metaphor is not in Deleuze's "style." And yet, as we shall see, Deleuze uses it regularly for three reasons: the first, as we have just seen, is that he is unaccustomed to this language and its surprisingly rich semantics; the second, and rightly so, is that he cannot stand the Heideggerian injunction to "return to the Greeks,"[3] to take the "step back" (*Schritt zurück*) toward the "Dawn of thought." And it is true that this refrain, presented as simple and in fact very pretentious, which seduced French intellectuals for a long time, was nothing but "tinfoil." The third would be a distortion of Nietzsche's formula: "The Greeks, superficial by depth," taken literally and modulated according to an unfavorable version. The full quotation from Nietzsche restores the exact meaning of "surface" and "depth":

> The Greeks did know how to live: which requires a courageous way of stopping at the surface, the fold, the epidermis, the adoration of appearance, the belief in forms, sounds, words, the whole Olympus of appearance. The Greeks were superficial—for depth's sake. (Nietzsche 1967, foreword, section IV)

Neither the term "superficial" nor the term "depth" here have the meaning that Deleuze seems to attribute to them. "Superficial" here means "in love with the surface," reversing the false metaphysical depth into an immanent surface and plane of consistency. It is this reversal that Deleuze advocates in *Difference and Repetition* (1994) and, with reference to Lewis Carroll, in *The Logic of Sense* (1990). As for the term "depth," it has no pejorative connotation in Nietzsche's formula. It indicates an ultimate subtlety of thought that goes further. In short, in this text, neither depth nor surface can be translated by the image of the first humans, drawn from deep caves, nor by the idea that they would remain at the level of superficial thought without any real content. And yet this disparaging comment is often repeated. In *Logique du sens*, Deleuze dismisses both anti-Platonists and Platonists:

> Presocratic philosophy does not come out of the cave; it considers that we are not sufficiently engulfed in it. The Presocratics installed thought in caves, life in the depths of the earth [...] Presocratism is the philosophical schizophrenia of absolute depth, the rediscovery of depth [...] Philosophy is neither the being of the caves nor Plato's bird, but the animal of surfaces. (Deleuze 1969, 153–8, my translation)

In one of the texts in *Deux régimes de fous*, again: "pre-Socraticism is a return to the subterranean" (Deleuze 2003a, 59). What is strange, and even incomprehensible, is that the thought of Anaximander and Anaximenes on the *apeiron*, the paradoxical thought of Parmenides on a being *ateleston* ("without end or goal, infinite") and *tetelesmenon* ("well finished"), Antiphon's even more elaborate thought on *arrhythmiston*, coincide on several points with Deleuzian thought on chaos and, beyond that, with the recent search for a scientific theory of chaos.[4]

Versions of the Indeterminate

The great specialist in Hellenic science, Paul Tannery, whose work has just been republished (Tannery 2022, see particularly 118–24), understands Anaximander's *apeiron* as the opposite of *peras*, meaning "limit"; it would therefore be above all "unlimited," with the possibility that this absence of *peras*, of the "formal envelope of bodies," is opposed to a Greek gaze that sculpts beings and things and brings them to light. Hölderlin had assumed, in the cross-fertilization of the Hellenes and the Hesperians (Westerners), a "natal reversal" in which the Greeks would have moved from their aorgic (inorganic) origins toward great organic subtlety, but would have compensated for this movement by a frantic "upstream return" toward the inorganic: Empedocles jumping into the volcano, leaving only his sandals as evidence (Villani 1986).

For Tannery, the semantics of this term are therefore centred on "indeterminacy" (2022, 122). The problem that is generally posed, namely whether the absence of formal differentiation refers to a mechanistic or dynamistic mixture, is resolved as follows by Tannery: If the Milesians assume an indeterminate primitive element, even if appearances plead for a mechanistic point of view, in their works (all entitled *Peri physeôs*) things and beings proceed from a dynamism,[5] namely "the *physis*":[6] "the power to bring into being, to launch into existence, to bring to light," a power that Plotinus (1968) will call "always young, indomitable." Spinoza (2005) aptly named it "nature naturante," and Johann F. Blumenbach (1792) defined it as *Bildungstrieb*, "aspiration to form," "informing impulse." For Tannery, these thinkers of origin are dynamists expressing themselves mechanistically. What I notice, however, is that Tannery cites Antiphon the Sophist only three times, and each time only in two words, whereas in recent writings, notably those of Gilbert Romeyer-Dherbey (1985), he occupies twenty pages for his theses, the squaring of the circle, his theory of *alupia* ("pain stop"), and his hypothesis on the indeterminate, which led him to change the term *apeiron* to *arrhythmiston*.[7] Is this just a lexical variation?

It should not be forgotten that at that time, opposites, while strongly opposing each other, were supported and held together by the energizing effect of their very disagreement. This contradicts modern thinking, which believes that harmony means overcoming all opposition. No, the opposite is true. Harmony (radical +*ar*– or +*har*– from carpentry) means the forceful interlocking of opposites. And it is the persistence of fundamental oppositions that ultimately produces an effect of calm, of serenity. At that time, fluidity guaranteed solidity. Because of the firm link between opposites, nothing is more stable than that which goes through a period of floating, nothing is more luminously laughing than the waves of the sea calmed after a storm, producing what the Greeks call *galênê*.[8] Harmony comes from the struggle of two opposites: *rhythmos* is Heraclitus' river, the same and always different, regulated fluctuation, stable movement, Cézanne's "modulation," the clash of *instress* (the slap of the sensible) and *inscape* (the intelligible structure) in the poet Hopkins, Calder's mobile changing form but retaining its unitive hold (*systema*).

I absolutely had to recall this trace in the Greek tragedians and in their heirs, the German Romantics, of this two-in-one thinking that sees its being only in the crackling conjunction of opposites, regenerating itself from their invisible and tenacious struggle. What, then, is to be understood in Greek by *chaos*, which certainly concerns the Milesians, but also Socrates (the obstetrical void and the *mêden eidôs*) and Gorgias, who claims that "there is nothing. If there were something, we couldn't know it, and if we could, we couldn't say it" (Pseudo-Aristotle 1955, 462–3, translation revised by me). First, it represents the semantic family of the gap, radical +*cha*–, "to gape, yawn, open the mouth," *hiare* in Latin, *inhians*, hence the "hiatus," originally abyssal. Van Helmont (1670), referring to amorphous matter, likens chaos to a "*gas*" (same word), said to be "perfect," in which no "preferential direction" is foreseen or assignable, or into which any of the watermarks that, as "preferential directions," prefigure organized matter have yet to creep. In short, where everything would be isotropic,[9] polytelic,[10] exploded in all directions.[11]

Thus *arrhythmiston*, without losing its constitutive indeterminacy, is fully positive in Antiphon, and it persists in its ambiguity or equivocation: it is at once and by nature, "fluid and formal, liquid and solid," and so on. If rhythm[12] is a notion that is both stable and labile, if it is made up of opposites (supple and rigid, fast and slow, laminar and swirling),[13] then this change of term, far from being a coquetry of language lovers, is a brilliant *invention of concepts* (Deleuze). We see the birth of a term that confounds thought, since it is equivalent to its opposite, and shows that form and non-form are opposed in it only by an "incision" present in each of them like a "fifth column." What attracts them reinforces their repulsion and increases their resistance to cohesion. This hesitation of form, through its active virtual points and the encounters between these points, certainly ends up producing a form, but leaves it with a taste of chaos, a hesitant, ambivalent, *non-finito* trace. This ability to say one thing and its opposite together is sophistry itself, too often confused with a procedural technique, when in fact it is the expression of true thought. Such is the gigantic advance of the Greeks on the *apeiron*, made possible by the existential sophist Antiphon. Just because Plato deliberately wanted to "kill" the Sophists (in other words annihilate their thought, which threatened the status quo of his own) does not mean that we should not rehabilitate them today. They have a lot to teach us about politics, society, discourse, and philosophy.

This hold of non-form over form, this driftwood that gets stuck at the foot of forms and prevents them from becoming substantial, this eraser that rises up to erase the form advocated the moment before,[14] makes the formal, the "substantial" as Plato and Aristotle would say, a random structure, a "stochastic cloud," where every position is statistical, and every clarity and distinction, every rectitude, is exchanged for an inaccurate blur. Antiphon makes us understand that the opposite of a position by oppositions is a new position by oppositions. And he makes us understand the weight in politics, society, and thought of the reciprocity of ancient peoples. When adversity is the law, reciprocal reversal is part of its power. Parmenides knew that the "hyper-metaphysical Being" of the future circus of Platonic transcendence was in fact the mystery of a unity without consensus, achieved by the powerful resistance of the opposites that define its reality. The indefinite dyad posits the one and the other instead of decimating the one by the other. I have given this process the name of "multiplunity," a combination capable of deflating the Platonic cathedral of aggressive Manichaeism that guards against opposites.

Here, then, is the irreversible[15] machine that has reversion as the very center of its movement, its relaunch.[16] However many hesitant choices the painter makes in the creative tunnel of the unconscious, where he advances by scanning, he will arrive, if he is "fluidly rigorous," at the finished work, without having consciously followed a *telos* of this work.[17] Ehrenzweig's proposal, which introduces a degree of vagueness/unconsciousness into our understanding of creation, is akin to Da Vinci's "serpentine line" and to the trial and error of painters and poets until they discover a form that hints at the persistence of a non-form. A thinker/painter like Paul Klee went furthest down this road when he evoked the "grey dot that leaps over itself" in *On Modern Art* (1945). Others have said the same about painting's interplay between visibility and invisibility. And it's possible that Balzac, in *The Unknown Masterpiece* (1831), already had this in mind. Propositions such as Nietzsche's, that you must "have chaos within you to give birth to a dancing star" (1971, 29, my translation), sum up the whole debate and become so intelligible that they shine through us with an overwhelming light. Antiphonian chaos is an abundance of opposites, like that indefinite "*ar* without death" that Heraclitus called *polemos*, and in which philological science today finds a relationship to *pallô* (radical +*pel*–, "to shake"), and to *palmos*, Verlaine's swinging palm, opening the sky from a prison (*Le ciel est, par-dessus le toit...*).

Visible forms are divided into gaseous, fluid, and solid states, with a whole series of intermediaries well known to Bachelard and his material imagination. The pre-Socratics all expressed this reciprocal exchange (*antamoibê*) between earth, air, fire, and water in the form of *tropai*, or "turns."[18] The link between the Milesians and the later pre-Socratics is metamorphosis, which, consistent with their mythological and social conceptions,[19] forms the basis of Greek thinking on the material/formal "given." These sudden versions of reality involve intermediaries between states (not yet states of being)[20] and within these states themselves (water turning to ice, earth becoming mud, cloud becoming rain). Greek chaos comprises all these forms and their states, as well as the intermediate stages from state to state and from form to form. This is how the meta- or para-morphic meaning runs beneath the substantial forms, which sometimes reveal their invisible underpinnings.

Chaos is no longer a mouth of darkness, as Hugo saw it, but rather a luminous, solar source, the orifice of all birth, a fountain in the village square, the origin, through its irrigation capacity, of the choice of its location. Chaos is the swallowing up of the old and the launching of the new which is waiting to happen. It is positive and negative, like the real unity of a ball of power born of these immediate, non-dialectical oppositions. The polymorphism of chaos includes a polyscopicity, a capacity to give itself this goal rather than that other, as demonstrated in the Egyptian or Greek language by the long persistence of antonymic terms: *deinon*, "marvellous/terrible," Latin *altus*, "high/deep," the verb *echein*, "to contain the order of the cosmos/to allow oneself to possess venally," the radical +*dap*–, "the joy of the banquet in equal parts/the regret of a sumptuary expenditure," the meaning of *kratos*, "endowed by benevolent mothering with a power that increases/exercising by force a power that diminishes," on whose meaning Karl Abraham and Freud visibly stumbled.

The In-between

If Deleuze makes the "in-between," as a "rule-fleeing," the foundation of his philosophy, it is because he is constantly confronted with chaos and chaoids, his true characters. Let us compare his type of chaos with that of the pre-Socratics. Refusing to take the concepts that articulate his philosophy as "concepts as big as hollow teeth" and merely "baptizing" the difficulties, Deleuze often uses the formula "the bottom rises up" (*le fond remonte*), for instance in various places in *Difference and Repetition* (Deleuze 1994). And he evokes the *fascum subnigrum* which, as a dark precursor, heralds a cosmic phenomenon. But if the bottom is rising, then we are at the right level to talk about it with full knowledge of the facts. The process of chaos is not only the tension between the invisible and the visible/formal,[21] but also the way in which these two domains penetrate each other, are tempered and rise to power, starting from a suspension of form in the formless, of the visible in the invisible.

Here Deleuze makes the leap to a major concept. For the virtual, which bears a strong resemblance to the invisible of traditional societies and which arises from the strong refutation of the category of the "possible" in Bergson's work, must absolutely be said to be as "real" as the actual or actualized. The possible is not yet real, and to pretend to think so is a deception. But both the actual and the virtual are real, the former formal and visible but slowed down and almost frozen, the latter informal and invisible, ultra-fast and elusive. The virtual is real, but we do not see it: it exists, it swarms between forms, but not as a recognizable, categorizable form. It is not by deifying the Forms, as Plato does, that we arrive at their essence, but by surrounding them with their "trains of virtuality," which accompany them as much as the traditional train accompanies the bride. Whereas common opinion sees beings only in forms that have reached their actualization, the virtual is the pulverization of fragments of reality, not aggregated into forms offered to the senses but constituting the median tissue, the connective cartilage, the in-between of these forms.

It has been the great misfortune of Western philosophy to add to this by attempting at all costs to substantialize these forms. The result is the world of stasis, of "frequent stops,"[22] a world in slow motion on the verge of coming to a halt, as Deleuze describes it. In this way, the time and space that accompany an actualized entity are suspended in an eternal present, an instant that will never again be an instant. This "present" time, which Deleuze eschews, must be overcome by a new time that bridges the past and the future and which Deleuze will call *aiôn*, capable, through the invisible entities of the virtual, of a perpetual movement that blends with all times and places in what Deleuze calls the "absolute flyover" (*survol absolu*) (Deleuze and Guattari 1991, 26). It is in this absolute, transhistorical overview, divorced from a chronic time of things eaten away by the past and thrown back into that past by an unknown future, that passages and exchanges take place. It is this Aionic time that virtualizes/actualizes interstitial fragments, indifferent to time and its prohibition of return, and never ceases to "move" everything. It is in this non-space and this non-time, billions of *stoicheia* ("little bits"), remarkable and brilliant singularities of a reality moving by the force transmitted to them by chaos which distributes them on flows ready for encounters with other singularities on other flows, where innovations slip in.

This is where the transcendental but immanent empiricism that leads Deleuze to a semantics comes into play. Following the model proposed by Gilbert Simondon, in which two distinct singularities give rise to an innovative third party that combines their contribution (binocular vision giving rise to the innovation of relief vision),[23] Deleuze evokes a *dispars*, which for Simondon is a *disparation*, or an operation of the difference of singularities and their propensity to meet and create the new. The result is the characteristically Deleuzian description of chaos. To shed light on the tightly woven formulas that he and Guattari forge, it is important to list a few key definitions, like so many inventions of concepts whose use must be grasped, since Deleuze's entire philosophy demands that everything take place not between concepts dripping with idealities but between acts, facts, gestures, situations, ambiences, and attitudes:

1) Folding. This "folding" is fundamental to two books that encircle the work, *Spinoza et le problème de l'expression* and *Le pli: Leibniz et le baroque*. As *Difference and Repetition* summarizes:

> We call complication the state of chaos retaining all the intensive series that affirm the ideal series. We call implication the state of intensive series that communicate through their differences. We call explication the state of the qualities and extents that cover the system. (Deleuze 1968, 359, my translation)

In other words, we ascend from the perfectly virtual and informal to what extends formally and is measured temporally, proposing sensible *qualia*, by passing through an intermediate, semi-chaotic state, where the flows of intensive singularities intersect. It is this intermediate plane that Deleuze will devote to speed, since it carries within it one or more of the characteristics of chaos and bears the mark of the chaotic.

2) The notion of "absolute speed." Deleuze has made it clear in the way he describes the "baroque" that his theory of the "line that continues ad infinitum" implies that in an instant, two flows, while remaining distinct, can merge together and form a single block. The thunderbolt of their meeting then produces the release of an absolutely new third which operates a vast two-way ascent in the flows. Nowhere is the polyscopy we detected in the pre-Socratics clearer in Deleuze than in this theory of flows. As soon as one of their singularities has noticed and accentuated another, which, provoking their encounter, opens the possibility for each of these well-singularized flows, and is suddenly placed in the spotlight of traversing the other flow and being traversed by it, this encounter triggers a formidable round trip at infinite speed. Such absolute speed, which is also that of the neuronal brain to which Deleuze devoted a few pages before his death, sums up the essence of what he retains from pre-Socratic chaos. It allows him to bridge the past and the future, to travel through history in instantaneous intuitions, and to create "crystals of time" that he calls, after Proust, "a little time in its purest state" (*un peu de temps à l'état pur*) (Deleuze 1985, 27, my translation). It enables him to describe the perceived world as an indefinite process of slowing down, as if an exhaustion were running through creation and forcing the dull regularities and repetitions to repress the minute differences and unpredictability of creative movements; everything gradually becoming chitinized, as women in Delvaux's work. It is the task of chaos to deterritorialize: it functions as an eraser, not of abolition, but of regeneration, threaded randomly between entropy and negentropy.

3) If we want to go from this principle to the realization of the "plane of immanence," also called the "metaphysical surface," and then respond to Deleuze's announcement: "I feel like a pure metaphysician" (Deleuze in Villani 1999, 130, my translation), we need to engage in a strategic process of "cutting through chaos" (Deleuze and Guattari 1991, 147), where virtuality goes from chaotic to consistent.[24] If, for example, a form of chaos can represent a primary layer of the unconscious, we can say that the plane of consistency extends and is articulated between the primary layer and the secondary layer.[25] This implies respecting the unconscious and not taking it for the sympathetic guest of

psychoanalysts who have bought a sofa from Darty. This then implies asserting the value of difference as a whole and "as such," that is, free from any subjection to subjectivity, emotionalism, or power of any kind.

4) What are these interstrata woven from? Free differences. If the micro characterizes Guattari's work better, as Deleuze told me in a personal letter, we already have, from *Difference and Repetition* onward and with these free differences, a *whole micrology*. The plane of consistency takes on substance, arranged as it is with divergent series, differential filigrees, flows of singularities like so many "salient points," or as Simondon put it, so many "strong points" (*points forts*) that demand to be noticed and matched (2012, 293–326). This is not chaos as such, but what can be positively extracted from chaos, preserving isotropy while bordering on nonsense, like Artaud getting too close to his *caldera*. It is at the end of *The Logic of Sense* that Deleuze abandons the "honest surveyor of surfaces" (Lewis Carroll) and assumes Artaud's happy suffering: "For all Carroll, I wouldn't give a page of Artaud" (Deleuze 1969, 114, my translation).

5) This reversal in his reading passions shows that the activity of the intermediate plane is based on the speed and freedom of singularities and the flows that carry them. This speed comes from the informal hole or abolition, into which, moreover, the "drugged wreck" is always in danger of falling (we can here guess the preoccupation of a Guattari in his various positions in psychiatry or psychotherapy). The philosopher becomes the alchemist who passively synthesizes this quality of absolute speed. In this way, he liberates the flows, that is, concentrates their power on calling singularities together, noticing them, receiving their signs, taking them as one, turning them around in all the flows concerned, triggering a "semiotic" circulation.

This transition to semiotics is exactly when the problem no longer concerns language or concepts, but facts, bodies, and their combinations. The real, we understand, is not at all in visible things and beings, but in these invisible differences, in and of themselves, which stand up on their own, and nourish what we call the world by giving rise to the tiny innovation that changes everything (a simple idea, which visits you at night, and redirects your research, or the conjunctive synthesis, which benefits from the process, immanent synthesis in "So it was you, the love of my life!," or transcendent in "So it was you, the traitor, the murderer!" (Deleuze 1969)).[26] But in this semiotics of singularities, their calls and their encounters, there are also words, concepts, and ideas. Are we then returning to a semantics to which we would be condemned?

No. This semiotics/semantics has nothing to do with that of scientific concepts or intelligible ideality. It also does not involve feelings and sufferings. Once again, it stands in between. It extracts from the sphere of affects and percepts blocks of sensations, sufferings, and passions that are independent of the person they attach to. It is this nonindividualized status that sometimes turns these sufferings and passions into a thing-in-itself, becoming a character, a "subject" detached from any subject, a "process of subjectivation" that can affect a tree, a flower, a poem, a word—it is this that allows us to say that the "concept" can "scream."[27] The *Logic of Sense* (1990) becoming a *Logic of Sensation* (2003) is the rupture that opens philosophy to a creative-active and tragic use of chaos. This discovery of a philosophical *interstitium*,[28] the discovery of an interstratum of interformal, a hippodrome of flows (the very Lawrence-like "great cosmic winds" evoked by Deleuze in this book), celebrating both da Vinci and Ravaisson's "serpentine line," Bergson's "intervals," Nietzsche's "grand style," and explaining the possibility of a "transcendental empiricism"—makes Deleuze the immense philosopher of the second half of the twentieth century.

A whole series of formulas become illuminated. We understand better what it means to be grafted by rhizomes. We hear these phrases: "these were extraordinary moments, those moments when philosophy made the bottomless speak" (Deleuze 1969, 130, my translation); "the divergence of affirmed series forms a chaosmos and no longer a world" (Deleuze 1969, 205, my translation); "[one must see] chaos as the power to affirm heterogeneous series" (Deleuze 1969: 301, my translation); "the problem of philosophy is to acquire consistency without losing the infinite" (Deleuze and Guattari 1991, 45, my translation); "the philosophical sieve as a plane of immanence selects the infinite movements of thought" (Deleuze and Guattari 1991, 112, my translation); "the bottomless is without differences even though it swarms with them, it is the limit-illusion" (Deleuze 1968, 355, my translation). We follow him when he reveals "the state of free differences when they are no longer subject to form" (Deleuze 1968, 148, my translation); when he discovers Pascal's "prophecy: to make chaos an object of affirmation" (Deleuze 1968, 313, my translation). We understand the larval swarming of small free differences (Deleuze 1968, 333; see also Deleuze and Guattari 1991, 155), the idea of a "neutral and displaceable energy" (Deleuze 1968, 149, my translation), the "abstract line coupled with chaos and producing the desert" (Deleuze 1993, 37, my translation; see also Deleuze 1968, 353).

To sum up, if we follow Deleuze's intuition, the reddened eyes and smoky faces[29] of the pre-Socratics are not a negative sign after all. They saw something that brought them closer to us. Deleuze, ill at ease in the Greek domain, also saw something of them, but he did not describe it well. For, in the immediate exchanges and metamorphoses that filled pre-Socratic studies and that Plato would so strongly condemn,[30] what they saw was structural freedom, the timelessness of qualitative multiplicities, the speed of liberation of a semiotics in formation, the existential content of an Antiphon that bridges the gap between the Ancients and Deleuze.

I'll use the following image. Everything happens as if there were a plane between pure chaos and stabilized forms. This plane has one side turned toward chaos, and the other toward the objects of nature and to *skeuaston holon genos*, "the whole genre of objects prepared by human technique." These objects could only come into being (and the same is true of living beings) through a completely random encounter between a system of flows with their own singularities (a man, a woman, two types of smile, two stores of secrets, but also two openings onto the world, two experiences, etc.) and another same flow. If we go down to the micro level, this encounter between two living, conscious beings is transformed into an encounter between a multitude of flows (the "hair" flow, with its general appearance and its micro details, the "body" flow, the "intelligence" flow, the "sensitivity" flow). In each flow, there are so many unique elements waiting for their moment to make an impression on other details of other flows, waiting for the moment to "speak," to engage in "encounter and conversation," until they produce the metamorphosis effect. This intermingling of flows and their singularities takes place at "absolute speed"; that's it for life. Something is born that, like the meeting between Deleuze and Foucault, Deleuze and Guattari, is a sharing of moods. Orson Welles said it best through the word "Rosebud" (in his movie *Citizen Kane*), which burns on the discarded sleigh. A discarded sleigh which represents the end of a life, the end of the creative separation between a child and this badge. Similarly, in Kafka, the gold teeth and noses of the participants at a boring party (Villani 2019b). You hold on to them, he says, to avoid falling. And then there's the grip of a novel hero (*Berlin Alexanderplatz*) and a plant or root, that jazz tune, poorly defined by Sartre at the end of *La Nausée*, or the white of the white whale, Moby Dick, in Melville's novel of the same name. It takes place between chaos and the chaotic singularities that flow at dizzying speed. That is the essence of all true poetry. And I'm back to where I started: is that why I got "hooked" on Deleuze's philosophy? Without my being able to analyze it for long, he would have given me the only profound definition of the poetic as an infinitesimal meeting of grains as vigorous as the "chaos" from which they come and which they commemorate.

Finally, between Deleuze and modern theories of chaos I would insist on the "sensitivity to initial conditions" that refers, even more than to the holism deployed by the pre-Socratics (Hippias, Antiphon) and then by Leibniz and Hölderlin, to a *theory of micro-multiplicity* that is characterized by 1) fragments and pure differences, 2) a resistance to a formalized spatiotemporality, and 3) a sense of indetermination (incidentally, from Anaximander to Antiphon, the notion of indetermination has continued to make progress in the concept)—three characteristics that, for me, define Deleuze's true contribution, and explain the accuracy of his "literary" readings. And so, "chaosmos," already used by Joyce and taken up with gusto by Deleuze and Guattari, this rather barbaric word-valise that I think we could have avoided, is nothing but the plane of consistency where everything happens, which no one sees but which in reality constructs the essence of reality, the rest being a bricolage adapting these strokes of genius of the creative unconscious to the realities of the environment.

Notes

1. And which also falls under the criticism of pre-formation or retrograde illusion that Bergson and Deleuze strongly denounce.
2. In *Esquisse d'un tableau historique des progrès de l'esprit humain*, written in secret during the French Revolution and while under threat of being guillotined.
3. This is the formula that, in France, commonly translates Heidegger's *Schritt zurück* or "step backwards." Heidegger says, in an annoyed tone that he very rarely uses: "There is something worse than objections and refutations, it is the 'return to.'" This is doubtless why, in Deleuze's reaction to my book about him, *La Guêpe et l'orchidée* (Villani 1999), which was still in manuscript and which he read toward the beginning of the 1990s, he wrote that, while scrupulously respecting his thought, I had "mediterranéised" it, without going so far as to "grecise" it.
4. See, for example Deleuze and Guattari (1991, 45), which states that the pre-Socratic Greeks were the first to discover the possibility of cutting through chaos to produce a plane of consistency. In a discussion with Alain Beaulieu, the latter suggested to me: "If Deleuze distanced himself from the depth of the Presocratics, it was in short to take credit for thinking on the surface, without seeing or admitting that they had achieved this before him." I find this remark very pertinent. It means that the first Greek philosophers would have understood the need to preserve the speed of chaos and its indeterminacy, but would not have been able to give their discovery its real importance, for want of conceiving—and this is one of Deleuze's great discoveries—*the idea of a plane of consistency*.
5. From Greek *dynamis*, that is, power.

6. Etymologically speaking, the original root is *+bhew–*, responsible for *physis* in Greek, and *fu-turus* (that which will be born), *fenus* (the interest produced by money) in Latin. The Latin *natura* (Lucretius' title, *De Rerum Natura*, is the strict equivalent of the Greek titles of the pre-Socratics: *Peri phuseôs*) is formed from the future participle of *nascor*, "that which will be born," "the action whose future is birth," whether we see things in terms of their result or their origin.
7. For the quadrature, see Romeyer-Dherbey (1985, 100); for the *echnê alupias*, which gives him so much advance, see Romeyer-Dherbey (1985, 110–15); for the question of the "free structure," see Romeyer-Dherbey (1985, 96–105).
8. *Galênê* is the laughter of the waves. So much so that Greek mythology dedicated to her the myth of Alcyoné, the wife who, after a storm, seeing the corpse of her husband return to the beach in the morning, transforms herself into the bird, the Alcyon, which makes its nest in the bonaces. The eye of the hurricane or the alcyonian, coined by Nietzsche in relation to this myth and characterizing the brightening of the sky over Nice, are images of two opposites inseparably linked by the power of their disagreement, their battle without death, their "discord."
9. *Tropos*: "to turn toward, to orientate according to a tropism." The verb meaning "to turn" in Greek is *trepomai*, where we see the play of the vowel alternation degree e, degree o, degree zero.
10. A multiple-aimed plan (*telos*, the contrary of *skopos*, is the long-term goal).
11. Like the limbs of the body, torn and scattered, in Parmenides, in Empedocles, in the dismembering of Pentheus in Euripides, and of Orpheus in Virgil, *Georgics* IV.
12. As is the case in all music and, more generally, in all art. This is what Plato persisted in saying, in the education of the lower guardians of the Republic, as if to clear himself of having long destroyed ambivalent, uni-dual or symbolic thought.
13. "Laminar" and "swirling" are the terms used to definite light in its paradoxical nature. Michel Serres, in his book *La Naissance de la physique dans le texte de Lucrèce*, sees in the Lucretian *clinamen* a "departure in swirling," which quite well resumes the *physics of fluid* in Archimedes' work.
14. Beckett made astonishing use of it in his "philosophical" novels: *Molloy*, *Murphy*, *Malone Dies*.
15. The ideal of the irreversible model was sketched out by Hölderlin, both in his "native reversal" and in his *Wirkungskreis*, his "circle of effectivity."
16. Hölderlin's rivers, which flow back to their source, are therefore welcome here. This process of inversion manifests itself perfectly when a chaotic disorder at the macrophysical level corresponds to a high level of organization at the microphysical level.
17. This model is the best that contemporary critics have come up with for understanding the genesis of the work of art, the obvious reference being Anton Ehrenzweig's unprecedented book, *The Hidden Order of Art*.
18. *Tropos* comes from the same root as the verb *trepomai*, "to turn in that way or in another way." Hence, the scientific terms isotropic or anisotropic, the first meaning "able to turn in every way," and the second "able to turn only in certain predetermined ways."
19. The late Marcel Hénaff demonstrated the essential importance of reciprocal exchange in traditional societies. See his book *Le Prix de la vérité*. I suppose the pre-Socratics were aware that through their thoughts on opposing principles they were engaged in political and societal work.
20. See the theme of the "unborn" in the poems of Trakl and in Samuel Butler's *Erewhon*, specifically the section titled "The World of the Unborn."

21. See Jean Servier's book *L'Homme et l'invisible* (1964), as well as its second volume, *Les Techniques de l'invisible* (1994).
22. This is the title (in French: *Arrêts fréquents*) of a collection of poems by Michel Deguy.
23. As demonstrated in his work *L'Individuation à la lumière des notions de forme et d'information.*
24. Chaosmos, as coined by Joyce as a portmanteau word, serves as an intermediary between chaos and cosmos. This plane of immanence is also referred to as "redoubling chaos" (Deleuze and Guattari 1991, 112).
25. Deleuze emphasizes the reciprocity between differences and the flows that circulate on the plane, as well as the constitution of the plane by the differences and flows that rely on it. Both operations are like a heads–tails, strictly contemporary. Hence the generally constructivist nature of Deleuze's philosophy. As we can see, this constructiveness does not require a precedence or superiority of the plane over its constitutive flows and differences.
26. Here, Deleuze reaps a rich harvest from Kleist's *Penthesilea* and Büchner's *Woyzeck*. The theory of syntheses, notably the conjunctive immanent one, are extensively studied in my book *La Guêpe et l'orchidée* (Villani 1999).
27. See this challenging passage in Deleuze and Guattari's *Qu'est-ce que la philosophie?* (1991, 21–31), in the chapter titled "Qu'est-ce qu'un concept?," and the notion of conceptual heterogenesis.
28. The interstitium is a medical and anatomical discovery that highlights an organ of organs, which is not the hand as in Aristotle, but maintains the "consistency" of all organs.
29. We can also reconsider that description as a revaluation of night. "The daughters of the night" abound in ancient thought. Even before Nicolas de Cusa and Novalis, Parmenides was the first to revalue the night.
30. Indeed, Plato only banishes poets from the city, after insulting them by awarding them a garland of flowers, because they constantly endow mythical gods with a thousand metamorphoses. It is this constant motion of things and beings that scares Plato. One could argue that he managed to invigorate his representation of immutable Ideas by comparing them to "statues by Daedalus" that move by themselves and need to be bound. However, neither this nor the spermatic aspect he attributes to them later can convince us that Ideas possess the motion of life, unless they belong to a body that guides them, as a dog guides the blind.

References

Balzac, Honoré de. 2001 [1831]. *The Unknown Masterpiece*. Translated by Richard Howard. New York: New York Review Books.
Blumenbach, Johann Friedrich. 1792. *An Essay on Generation* [*Über den Bildungstrieb*]. Translated by Alexander Crichton. London: T. Cadell.
Butler, Samuel. 1970 [1872]. *Erewhon*. New York: Penguin.
Condorcet, Marquis de. 1988 [1795]. *Esquisse d'un tableau historique des progrès de l'esprit humain*. Paris: Flammarion.
Deguy, Michel. 1990. *Arrêts fréquents*. Paris: A. M. Métaillé.
Deleuze, Gilles. 1968. *Différence et répétition*. Paris: Presses Universitaires de France.
Deleuze, Gilles. 1969. *Logique du sens*. Paris: Éditions de Minuit.
Deleuze, Gilles. 1985. *Cinéma 2. L'Image-temps*. Paris: Éditions de Minuit.
Deleuze, Gilles. 1990. *The Logic of Sense*. Translated by Mark Lester and Charles Stivale. New York: Columbia University Press.
Deleuze, Gilles. 1993. *Critique et clinique*. Paris: Éditions de Minuit.

Deleuze, Gilles. 1994. *Difference and Repetition*. Translated by Paul Patton. New York: Columbia University Press.
Deleuze, Gilles. 2003a. *Deux régimes de fous. Textes et entretiens, 1975–1995*. Paris: Éditions de Minuit.
Deleuze, Gilles. 2003b. *Francis Bacon: The Logic of Sensation*. Translated by Daniel W. Smith. Minneapolis: University of Minnesota Press.
Deleuze, Gilles, and Félix Guattari. 1991. *Qu'est-ce que la philosophie?*. Paris: Éditions de Minuit.
Ehrenzweig, Anton. 1971. *The Hidden Order of Art*. Berkeley: University of California Press.
Hénaff, Marcel. 2002. *Le Prix de la vérité*. Paris: Seuil.
Klee, Paul. 1975 [1945]. *On Modern Art*. Translated by Paul Findlay. London: Faber and Faber.
Nietzsche, Friedrich. 1967 [1882]. *Gai savoir*. Translated by Pierre Klossowski. Paris: Gallimard.
Nietzsche, Friedrich. 1971 [1883–85]. *Ainsi parlait Zarathoustra*. Translated by Maurice de Gandillac. Paris: Gallimard.
Plato. 2012. *Republic*. Translated by Christopher Rowe. New York: Penguin.
Plotinus. 1968. *Enneads*. Translated by Anton H. Armstrong. Cambridge, MA: Harvard University Press.
Pseudo-Aristotle. 1955. "De Melisso, Xenophane, Georgia." In *Aristotle. Minor Works*. Translated by Walter S. Hett, 461–507. Cambridge, MA: Harvard University Press.
Romeyer-Dherbey, Gilbert. 1985. *Les Sophistes*. Paris: Presses Universitaires de France.
Sartre, Jean-Paul. 1938. *La Nausée*. Paris: Gallimard.
Serres, Michel. 1977. *La Naissance de la physique dans la tête de Lucrèce*. Paris: Minuit.
Servier, Jean. 1964. *L'Homme et l'invisible*. Paris: Robert Laffont.
Servier, Jean. 1994. *Les Techniques de l'invisible*. Paris: Plon.
Simondon, Gilbert. 2005. *L'Individuation à la lumière des notions de forme et d'information*. Grenoble: Jérôme Millon.
Simondon, Gilbert. 2012 [1958]. *Du mode d'existence des objets techniques*. Paris: Aubier.
Spinoza, Baruch. 2005 [1677]. *Ethics*. Translated by Edwin Curley. New York: Penguin.
Tannery, Paul. 2022. *Pour l'histoire de la science hellène, de Thalès à Empédocle*. Paris: Les Belles Lettres.
Van Helmont, Jean Baptiste. 1670. *Les Œuvres de Jean Baptiste van Helmont traitant des principes de médecine et de physique*. Lyon: Jean Antoine Huguetan.
Villani, Arnaud. 1986. "Figures de la dualité. Hölderlin et la tragédie grecque." In *Cahier de l'Herne: Friedrich Hölderlin*, edited by Jean-François Courtine, 277–96. Paris: L'Herne.
Villani, Arnaud. 1999. *La Guêpe et l'orchidée*. Paris: Belin.
Villani, Arnaud. 2019a. "Le vrai-fuyant, la poésie, les Grecs." In *Agencer les multiplicités avec Deleuze, Actes de la Décade Deleuze à Cerisy, 2015*, edited by Anne Querrien, Anne Sauvagnargues, and Arnaud Villani, 300–10. Paris: Hermann.
Villani, Arnaud. 2019b. *Kafka. L'Homme en chute libre*. Paris: Kimé.

Chapter 3

Cosmic Triptych: A Deleuzo-Guattarian Exploration of the Transistantial, the Fleeting, and the Akoúômorphical

Antoine Renzo

For a long time I dreamed of devising my own movement and rhythm.
—Hubert Aquin, *Next Episode*

If milieus and rhythms fall within the domain of very ancient cosmogonies, it is because, as Pierre Schaeffer implies in his work "In Search of Concrete Music," there would have been a kind of emergence, or rather, an original cosmogenetic chant (Schaeffer 1952, my translation; see also Peinaud 1998), which is to be understood as a hybridization or radical multiplicity at the genesis of bridges and passages between matter and sensory experience: "music itself is a hybrid language, a suspension bridge between matter and our sensibility, responsive to the same vibrations, musical acoustics perhaps offers the mind an unhoped-for pathway between the inner and the outer, the cosmos and man" (Schaeffer 2017, 231).

The cosmological nature of the Deleuzo-Guattarian project becomes evident from the very beginnings of Deleuze's work, notably through *Difference and Repetition* (1969), as well as Guattari's early psychiatric work at La Borde. It extends through their final collective writings, particularly *What is Philosophy?* (1991), encompassing *A Thousand Plateaus* (1980), and finds expression in a wide range of commentaries across multiple fields, including ethics (Beaulieu 2006), astrophysics (Barrau and Rosenvallon 2010), aesthetics (Villani 1988; de Assis and Giudici 2017), technology (Yuk Hui 2016; Stiegler 2017), anthropology (Viveiros de Castro 2019; Meziane 2023), semiotics (Genosko 2002; Berardi 2014), politics (Sibertin-Blanc 2010; Montebello 2017), ecology (Morton 2012; Hörl and Burton 2017), and metaphysics (Lapoujade 2014; Montebello 2015a).

While these disciplines and territories might appear fragmented and disconnected, it is essential to recognize that the cosmological character of Deleuzo-Guattarian thought consistently weaves connections and relationships between these different "disciplinary" territories, forming an intricate interweaving of theoretical galaxies. Guattari referred to this as "transcoding," which is the way one milieu serves as a basis for another, establishing itself on another, dissipating, or constituting itself in another milieu (Deleuze and Guattari 1980, 385). Therefore, it is reasonable to assert that their thought allows multiplicities to move. Indeed, they traverse boundaries, transcending time and space, facilitating the passage of thought between different propositional circles, weaving bridges and passages on either side of the *planomenon*,[1] dissolving through an *aformal* exterior.

Therefore, the movements of matter and thought, as the intensive forces of life, are directed, hence rhythmic, and carried by a particular refrain, or, more precisely, by an undulatory and reverberating acoustic. This enables their navigation both below and beyond the bridges and passages woven through assemblages and between the two limits of any possible assemblage, namely "between the system of strata and the plane of consistency" (Deleuze and Guattari 1980, 415, my translation).

If Deleuze grants a cosmological, thus aesthetic and political, privilege to music (Bogue 1991; 2003; Montebello 2015b), it is because it "expresses both the doubling of time between past and present and the division of space between the Earth and the deterritorialized [... it] is listening to the inhuman outside [...] that the forces of art compose" (Montebello 2015b, 14, my translation). Right from the early pages of *A Thousand Plateaus*, Deleuze and Guattari, in the background of Pierre Boulez, assign both an aesthetic and a cosmological priority to music "against the overly significant cuts that separate structures or run through them" (1980, 16, my translation). In this departure, bifurcating from structuralism, symmetry, and binaries, they attribute to music a character of passage or transference, even a logic of passage, as music acts as a ferryman, allowing "its lines of flight to pass, as so many multiplicities in transformation" (1980, 19, my translation).

Pierre Montebello points out that Deleuze conceives of music (serial music) as "two ways of connecting with the world," with one being political, instituting nomadic and sedentary elements, and the other being cosmological, instituting the mobile and the static (Montebello 2015b, 13–14). Therefore,

> In a smooth spacetime one occupies without counting, whereas in a striated spacetime one counts in order to occupy. One makes palpable or perceptible the difference between non-metric and metric multiplicities, directional and dimensional spaces. One renders them sonorous or musical. (Deleuze and Guattari 1980, 477, my translation)

Here, two fundamental points of interest become apparent: 1) smooth spacetime is a directional space, hence inducing movement and rhythm (Goddard 2008, 112), while striated spacetime is a dimensional space, thus schematic[2] and structural; 2) smooth spacetime becomes sonorous and musical; it becomes music. In other words, music ontologically[3] precedes spacetime.

Thus, echoing the inquiries of Élie During and Jean During, the nomadic character inherent to music can be found within what they call "scientific music" (which is to be understood as an auto-referential movement), as well as in "nomadic music" (performative tradition-based music). Henceforth, drawing on the performative, thus "nomadic" or moving character inherited by a pre-philosophical plan or pre-individual plan, one could ask: what are the possibilities for a "tradition to establish itself, so to speak, from the outset as nomadic music, unfolding its effects in a smooth musical space" (During and During 2015, 13, my translation)? By acknowledging the constitutive and performative nature of music, particularly within its intrinsic nomadism, this inquiry explores how music emerges as a pivotal agent in the construction of a relational field, as a relational force. Moreover, it delves into the intricate ways in which music comes to rhythmically orchestrate or pulsate our interconnections, thus shaping the ontology of our relations—that is to say, the temporality or milieu of our relationality. In light of the constitutive and performative nature of music, given its intrinsic nomadism, how does music evolve into a key factor in shaping the relational field? Furthermore, how does it intricately weave itself into the rhythmic orchestration or pulsation of our interconnections, thereby influencing the ontology of our relations—specifically, the temporality or milieu of our relationality? That is, music practiced by nomads (or, furthermore, nomad music); "not the principle of a general process of deterritorialization affecting all music, but music that actually organizes its activity based on operations of connection and selection [...] from heterogeneous musical flows distributed in a smooth space" (During and During 2015, 13, my translation). Therefore, revisiting the archetype of the planomenon; *milieus* are not places of the passages but are what passes through locations (*lieux*).

What about the passage? What is the importance of granting an ontological, metaphysical, and cosmological privilege to the passage and the fleeting? What is the significance of a sonic framework, a musical force, an acoustic for the notion of passage? To what extent is our ability to make audible non-audible or akoúômorphic forces related to our ability to *make-world*? Our work will be divided into three passages: the logic of passage and transistantiality; the passage, the passer/passerby, and the fleeting;[4] and, finally, akoúômorphic thought: toward a metaphysics of the ephemeral (*métaphysique passagère*).

Logic of Passage and Transistantiality: Movement and Limit

> Waves are vibrations, shifting borderlines inscribed on the plane of consistency as so many abstractions [...] Each is simultaneously in this multiplicity and at its edge and crosses over into the others. (Deleuze and Guattari, 1987, 252)

Here, beneath the proposed aberrant movements of Lapoujade (2014), what seems crucial is to return to the sonic framework, the Ariadne's thread of Deleuzo-Guattarian thought, and singularize it as a logic of passage, a metaphysical passage, or a necessary passage for metaphysics. We draw upon this Deleuzian *logic* and Guattari's logic of transistantiality.

In line with Lapoujade, we posit that philosophy in the Deleuzian context is established through and within a logic, an *irrational logic of aberrant movements*.[5] As Lapoujade notes, numerous commentators on Deleuze and Guattari weave the common thread of their thought based on the concepts they created. In this context, Lapoujade argues that Deleuze's thought is subject to aberrant movements and irrational logics. According to him, the genesis of Deleuzian thought—which can be applied to what is now coined "Deleuzo-Guattarian thought"—is established through an irrational logic of aberrant movements, focusing on a constant quest for a logic: the logic of meaning, the logic of desire, the logic of multiplicities, and so on. Following this logic, the uniqueness of Deleuze's work is built on its capacity to escape the principle of reason without being illogical. There is almost always a coupling in operation in Deleuze's work: the coupling of aberrant movements and irrational logics asserting themselves beyond all principles of reason (Lapoujade 2014).

Quid juris? What is it in law, that is not? In this context, Deleuze's task is to overturn the foundations of the philosophy of judgment and representation. It is in this sense that Deleuze seeks to give meaning to the foundation of the groundless, the philosophical nonsense, the groundlessness of representation; that is to say, the right to discord (Bensmaïa 2006). From then on, we find ourselves confronted with the smooth space: the philosophy of judges, courts, and foundations dissolves in favor of cartographers, amoral judgment, and the powers of being. This subterranean logic continually forms itself through the struggle of two major forces: movements and limits, or rather, through the mobilization of limits. It transcends the limits: aberrant movements or what is unthinkable in thought (the "unthought") "exceed the empirical exercise of each faculty and force to surpass itself toward an object that concerns it exclusively, but which it only reaches at its limit" (Lapoujade 2014, 11, my translation). This logic ultimately finds its exterior through the depths of thought. It is only when confronted with these limits that it unfolds, setting itself in motion through its own transitive apparatus. It bears witness to "an inorganic life that traverses [breaks through] organisms and threatens their integrity, a life indifferent to the bodies it traverses as well as the subjects it disrupts" (Lapoujade 2014, 14, my translation).

From this "encounter" between movement and limit—the in-between created by the interstice of thought—the passage beyond the outside is shaped. Thus, "one is no longer on one side or the other of the limit but straddles it, torn apart" (Lapoujade 2014, 350, my translation), like a tightrope walker. In this context, the limit is always to be crossed and impassable, as it is underlying to our own becoming; making us acknowledge that something happens, that something must be passing. Like the Deleuzian gesture, this encounter is twofold: through thought and within oneself. Hence, the importance of combining this logic with Guattari's notion of transversality to fully grasp the *logic of passage*.

The logic of "transistantiality" in the work of Guattari, which he developed in his seminars in 1981, following the authorship of *Anti-Oedipus* (1972) and *A Thousand Plateaus* (1980), is intimately linked to the convergence and interweaving of his ideas with those of Deleuze. This concept's origins can be traced back to Guattari's earlier work on "transversality" in 1964, and reached its culmination with the publication of his books *Psychoanalysis and Transversality* (1972), *The Molecular Revolution* (1977), and *The Machinic Unconscious* (1979). It raises questions about unconscious transfers between groups and structures, particularly in terms of language and communication: "does the group and its non-sense engage in a sort of secret dialogue that is productive of potential alterity?" (Guattari 1972, 52, my translation). Guattari questions binary and dualistic thinking, advocating for an opening to the outside by unfolding the institution or group-subject from the inside, thus confronting alterity. This perspective directs our attention to the mediating object, which acts as a transitory agent: "we move within the realm of transfer only to the extent that something can move. Something that is neither the subject nor the other. There is no intersubjective relation that suffices to establish a system of expression, that is, alterity" (Guattari 1972, 52, my translation). Even through clinical experiences and psychoanalytical research on the unconscious, Guattari perceives the imperative of non-sense, which sets in motion a transitive logic, resulting in a reality that can only be characterized as mobile and elusive:—"we are part of the group not to hide from desire and death, but due to a specific problem, *not for eternity, but on a transitory basis. This is what I have called the structure of transversality*" (Guattari 1972, 52, my translation).

While Guattari never had the opportunity to precisely define the notion of "transistantiality," it is reasonable to assert that it pertains to a certain logic which bears the notions of "transistant objects" and "transistance." Within an interwoven analysis of language and the unconscious, Guattari observes that in classical analysis and classical logic,[6] there are objects that have been "foreclosed, forgotten, or denied." Nevertheless, these objects constitute unconscious mechanisms that need to be considered. According to Romain Sarnel's analysis, in contrast to classical logic and structuralism influenced by a phallic monolithism oriented toward an already constituted reality, transistantial logic is bifocal. It has a generative facet that encompasses persistent objects like an arching rainbow, and a transformative facet that illuminates transistant objects like a lightning bolt, those which displace lines and undergo metamorphosis. This implies the existence of a space between the conscious and the unconscious, where several "transistant objects" temporarily escape habitual capture. Transistance is the process of setting these transistant objects in motion beyond (they pass through but do not dwell there) machinic phylums and heterogeneous deterritorialized universes, on either side of the plane of consistency and beyond its limits. Transistance is what holds together entirely heterogeneous universes, and the transistant object, through its mobilization, is what enables a "gravitational imbalance" among these different universes, such as mathematics, chemistry, or computer science. Thus, it is simultaneously independent and parasitic with respect to persistence[7] and evolves through a dual modality of existence: one "confined to spatiotemporal, substantial, and energetic coordinates, and [the other] pertaining to systems of existents beyond coordinates, resembling a form of antimatter that duplicates all matter confined within coordinates" (Guattari 1981a, 8, my translation).

In contrast to assemblages that function to articulate material or energetic flows, relationships of segmentarity, territory, coordinates, and references to the machinic phylum, transistant objects operating under a transistantial logic move between the different lines of universes, phylums, and assemblages, existing on the "smooth" side of the line:

causing "a series of possibilities carried by the universes in question to articulate with each other" (Guattari 1981b, 6, my translation). An assemblage articulates different universes by surrounding them, while transistance articulates the possibilities of these universes in a way that traverses them and yet hovers above them, both horizontally (phylum) and vertically (rhizome). Guattari illustrates this through an example when discussing Italian neorealism:

> It transists in the sense of a phylum: in that Italian neorealism precedes a certain type of films, announcing others; it has a place in the history of cinema, within the phylum of films. But, at the same time, it is involved in rhizomatic relations with things that are entirely external to cinema. (Guattari 1981a, 9, my translation)

From incorporeal[8] to transistantial, time, space, and the exprimable are taken up by Guattari to shift the transcendental field into a transversal field that is "oriented both toward the various places of the real and the different vectors of the imaginary" (Sarnel 2007, 6, my translation). The logic that these transistant objects engage with is vastly different from the logic of degrees—a continuum of existence. This leads to an absolute radicality: abstract machines, through abrupt levels of transistance, begin to exist in terms of all or nothing; moreover, these levels of existence and nonexistence coexist. At the same time, abstract machines persist locally in a field, appearing on occasion through one event or another, and at the same time they transist forever once they have started transisting (Guattari 1981a, 7, my translation).

This logic of passage can be considered at times as a theoretical-conceptual alloy between this transistantial logic and that of aberrant movements, and at other times as a sonic framework, a dual act concomitant with thought put into a relationship state, ultimately, with their nominal function: "connection between the field of immanent experimentation and subjectivity in becoming within space and time" (Guattari 1981a, 2, my translation). The Deleuzo-Guattarian philosophical operation distributes thought, "unfolds its own original topology according to the battle that takes place within it and from which it proceeds" (Guattari 1981a, 2, my translation). Deleuze AND Guattari say what they *do* AND *do* what they say: they say what they are made of: a true movement—in-between;[9] "I am no longer myself, but an ability of thought to see itself and develop through a plane that crosses me in several places" (Deleuze and Guattari 1991, 62). This metaphysical activation is not peopled with dreams, fantasies, or projects, but with encounters which may be imaginary, delirious, fleeting (Deleuze 1985).

Triptych, or the Three Aspects of the Exprimable: The Passage, the Passer/Passerby, and the Fleeting

As Pierre Montebello points out, the Deleuzo-Guattarian philosophy can be considered an unlimited plane of nature (2015a), already beyond the Nature/Culture dilemma. The role of philosophy, with the help of the sciences, is to offer us a "Nature-Thought" (*Pensée-Nature*) that gives substance to the virtual and the power of the infinite (Deleuze and Guattari 1991, 112–13). This plane of nature[10] allows us to move from a Platonic intransitive thought to a transitive, fleeting metaphysical thought (Montebello 2015a, 33). Indeed, it is a matter of (re)connecting, of creating a passage between two modes of access to being, namely the ontological and the metaphysical. Charged with filtering the "outside," they are inseparable, since it appears to us that "to think being without beings is impossible" (Montebello 2015a, 49, my translation).

In the background of Montebello's thought it is essential to reclaim a boundless world, to let oneself be carried by music, to enter the dance and plunge into it: "not so much between assemblages and something else as between the two limits of every possible assemblage, that is, between the system of strata and the plane of consistency" (Deleuze and Guattari 1980, 415, my translation). In this context, we consider the passage as the "connection between the field of immanent experimentation and subjectivity in the process of becoming in space and time" (Sarnel 2007, 2, my translation), that is to say, the "field concerning the conditions of possibility of experience." It is in this sense that I want to explore the third prerogative of transistantial logic: that of the exprimable, namely, the passage, the passer/passerby, and the fleeting.

What about the exprimable? What about our proposal regarding the passage? Does this logic of passage not also call for a reversal of universalism in thought, experience, and the sensible? The term "exprimable" is to be apprehended as the disjunction of the consubstantial forces intrinsic to the realms of chaos, terrestrial existence, and the cosmos. This disjunction constitutes a corporeal mobilization and an unequivocal transcendence that extends beyond the confines of cogitation, thereby permitting a unique and multifaceted intertwining of the vital forces of existence. The Guattarian "exprimable" or Deleuzian "expression" is of a corporeal nature; it forges a nexus between the singular individual and the cosmos, thus unveiling its cosmic dimension (Schérer and Passerone 2006, 176). The conventional subject, akin to a fleeting phenomenon, gives way to the process of individuation.

Three distinct movements, or what Klossowski aptly characterizes as "power-qualities" (Pinhas 2006, 215), emanate from the concept of the "exprimable": the passer, the passerby, and the fleeting. The passer, reminiscent of the obscure precursor, is the agent that establishes correlations among dissimilar potentials (Deleuze 1996). The passerby, the lightning, akin to a sudden flash of insight, embodies the visible event, signifying the very process of differentiation (Deleuze 1996). The fleeting assumes the role of

> What passes through a series of states, and never exists at the center, but always on the periphery of the disjointed circle around an elusive, unnamable center [...] Hence, identity is essentially contingent, and a series of individualities must be traversed by each one, so that the contingency of this one or that one makes all of them necessary. (Pinhas 2006, 216–17, my translation)

A true disjunctive-synthesis,[11] the exprimable, through its entanglement of power-qualities, forms a major synthesis that transforms Deleuze's thought into both a musical and philosophical synthesizer (Pinhas 2006, 214). It is a synthesis of the *passer* to the *passerby* (Sarnel 2011, 8), translating from the *Is* to the *And*, from connective syntheses to conjunctive syntheses, and vice versa. This marks a double reversal of the philosophy of desire—a transformation of the philosophy of the exprimable—in favor of a transient synthesis (synthetic passage) in which the fleeting, or in Deleuzian terms, the figuration, is an "elaboration, a transformative labor of energies whereby the circulation of metamorphoses provokes the transvaluation associated with Becoming [...] the figurative is to figuration what feeling would be to sensation" (Pinhas 2001, 71, my translation).

Entanglement or fragmentation, terrestrial tremors or stellar twinkles, the fleeting—this people yet to come—is merely a sporadic and ephemeral effect. It only harmonizes at specific times and places in response to celestial calls, much like the tide dances with the moon until its penultimate ebb. Both close and distant simultaneously, this philosophy of Nature (of the natural event) developed by Deleuze and Guattari (1980; 1991) demands a kind of natural passage, a "philosophical javelin throw" that pierces and orbits all at once. In short, it necessitates a fleeting philosophy, and metaphysics, that is akoúômorphic.

Akoúômorphic Thought: Toward a Metaphysics of the Ephemeral

What is the significance of a soundtrack, a musical force, acoustics for the notion of passage? To what extent is our ability to make audible the inaudible correlated with our capacity to create a world? As mentioned earlier, in the context of the insights of Deleuze and Guattari, artfully articulated by During and During[12] as well as Montebello,[13] we consider that music ontologically precedes spacetime.

As Beaulieu and Sholtz point out, it is modern music that intrigues Deleuze and Guattari, as it allows for "a deterritorialization of the refrain through music, and a deterritorialization of music itself" (2022). It is this dual movement that interests us, as it enables the capture of non-terrestrial forces: music of the world *is not of this world*. Here, it is music that enables a real connection with the outside. More precisely, it is acoustic reverberation that allows for both its generation and relay. The phenomenon of reverberation, which is sometimes seen as the persistence of sound after its source has ceased emitting it, due to the reflection of sound waves returning to the listener's ears with a certain delay (Schaeffer 1952, 216), and at other times as the persistence of sound in an enclosed or semi-enclosed space after the interruption of the sound source (Schaeffer 1952), expresses itself through the discovery of singular objects and is understood as the rediscovery of familiar-unfamiliar material, or unknown-known matter. That is to say, the "alleyway" or *allée-retour*[14] is the transistantial movement that the reinjection of singular objects takes within the planomenon (Renzo 2023).

Here, it's precisely this acoustic reverberation that gives passage, *allée-retour*, and abstract objects their transient nature and their persistence for both thought and sensory experience. However, this "connection" with the outside,[15] this *allée-retour* through this interstitial passage, is not sufficient to establish consistency: the Earth's boundaries become porous, while the movements of the cosmos even transcend gravitational forces; "this world has had different aspects, it will have others, and there are already others on other planets" (Deleuze and Guattari 1980, 416, my translation). To establish consistency, it is necessary to acknowledge the genuinely transitive nature of the constitutive relations that bind us to the world, and, above all, to consider our relationships with the world as transitory. If Deleuze and Guattari's major challenge is to conceptualize life as the power of the outside, it is equally important to apply their own logic, namely that this outside is only a passage giving way to "other" passages in what Vincent Jacques (2005, 2) called an existential logic of sensations. Thus, the process of exteriorization[16] can only be transient, not in terms of time or space, but in terms of life, whether organic or inorganic. Therefore, all life is fleeting because it is transitive, and vice versa. By acknowledging our transitory nature and the transitory nature of things and the world around us, it becomes possible to establish consistency, to create a world. Music is transient, and we are merely passengers of it.

The importance of this acoustic reverberation for life is crucial in the formation of a sound framework that rises as a fleeting, hence cosmic, sensation in that it resists finitude, "as if the powers of life were caught between two self-destructive processes; on one side, the death from within as a process of transmutation, of metamorphosis; on the other, the death from without that sweeps us away and destroys us" (Lapoujade 2014, 300, my translation). This might be what Frédéric Neyrat meant when he called for a "new Copernican revolution" (2016), or what Pierre Montebello understands when he exhorts metaphysics to find something *to apply to again* (2015a, 198); namely, to invent a "common world, to produce an unlimited plan of nature, so that anthropocentrism, dogmatic humanism, reductionism [...] can no longer be taken for granted [:] the weaving of being and thought is to be endlessly repeated, without precondition, without a proper combination" (2015a, 232, my translation) by embracing the fleeting nature of this entanglement, which is to say, dazzling and ephemeral. Thus, to the Deleuzo-Guattarian intensive self-relation and expressive relation to the world, we want to extend the argument to a passing relation to the cosmos; and this, by admitting the consistency that cosmic otherness takes on—or the reverberating intensities of telluric and cosmic chants—in our relationship with the universe (Renzo 2023); that is, to acknowledge that the "mechanisms of localization" are transitive. The telluric forces that make up our world are the same transitive forces, "recovered or liberated from a deterritorialized cosmos" (Deleuze and Guattari 1980, 403, my translation). We transist; the world transists; our relations transist as well as the relations between them; what allows us to gain consistency is to admit that the real is passing. Consistency is a process

> that comes to combine differences and separate heterogeneous states, a process of active connectivity always precarious, fragile, transitional. It allows us to think about how new arrangements are created through crossings, marriages, catalyses, symbioses, polymerizations, "entity compositions." Taking on consistency is hybridization, conjugation, new symbiosis. (Montebello 2018, 566, my translation)

To do this, it is necessary to develop an "akoúômorphic" relationship with the world to make audible the telluric forces as well as those of the cosmos.

While Montebello's cosmomorphic[17] thought allows us to "connect" with the cosmos, to re-anchor human beings in their vital continuum and thus give consistency to the relationships that constitute them (Montebello 2015a, 249), I propose to circulate an aesthetic and a haptic—hence political—proposition that aims to be akoúômorphic, and, *ipso facto*, to participate in a cosmomorphical metaphysics. Deleuze's aesthetic is grounded in the transcendence of the Kantian cleavage between aesthetics as a theory of sensible intuition and aesthetics as a theory of art, wherein art no longer constitutes merely a representational activity and where the subject is not confined within the artist's subjectivity but rather emerges as a genuine practice of experimentation and problematization of the real, suturing the unified subject in favor of subjectivity, henceforth subjectivities. Masterfully expounded by Réda Bensmaïa (2006), the Kantian "illusion" consists in folding the dynamisms aroused by the violence of the conflict of faculties onto a singular instance of subjectivity, onto the singularity of a "molar" subject. In his discourse, Deleuze adopts the term "haptic," initially coined by Aloïs Riegl, to signify an aesthetic resistance whenever there is no longer strict subordination in either direction between senses, nor a loose subordination or virtual connection. Instead, this occurs when vision (or any other sensory category) reveals within itself a tactile function intrinsic to it alone, distinct from its primary function. Thus, the Deleuzian aesthetic serves to reconfigure the Kantian subject, previously fixed in orbit. The haptic function facilitates sensory passage (from sense to out-of-sense) by apprehending faculties that are not merely reallocated within a framework of reason or truth, but rather of re-cognition, thus enabling the agency of the otherness.

Akoúômorphism, or what we could describe as akoúômorphic aesthetics applied to a cosmomorphic metaphysics, is understood as the "final contraction that makes audible the density of the heavens, an expression of difference, a forced movement toward an unspeakable universe" (Pinhas 2001, 70, my translation). The term *akoúô* originates from the ancient Greek *ἀκούω*, meaning "to hear," to perceive by the ear what is announced before us, to learn through listening, something that reaches our ears, and it poses as an auditory channel, a faculty related to hearing, much more as a tinnitus or what Deleuze aimed at when he was searching for the "outside of the inside" (Deleuze 1986).

Regarding the precedence I accorded to music[18] over time and space, I consider that this music—this acoustic reverberation—emanates from cosmic chants:[19] "music embodies the act of attuning to the inhuman outside, forging a linkage to a plane of primordial nature that the art's forces configure" (Montebello 2015b, 14, my translation). In this manner, music acquires its cosmic character as it rhythms what Deleuze characterizes as "pure past," subtly interwoven with Bergson's notion of duration, connecting and forming "a plane of coexistence, so that our own duration becomes entangled with other durations, such as the duration of the universe" and "the duration of life" (Montebello 2017, 39). It is in this way that reverberational acoustics puncture our traditional conception of a world-as-fact, transforming it into a world of becoming: "it is no longer subjectivity that temporalizes, but rather [...] it is subjectivity that is temporalized by the timelessness of time" (Montebello 2017, 39). Consequently, we witness not a recentralization of the infinite within finitude, but rather the decentralization of the finite within the infinite (Montebello 2017).

This sonic material that expresses the totality of the world (*le grand plan* or big picture) inexorably leads us, through rhythm and pulsation, into various experimental intertwinings and infuses us with the sensation of modes of habitability of the Earth through the plurality of living beings (Escobar 2020). It is this floating matter that Deleuze mentioned during his reflections on art at IRCAM, as Pierre Boulez points out the fluidity of these "sonic beings whose becoming generates [...] a time of suspension in uncertainty where everything appears in perpetual transition" (Pinhas 2001, 177, my translation).

In this context, the significance of this material lies not only in its inherently mesomorphic composition but also in the revolutionary nature of the "akoúômorphic passenger." In contrast to the connections that give rise to acts of resistance, the revolutionary character of the sonic material is established through a haptic capacity that moves as a "transversal vector [...] motivated solely by its own wandering, [abstract but primarily] independent of any pre-established form" (Parret 2009, 2, my translation). Deleuze defines this haptic force as a form of spatialization in which "there is no longer a strict subordination in either direction, neither a relaxed subordination nor a virtual connection, but when sight discovers within itself a specific function of touch that is uniquely its own, distinct from its optical function" (Deleuze 2003, 155). This haptic spatialization or "hearing window" is what opens the world to revolutionary forces, inevitably propelling us into other modes of inhabiting the world. Here, the "haptic touch" does not occur synesthetically by invoking a fusion in the passage or between the outside and the inside, but by admitting a (synesthetic) moment of doubled "elevation, a redoubling of vital forces" (Parret 2009, 6, my translation).

Hence, akoúômorphism is not ascribable to the subject, cognition or thought, but rather emerges from the aesthetics (sensible) of the universe-cosmos or *grand plan*. It reveals itself as a form of political emancipation, exercising itself negatively by "undoing the givenness of the given" (Grosz 2011, 3). This implies that becoming is but a process of un-becoming, disrupting the stability of identities, knowledge, situations, and being. Through disjunctions or perforated spaces, akoúômorphism causes "the social to escape through the multiplicity of gaps that erode and puncture it, like a genuinely revolutionary force" (Culp 2020, 66, my translation); in doing so, it makes audible new threads of inhabiting the world. Indeed, as Montebello points out: "to perceive is not to create a representation of the world but to fully engage with the light of the world; perception would thus be an endpoint, not a starting point" (2017, 33–4, my translation).

Thus, if the metaphysical and cosmological apparatus seems to be in place, passing from time to time as an asynchronous back and forth, all that remains after all is this "earing window" of potentialities that enables us to make-world. It is the very condition of transistantiality and also composes these modes of habitability. Nomads roam, while in addition to their mobility, the akoúômorphs are carried by the ephemeral of a becoming in action. Cosmic fugitives, these modes of habitability of the world—ultimate breaches of emancipation—merely pass, and from this fleeting consistency, *make audible non-audible forces.*

For the Penultimate Time: Toward a Metaphysics of the Ephemeral?

Deleuze and Guattari's philosophy presents a unique perspective on the fleeting nature of experience and the sensible. This fleeting nature implies a gateway and hence a passage which forms the heart of this philosophical exploration and goes beyond and under spatial or temporal transitions. It represents a dynamic connection between diverse modes of being, bridging the ontological and the metaphysical realms. The forces of the outside thrust thought into a realm of informal connections and diverse perspectives, fostering a constant interplay between heterogeneities. These forces provide thought with an "absolute rapport to the outside" and infuse it with a sense of vitality. Henceforth, we embrace a transitive, ephemeral, and ever-passing philosophy that invites us to recognize our own passageway in the world, acknowledging that everything is transistant. It is by acknowledging this fleeting character of existence that we can attain consistency, create a world, and fully engage with the forces of the cosmos and the Earth.

Akoúômorphism is an aesthetic and philosophical gesture focusing on how sound and acoustics can be used to understand and express invisible forces and dynamics of the world that escape ordinary hearing. This approach emphasizes that music and sounds can reveal hidden aspects of reality by making subtle dimensions of existence audible. It is philosophical because it enables us to hear the object (the other), and aesthetic—hence political and critical—because it provides a material gateway to the outside, allowing thought and experience to pass through it. By reintegrating the sensibility of matter, it invites us to reconfigure our relationships with the world, emphasizing the importance of ever-evolving and constant relationships; as if the only thing that fixed us to the world was fleeting.

Notes

1. "Planomenon," a term encapsulating the idea of erasing the rift between human and nature while reshaping territorial forces into nomadic entities, stands as the central theme of this discussion. It sheds light on the interplay between life and thought, highlighting concepts of circulation, proliferation, and expansion as delineated by Villani and Sasso (2003, 276).
2. The schematic is defined as of a fixed form posited as an object (Goddard 2008, 112).
3. In the strong sense as a mode of pre-reflective action; we could refer to Descola's concept of worlding, or furthermore to Montebello's situated metaphysics (*métaphysique située*), in order to account for it.
4. Fleeting or *passager* in French. By fleeting, I mean the one who is just passing through, who is just passing by and/or for a short while, for a brief moment.
5. According to Lapoujade (2014), aberrant movements have a dual characteristic, meaning they are both unexplainable and simultaneously obey an imperative logic.
6. Implicitly influenced by René Schérer, I hold that transitive thought, or the logic of passage, is established through "three basic principles: the principle of alterity, the principle of transition, and the principle of the admitted third, in contrast to the three basic principles of formal logic: the principle of identity, the principle of non-contradiction, and the principle of the excluded third" (Sarnel 2011, 106, my translation).
7. Like striated spaces, persistence establishes the relationships between "matters of expression and substances" (Guattari 1981b, 3, my translation).
8. Drawing upon the Stoics' three incorporeals: time, space, and the exprimable or expressible.
9. "The proposition; the language enunciates the potential, yet it prepares it for actualization: language does not articulate what is, it articulates what might be [...] When I say it is day, it is not at all because it is day [...] but because I have my own intention to fulfill, and to which the day merely serves as a pretext or argument" (Deleuze 1992, 58).

10. “Deleuze and Guattari have made expression the foundation of their philosophy of nature. Everything in nature is expressed, everything is traversed by the forces of the Earth and the forces of the cosmos” (Montebello 2015a, 143, my translation).
11. The disjunctive synthesis “designates the system of potential permutations among differences that consistently return to the same, through movement, through sliding, or rather, differences that recurrently converge without ceasing to be differences” (Deleuze and Guattari 1972, 18, 82).
12. “If one immerses themselves in nomadic music to observe its functioning, it all happens as if the process of becoming-music were reversing its course” (During and During 2015, 6, my translation).
13. “The new conception of time, according to Deleuze, essentially lies in the possibility of generating subjectivity through ontological planes that precede it” (Montebello 2017, 38, my translation). “It’s not that music doesn’t exists in space and time, but rather that space and time fall under a form of musicality” (Montebello 2015b, 9, my translation).
14. The *allée-retour* is conceptualized as a movement for passage—the inextricable mark left by the crossing—and describes a moment where only the return is contingent.
15. Deleuze recognizes the synesthetic and even anamorphic nature between thought and the outside world. Without reducing one to the other, he posits that “thinking always comes from the outside, is directed toward the outside, belongs to the outside, and is an absolute relationship with the outside” (Pelbart 2008, 23, my translation).
16. The forces from the outside “place thought in a state of externality by casting it into an informal field where heterogeneous perspectives, corresponding to the heterogeneity of the forces at play, come into relation” (Pelbart 2008, 24, my translation).
17. “The substitution of a cosmomorphic thought for the anthropomorphic schema plays a decisive role in the recomposition of the world. By cosmomorphic thought, we mean any thought capable of transcending correlation, not in an abstract or formal way, but by integrating the focal points of heteromorphic alterity, whatever they may be, whether vital, aesthetic, cosmological. What we call the consistency of relations is the way in which participation in the world of beings, in whatever form (aesthetic, political, desiring), colors activities and behaviors, giving them an existential gravity” (Montebello 2015a, 253, my translation).
18. We can consider this precedence as an essence, but only if we acknowledge that the meaning of the sign does not find its foundation in either the object or the subject. Indeed, it pertains to an “intermediate zone” (Montebello 2017) or what I characterize as a passage; thus, the sign reveals, neither purely objectively nor purely subjectively, an inception of a world, an emerging world: “neither percept, nor affect, nor subject, nor object, the sign thus leads us elsewhere [makes us pass elsewhere]” (Montebello 2017, 56, my translation).
19. In this context, I am referring to what Deleuze and Guattari (1987) have termed “material forces,” in contrast to the notion of “matter-forms” (substances-attributes), which captures the forces of the cosmos as molecularized matter: “aligned with Deleuzian thought regarding philosophy as the art of exploring the genesis of things and thought—prioritizing the process over the outcome” (Jacques 2005, 18–19, my translation).

References

Aquin, Hubert. 2010. *Next Episode*. Translated by Jean-Louis Major. Toronto: McClelland and Stewart.

Barrau, Aurélien, and Jérôme Rosanvallon. 2010. *Cosmogenèses (à partir de Deleuze et Guattari)*. http://www.rehseis.cnrs.fr/IMG/pdf/philmaths_2010_05_21_Cosmogeneses.pdf. Accessed September 8, 2025.

Beaulieu, Alain. 2006. "Introduction to Gilles Deleuze's Cosmological Sensibility." *Philosophy and Cosmology* 16: 199–211.

Beaulieu, Alain, and Janae Sholtz. 2022. "Introduction to the Virtual Art Exhibition *Gilles Deleuze, Félix Guattari and Cosmic Art*." https://deleuzeguattariart.wixsite.com/deleuze-guattari-art/introduction. Accessed September 8, 2025.

Bensmaïa, Réda. 2006. "Le sujet de l'art: Prolégomènes à une esthétique deleuzienne future." *Cinémas. Revue d'études cinématographiques* 16, no. 2–3: 208–37.

Berardi, Franco. 2014. *And. Phenomenology of the End*. New York: Semiotext(e).

Bogue, Ronald. 1991. "Rhizomusicosmology." *SubStance* 20, no. 3: 85–101.

Bogue, Ronald. 2003. *Deleuze on Music, Painting, and the Arts*. London: Routledge.

Culp, Andrew. 2020. *Dark Deleuze*. Paris: Éditions Divergences.

de Assis, Paulo, and Paolo Giudici. 2017. *The Dark Precursor: Deleuze and Artistic Research*. 2 vols. Leuven: Leuven University Press.

Deleuze, Gilles. 1985. *L'image-temps*. Paris: Éditions de Minuit.

Deleuze, Gilles. 1986. *Foucault*. Paris: Éditions de Minuit.

Deleuze, Gilles. 1992. "L'épuisé." In Samuel Beckett and Gilles Deleuze, *Quad et autres pièces pour la télévision*, 55–106. Paris: Éditions de Minuit.

Deleuze, Gilles. 1996. "Z comme zig-zag." In *L'Abécédaire de Gilles Deleuze avec Claire Parnet* (video). Paris: Éditions Montparnasse.

Deleuze, Gilles. 2003. *Pourparlers (1972–1990)*. Paris: Éditions de Minuit.

Deleuze, Gilles, and Félix Guattari. 1972. *L'Anti-Œdipe*. Paris: Éditions de Minuit.

Deleuze, Gilles, and Félix Guattari. 1980. *Mille Plateaux*. Paris: Éditions de Minuit.

Deleuze, Gilles, and Félix Guattari. 1987. *A Thousand Plateaus: Capitalism and Schizophrenia*. Translated by Brian Massumi. Minneapolis: University of Minnesota Press.

Deleuze, Gilles, and Félix Guattari. 1991. *Qu'est-ce que la philosophie?*. Paris: Éditions de Minuit.

During, Élie, and Jean During. 2015. "De l'espace lisse au temps troué à propos des musiques nomades." In *Deleuze et la musique*, edited by Pascale Criton and Jean-Michel Chouvel, 9–23. Paris: CDMC.

Escobar, Arturo. 2020. *Pluriversal Politics: The Real and the Possible*. Durham, NC: Duke University Press.

Genosko, Gary. 2002. *Félix Guattari: An Aberrant Introduction*. London: Continuum.

Genosko, Gary. 2012. *Félix Guattari in the Age of Semiocapitalism*. Edinburgh: Edinburgh University Press.

Goddard, Jean-Christophe. 2008. "Henri Maldiney et Gilles Deleuze. La station rythmique de l'œuvre d'art." *Revista Filosófica de Coimbra* 33: 109–24.

Grosz, Elizabeth. 1994. *Volatile Bodies: Toward a Corporeal Feminism*. Bloomington: Indiana University Press.

Grosz, Elizabeth. 2011. *Becoming Undone: Darwinian Reflections on Life, Politics, and Art*. Durham, NC: Duke University Press.

Guattari, Félix. 1972. *Psychanalyse et transversalité*. Paris: Maspéro.

Guattari, Félix. 1977. *La révolution moléculaire*. Paris: Éditions Recherches.

Guattari, Félix. 1979. *L'inconscient machinique. Essais de schizo-analyse*. Paris: Éditions Recherches.
Guattari, Félix. 1981a. *Transistancialités*. Toulouse: Chimères.
Guattari, Félix. 1981b. *Agencements. Transistances. Persistances*. Toulouse: Chimères.
Hörl, Erich, and James Burton. 2017. *General Ecology: The New Ecological Paradigm*. London: Bloomsbury Academic.
Jacques, Vincent. 2005. "Le monde de la musique et la musique comme monde selon Deleuze." *Horizons philosophiques* 16, no. 1: 1–23.
Lapoujade, David. 2014. *Deleuze, les mouvements aberrants*. Paris: Éditions de Minuit.
Meziane, Mohamed Amer. 2023. *Au bord des mondes: Vers une anthropologie métaphysique*. Brussels: Vues de l'esprit.
Montebello, Pierre. 2015a. *Métaphysiques cosmomorphes: La fin du monde humain*. Dijon: Les presses du réel.
Montebello, Pierre. 2015b. "Deleuze et Boulez." *Liminar* 2, no. 4: 9–22.
Montebello, Pierre. 2017. *Deleuze, esthétiques—la honte d'être un homme*. Dijon: Les presses du réel.
Montebello, Pierre. 2018. "Métaphysique située: La consistance du réel et la Terre." In *Choses en soi: Métaphysique du réalisme*, edited by Emmanuel Alloa, 563–75. Paris: Presses Universitaires de France.
Morton, Timothy. 2012. "The Œdipal Logic of Ecological Awareness." *Environmental Humanities* 1: 7–21.
Neyrat, Frédéric. 2016. *La Part inconstructible de la Terre: Critique du géo-constructivisme*. Paris: Seuil.
Parret, Herman. 2009. "Spatialiser haptiquement. De Deleuze à Riegl et de Riegl à Herder." *Actes Sémiotiques* 112. https://www.unilim.fr/actes-semiotiques/2570. Accessed September 8, 2025.
Peinaud, Michel. 1998. "Le chant du virtuel." *Littérature* 111: 29–36.
Pelbart, Peter Pál. 2008. "Cartographies du dehors." *Rue Descartes* 59: 20–30.
Pinhas, Richard. 2001. *Les larmes de Nietzsche: Deleuze et la musique*. Paris: Flammarion.
Pinhas, Richard. 2006. "Circulus Vitiosus Deus." In *Pierre Klossowski*, 209–21. Paris: Inculte.
Renzo, Antoine. 2023. "Le passage interstitiel: une pensée cosmologique de la philosophie deleuzo|guattarienne." *La Deleuziana* 15: 179–92.
Riegl, Aloïs. 1985. *Late Roman Art Industry*. Rome: Giorgio Bretschneider.
Sarnel, Romain. 2007. "Lieux de passages et transversalités: Pour une dynamique Deleuzienne." *Le Portique* 20: 1–9.
Sarnel, Romain. 2011. "Transversalisme de la pensée: pour une dianoétique schérérienne." *Cahiers critiques de philosophie* 10: 97–114.
Schaeffer, Pierre. 1952. *À la recherche d'une musique concrète*. Paris: Éditions du Seuil.
Schaeffer, Pierre. 2017. *Treatise on Musical Objects: An Essay across Disciplines*. Berkeley: University of California Press.
Schérer, René. 1996. "L'usine de l'âme." *Chimères. Revue des schizoanalyses* 27: 21–34.
Schérer, René, and Giorgio Passerone. 2006. *Passages passoliniens*. Villeneuve d'Ascq: Presses universitaires du Septentrion.
Sibertin-Blanc, Guillaume. 2010. *Deleuze et l'Anti-Œdipe: La production du désir*. Paris: Presses Universitaires de France.
Stiegler, Bernard. 2017. "General Ecology, Economy, and Organology." In *General Ecology: The New Ecological Paradigm*, edited by Erich Hörl and James Burton, 129–50. London: Bloomsbury Academic.

Villani, Arnaud. 1988. *Gilles Deleuze: La guêpe et l'orchidée*. Paris: Rue d'Ulm.
Villani, Arnaud, and Robert Sasso. 2003. *Le Vocabulaire de Gilles Deleuze*. Nice: Les Cahiers de Noesis.
Viveiros de Castro, Eduardo. 2019. *Politique des multiplicités: Pierre Clastres face à l'État*. Bellevaux: Dehors.
Yuk, Hui. 2016. *The Question Concerning Technology in China: An Essay in Cosmotechnics*. Windsor Quarry: Urbanomic.

Chapter 4

Following a Plastic Vortex of Spinning Waste[1]

Macarena Rioseco

> From chaos, *Milieus* and *Rhythms* are born. This is the concern of very ancient cosmogonies.
>
> (Deleuze and Guattari 1994, 364)

Following a Plastic Vortex of Spinning Waste is a project of artistic research that has been developed through an ongoing dialogue with Gilles Deleuze and Félix Guattari's cosmological thinking, which they present as an immanent, dynamic, and ever-changing process based on constant relation and continuum between order (cosmos) and disorder (chaos). This chapter is an essential part of the project, which proposes that the cosmological visions of Deleuze and Guattari, as well as Karen Barad, which conceive matter as active and with *agential potential*,[2] provide a view of formation processes in the universe that offers a new perspective for thinking about and guiding creative processes as nonrepresentational practices. Refocusing our attention on the affective potential or agency of materiality in dialogue with the agency of the artist, through the ontological category of immanence, offers new understandings of creativity. Creativity is described in this text as the generation of *difference* and a *potentiality* horizontally shared by *agencements*[3] composed of multiplicities of human and nonhuman entities, which is activated within the making processes and as the result of all entities' reciprocal interactions. Regarding Deleuze and Guattari's thinking about the processes of creation of the Earth, encountering *difference* involves *following matters* (materials and issues), to be taken away by non-terrestrial chaotic forces of the cosmos, and being able to identify the new forms of order that emerge from chaos.

The art project follows the environmental problem of plastic contamination of the oceans, and so-called "plastic islands" or "trash vortexes," by forming an *agencement* in between multiple and multidimensional matters. This text explores the oceans through Deleuze and Guattari's cosmology as archetypes of the smooth space, a virtual plane of immanence populated by material–forces relations and material–flows distributions with the potential to generate dynamic formations such as gyres. Today gyres are attracting the plastic that contaminates the oceans to their stable centers, where it gets trapped, resulting in trash vortexes. These vortexes offer us an uncanny materialization that allow to visualize non-visible and unstoppable flows and forces of the universe. These are sorts of monstrous wonders, in between a body without organs and strata, which so brutally illustrate the cosmos's potentiality for self-creation and self-organization, to continue changing, thriving, and evolving: *for becoming itself*.

This is an experimental text that reflects on Deleuzo-Guattarian cosmological thinking, combining and interlacing the voices of many different authors and my own practice. The intention of this text is to work together with these authors' cosmological ideas and unfolding vision of life, seeking to "materialize," by means of knitting and language, affective matters related to the processes of formation leading to monstrous trash vortexes. These are matters that are extremely difficult to translate into words, but which the art practice explored in this chapter clearly and simply *puts-materially-into-action* (materially enacts) and embodies as a *block of sensations*.

The first section develops a nonrepresentational approach to art practices, as the generation of difference embodied in material-discursive pieces, drawing from the Deleuzo-Guattarian understanding of cosmological flows and forces and their vision of the unfolding of life. The second section explores the agency of water as a morphogenetic force in the cosmos. The world's oceans are analyzed as interconnected virtual smooth spaces, in which the dynamics and flows of their circulating currents around the globe transport water and redistribute heat around the planet. The third section discusses the creation of oceanic trash vortexes, through Deleuze and Guattari's cosmological ideas considered in previous sections, as hybrid (natural/artificial) monstrous formations. The entanglement between water and microplastics is explored to propose trash vortexes as events that *actualize new potentialities* of the ocean as a smooth space. The final section presents *Following a Plastic Vortex of Spinning Waste*, an artmaking project-in-progress that follows the problem of trash vortex formations. The central piece is a crocheted plastic formation made by the re/collection of recycled plastic and through manual actions that combine *striated* and *smooth* making strategies. The process of formation of this piece is explored through the cosmological ideas presented herein, as processes of following singularities emerging from the interactions between material forces and flows in the cosmos and the oceans.

1. Cosmological Features as Nonrepresentational Practices

Deleuze and Guattari's immanent ontology and dynamic cosmological vision of the unfolding of life and the material world proposes that the cosmos is not created as an image of something else which transcends it, but through dynamic and ongoing interactions between matter, forces, and flows that are immanent to the processes themselves. Their conviction that "matter contains its own immanent possibilities for dynamism and change" (Gray et al. 2021, 203) has further enabled the development of a current of thought that problematizes matter as entities with potential capacities of agency. This ontological perspective corresponds to a relevant innovation for thinking about the universe, the material world, and the processes of formation of phenomena where, instead of a duality "matter–form ensemble," these can be thought of and interrogated as a "material–force interaction" and "material flows" (Deleuze and Guattari 2014, 111, 24). These interactions and relations are not predetermined or given but develop and transform as they occur. Cosmological features are nonrepresentational because, in their becoming, they cannot be reproduced (represented), but as proposed by the problematic model of nomad science, only be followed. In nomad science—a theory of flows, fluids, and change that seeks to understand earthly processes and that has as model "becoming itself"—matter or materials should not be understood as homogeneous, but "essentially laden with singularities" (Deleuze and Guattari 2014, 421, 430). These influence the forms, or "traits of expression" (Deleuze and Guattari 2014, 430), that matter can take. Nomad science sees matter (singularities of matter) and forms (traits of expression) as indivisibly braided together and is concerned with the forms of reciprocal connections and influences between them, and within this indivisibility. For that reason, this science does not endorse the *matter–form duality*, but *material–forces* processes; "it *follows* the connections between singularities of matter and traits of expression" (Deleuze and Guattari 2014, 430), looking to understand their nature (natural or artificial) and the level of these connections. As a nonrepresentational practice, nomad science has as a model the problematic and the hydraulic (fluids) of becomings and heterogeneities. This can be contrasted to the model of the theorematic and solids, of royal science, of "the stable, the eternal, the identical, the constant" (Deleuze and Guattari 2014, 421).

Nomad science seeks to understand the emergence of difference in earthly processes by putting the variables "in a state of continuous variation," instead of repetition, which is done by "extracting constants from variables" (Deleuze and Guattari 2014, 430). The specificity (individuations) of differences that emerge from these variations are events that capture singularities of a specific matter. In contrast, the specificity of repetition—extracting constants (essences) —is achieved through the composition of objects (forms) led by a dual relation of matter–form. Deleuze and Guattari argue that creativity, "the placing-in-continuous variation of variables, is in opposition only to the actual determination of their constant relations" (2014, 115). Regarding the distinction between royal science (seeing the universe as static, eternal, and out of time) and nomad science (seeing the universe as dynamic, becoming, and in time), Ilya Prigogine and Isabelle Stengers argue there is an ever-increasing understanding that "there is a more subtle form of reality involving both time and eternity" (1986, xxx). They also discuss the transition between paradigms of dynamics, from a static one (being) to a dynamic evolutionary one (becoming), leading to "a new concept of matter, matter that is 'active,' as matter leads to irreversible processes [disorder] and as irreversible processes organize matter [order]" (1986, xxiv). As I will discuss in section 3, these authors are describing scientists' realization that order (cosmos) emerges out of disorder (chaos).

Thinking of materials through the flat ontological category of immanence as active and as entities with potential agential "capacities to affect and be affected" (Deleuze and Guattari 2014, 304) has relevant consequences for artmaking. Also, the approach to *matters* (materials and issues) during artmaking processes with a sense of a problem to work with and to follow brings meaningful insights into these creative processes. In this approach problems are principally intuitive creative boosters that trigger new ways of thinking and allow us to imagine new ways of doing. Indeed, as Williams stresses "when thinking emerges and changes, it is necessarily accompanied by problems" (2003, 131). Affective and problematic approximations to matters are nonrepresentational approaches that decentralize the responsibility of creativity, which has traditionally rested in the human being who is expected to deliberately manipulate materials and fully guide the creative processes. Instead, nonrepresentational approaches involve the artist being dragged into *agencements* and being plugged into the chaotic forces of the cosmos in search of the emergence of new order. Thus, traditionally thought, singular human agency is fragmented and proliferated as a multiplicity that belongs to an open, fluid, dynamic, hybrid, and relational *agencement* plugged into the becoming of life, and brings the realization that creativity is part of an ongoing cosmological process. This opens the becoming of the creative processes (its possible directions and outcomes) to the *agencement* that makes each process possible. Consequently, creativity ceases to appear as an only-human endeavor or agency, to become the emerging and unexpected outcome of a multiplicity of interrelated and multidimensional agencies involved in the ongoing, ever-changing and undetermined process of creation of the world.

Karen Barad's (2007) notion of "intra-action" explores a similar idea. In her understanding of the creation of phenomena, she claims that entities involved in earthly processes of formation do not preexist their interactions but emerge from them. The specificity of their emergence is defined in relation to the conditions of possibility of agency that are "afforded" (Gibson 2015) by and shared between the entities involved during and within the very processes. Highlighting the dynamic nature of the world, Barad argues that, in its creation, instead of "an assemblage of agents"—as preexisting entities with already defined conditions of possibility for action—"there is an entangled state of agencies" (Barad 2007, 23) whose potentials emerge and are defined within, because of the processes, and while they unfold. Regarding nonrepresentational creativity, Barad's thinking endorses the view of creativity as multiple and as part of wider earthly processes. Intentions are not "preexisting states of mind" of individuals as isolated human beings but are conditioned by the "complex material conditions" (Barad 2007, 23) afforded by a specific context that allows their articulation. Deleuze and Guattari support a contextual and dynamic conception of thinking, arguing that ideas originate with sensations through people being literally in touch with the material world. Hence, ideas and intentions are immanent, contextual, and evolving because they depend on every individual's thought and affective experience. Lisa Åkervall endorses this by claiming that "[a]ffects exceed ordinary experience and precipitate thought, taking part in a hybrid configuration of thought and the body" (2021, 577).

In terms of artmaking, this challenges the traditional representational model of agency based on thinking and stable subjects, who act by imposing methods and forms on inert and stable objects. It opens the possibility of objects also acting upon subjects, changing their intentions or provoking new intentions in them. This expands the view of creativity as an *agencement*'s collective production and as the result of an agential co-responsibility and of being part of worldly processes in progress.

Consequently, this chapter proposes to understand and guide nonrepresentational artmaking as intra-acting processes evolving within *agencements*, which are based on reciprocal operations of following between heterogeneous and entangled states of agencies hoping to be dragged by chaotic forces of the cosmos, to witness and participate from the emergence of a new order (difference). In being carried away by forces of the cosmos, attention should be paid to the problems that arise, because these are "accidents that condition and resolve" (Deleuze and Guattari 2014, 420) the processes themselves. Hence, problems are principal nonrepresentational creative agents that can lead in unimaginable directions. The dynamics of artmaking *agencements* are uncertain and emergent; in their becoming, "material-discursive" knowledge is co-constructed by a multiplicity of co-emerging entangled and intra-acting states of agencies (Barad 2007, 132). Material-discursive refers to hybrid constructions resulting from the synergy of conceptual and material practices where the boundaries between them cannot be determined.

Nonrepresentational artmaking strategies draw from cosmological operations of generation and creation of difference, based on following the matters and the things that affect them, with the intention of discovering their singularities. The goal is to produce continuous variation of these singularities by putting them into action and to generate the difference embodied in material-discursive pieces. Moreover, the complexity of the knowledge embodied in these pieces, as compounds of sensations (Deleuze and Guattari 2011), can only be fully received by people through their direct embodied experience of them. Consequently, I propose that artworks are *material-discursive-affective* pieces, whose signification emerges from the entanglement of these three levels of experience.

2. Water as a Morphogenetic Earthly Force

The fluidity of water rarely puts up resistance, making it particularly vulnerable and almost inseparable from the effects of forces in the cosmos. But, at the same time, the same condition, as "direct relation *material–force*" (Deleuze and Guattari 2014, 398), makes this element a principal shaping agent of the Earth. Forces cannot be grasped directly as forces but can only be traced through the analysis of formations understood as processes, meaning that forms are effects of the specific processes that generated them. Hence, in natural and/or artificial formations the manifested relations between matter and form should be scrutinized as clues that can reveal *morphogenetic processes* (DeLanda 2013; Ingold 2013)—that is, a confluence of forces and matter with the agency to trans/form other materialities—which led to those formations. Deleuze and Guattari assert that, instead of forms organizing matter, "materials signal forces and serve as symptoms for them" (2014, 557). Water has a multiple role in life's genesis/destruction cycle on the planet, endlessly flowing and circulating through nature. Water allows the surge and maintenance of all life forms; as "a vortical body in a nomad space" (Deleuze and Guattari 2014, 427) it enables the actualization of the virtual, that is, life in all its potential. Philip Ball comments: "Water is the agent of geological, environmental and global change [...] water: the agent of planetary-scale life and death" (2015, 36–7).

The sea is a plane of continuous liquid matter, constantly and unceasingly moving and flowing, which Deleuze and Guattari signal as "the archetype of smooth space" (2014, 558). The waters in the world's oceans do not have forms or rigidities; they embody "an *amorphous*, nonformal space" (Deleuze and Guattari 2014, 554), "a shapeless mass" (Nietzsche as cited in Deleuze 2009, 123), that is, not formless but rather "of the pure unformed" (Deleuze 2009, 123). Indeed, the three oceans in the world (the Atlantic, Pacific, and Indian)[4] are interconnected; hence, masses of water are constantly circulating around a "global conveyor-belt," "both vertically and horizontally" (Ball 2015, 44) between surface-water and deep-water circulation. Surface-water circulation involves the water from the surface down to about one hundred meters and is principally affected by forces of wind. Deep-water circulation connects all the world's three oceans and involves the water from a hundred meters depth and below, and it is affected by unidirectional forces of differences of temperature moving warm masses into colder seas and redistributing "heat around the planet" (Ball 2015, 48). The density of water also changes with temperature variations and salt concentration; cooler water is denser than warmer water, and if salinity increases, density increases too. The denser the water, the heavier it is. Forces of differences in density and weight prompt cool, salty, and dense masses of water to sink to the depths of the oceans, and warm, less salty, and less dense water to rise toward the surface.

In addition, ocean currents present a localized type of motion related to their potential as a "vortical body in a nomad space" (Deleuze and Guattari 2014, 427) which is manifested in the form of gyres or "slow and elegant gyrations" (Ball 2015, 38). These are produced, in part, due to the distribution of the continents, whose land masses interfere in the circulation of the currents, cornering and obstructing them. This makes the masses of water "trace out huge, closed loops called gyres," which are amplified and modified by the Coriolis force, which is produced by the rotation of the Earth and acts on entities that move "within a rotating system" (Ball 2015, 46). In addition, inconstancies in the flows of the water in the oceans and within the circulating gyres can cause these latter to develop vortex-like whirlpools, with unpredictable rotating patterns, which can easily change their rotational direction throughout the year.

Regarding Deleuze and Guattari's cosmological thinking, they reflect on vortical formations by analyzing the differences between *rectilinear striated* and *spiral smooth* spaces. They discuss atomic interactions based on Lucretius' *clinamen*, which theorizes that if during atoms' downward fall there were no deviation in their course, then there would be no coalitions between atoms and thus no creation in the universe (molecules, compounds, substances, strata, etc.) would be possible. As Deleuze and Guattari note, "the simplest considerations of velocity immediately introduce the difference between vertical descent and curvilinear motion, or more generally between a straight line and the curve, in the differential form of the clinamen, or the smallest deviation, the minimum excess" (2014, 432). They further state that great vortical organizations and spirals are formed by "curvilinear declination on an inclined plane," and that, as a model, the vortical operates in open smooth spaces, "throughout which things-flows are distributed" (2014, 421). In striated spaces, the vertical descent of atoms would not imply any kind of motion other than a vertical descent, nor any kind of distribution between things other than "onto parallel layers" (2014, 431). For that reason, the force of gravity is the law of this space, because it is striated and homogeneous "by the fall of bodies, [in line with] the verticals of gravity" (2014, 431).

Rhythm, however, is the law in smooth spaces because it is heterogeneous through continuous variation and development of form. In smooth spaces vortical formations are provoked by "rapidity, celerity, [which] applies only to the movement that deviates to the minimum extent and thereafter assumes a vortical motion, occupying a smooth space" (Deleuze and Guattari 2014, 433). In addition, in vortical formations such as ocean gyres, the indivisibility of material–force and material–flow relations is manifested as a synergy between them, drawing a smooth space where they cannot be differentiated and where the "matter–flow [duality] can no longer be cut into parallel layers" (Deleuze and Guattari 2014, 433). Within the circulating currents and swirling dynamics of ocean gyres, water is put into a state of continuous variation, allowing us to seize the singularities of gyre formations as a sort of *swirlingness*. Singularities are variations of "problematic figures," which are forces that affect forms (e.g., roundness), as contrasted with "theorematic figures" (Deleuze and Guattari 2014, 428, 427), which are forms—that is, essences—as such (e.g., a circle) (Rioseco 2018). For that reason, singularities reveal that nature "presents to us the heterogeneity of the diverse with itself, and also the resemblance of the diverse with itself" (Deleuze 2009, 308). Nevertheless, as forces, problematic figures are "transformations, distortions, ablations, and augmentations" (Deleuze and Guattari 2014, 427) of essences, and for that reason, the authors argue that *variations of problematic figures* (singularities) are "vague essences" (2014, 428) seized from a matter put into a state of continuous variation. The reason is that singularities are "determinations, or constants [e.g., roundness], that are extracted from things that are more abstract than any characteristic or essential trait [e.g., round] of the thing itself" (Rioseco 2018, 4). Hence, singularities express a matter's multiple and heterogeneous potentialities for actualization, which are expressed as a variety of degrees of possibilities and capacities for existence, to affect, to be affected, and/or to connect to other multiplicities.

Deleuze highlights the inherent different and multiple character of nature, noting, following Lucretius, that nature "must be thought of as the principle of the diverse and its production" (2009, 303), whose heterogeneity of elements cannot be thought of as an indivisible whole. "Following" is proposed as a nonrepresentational approach to be plugged into the chaotic forces of the cosmos and the search for the singularities of a matter. To follow compels us to leave the representational contemplating viewpoint, which is external to the phenomenon—"watching the flow from the bank"—and "to be carried away by a vortical flow; when one engages in a continuous variation of variables" (Deleuze and Guattari 2014, 433–4). The necessity to be dragged into an *agencement* with the process one is committed to comprehend is endorsed by Hanjo Berressem, who argues that "nature is an infinitely complex abstract machine that surveys its planes of consistency and composition from within those very planes" (2021, 499). The only way to grasp them is to be carried away by and within those very planes. The universe is composed of a "multiplicity of heterogeneous singularities [...] that ceaselessly enter into new constellations" (Berressem 2021, 499–500). That dynamism does not allow it to compose a totality (a whole) with its parts; therefore, it can always remain multiple and fragmented.

3. Trash Vortexes: A New Order in the Chaos

Today, due to plastic contamination of the waters, oceans are developing garbage patches: areas for the collection of marine debris, mostly composed of non-biodegradable plastic. Because of rotating ocean currents, debris is drawn into the gyres' calm and stable centers where it gets trapped. The rotating dynamic of the gyres makes the waste spin, and the patches acquire vortex forms; for this reason, they are known as trash vortexes (Evers 2023). The idea of a garbage vortex evokes images of a floating island of trash, but in reality, because the plastic thrown into the ocean is constantly being broken into smaller pieces, "these patches are almost entirely made up of [...] microplastics" (Evers 2023) mixed with some larger pieces. Hence, the image of a floating island of trash is fictional. Microplastics are not easily observable, not by the naked eye nor by satellite imagery, but they make the water look cloudy. In addition, some of this marine garbage floats on the surface, but a large part of it (about 70%) sinks. Many pieces sink only a few centimeters or meters beneath the surface, but many others sink several meters or even all the way down to the bottom. Hence, the seafloor is most likely also filled with garbage. This makes it almost impossible to measure the vortex's full area.

Thinking of this matter through a Deleuzo-Guattarian lens, this multiplicity of *unformed* microplastics composes a smooth space that, when mingled with the waters in the ocean, forms garbage vortexes, an *agencement* that *actualizes* a new differentiated smooth space and whose corporeality renders visible the nonvisual dynamics between forces and matters forming the oceans. Discussing Lucretius' work, Thomas Nail argues that "the material flows of being are not necessarily observable as such [...] never appear as discrete, observable or empirical particles [...] are always just below the level of observation" (2018, 11). Trash vortexes embody the ocean's material flows differently, offering them a monstrous body whose materiality renders visually sensible these virtual oceanic forces and dynamic gyre formations, their rotating currents and gently rolling vortexes with stable centers, "which are not [visually] sensible in themselves" (Beaulieu 2016, 202)—that is, not visible to the naked eye.

This new monstrous body is hybrid because it is composed of water, an "absolute fluid" element, characterized by the complete "absence of coherence or cohesion" (Deleuze 1993, 5), and, for that reason, it easily separates into parts. Absolute hardness is pointed out by Deleuze as the opposite of an absolute fluid, and is characterized by "the inseparability of its parts" (1993, 6). Despite water being fluid, he notes that it can have a simultaneous dual (fluid/hard) nature: if "a boat reaches a certain speed a wave becomes as hard as a wall of marble" (1993, 6). On the other hand, the monstrous body of plastic vortexes is composed of plastic, which is affected by its contact with the sun and the water, breaks into tinier and tinier pieces, and unavoidably separates into micro-grains (microplastics). Microplastics also have this dual nature; they are "granular and fluid" (Berressem 2021, 504). At the micro level they are absolute hardness, and at the macro level they are absolute fluid. Consequently, as microplastics accumulate in gyre formations, these centripetal oceanic dynamics provide coherence and cohesion to this mass of smooth, grainy, microplastic multiplicity. Hence, in their entanglement, ocean and debris as "differential modes of intensity [...] plunged into a creative flux of becoming" (Beaulieu 2016, 200) embody an *event* that results in a new monstrous, multiple and heterogeneous *agencement* that actualizes a new potentiality of the ocean as a smooth space. In these new formations, emerging amid water and plastic, the latter is irreversibly integrated in the "chaosmos" of the oceanic ecosystem and becomes an agent of a particular "chaosmological experience" (Beaulieu 2016, 201).

Microplastics and water are mingled in such tight and homogeneous ways that separating them is almost impossible. These large areas of spinning microplastics are dangerously close to the fictional scenario of a fully entropic (chaotic) system, achieving its maximum possible equilibrium. The reason is that systems that have achieved a maximum of entropy are characterized by a complete homogenization of their matter–energy components, by a full equilibrium and a complete seizure of their work expressed as matter and energy exchange (transformation). Indeed, Prigogine and Stengers claim that scientists have come to an agreement that "nonequilibrium, the flow of matter and energy, may be a source of order" (1986, xxix), whereas Deleuze and Guattari argue that there is "a calming and stabilizing center in the heart of chaos" (2014, 362). Nonetheless, the potential of this shapeless mass of water of flows and movement whose currents behave as a vortical body in a nomad space makes this system an open one, meaning that matter (water and debris) travels and circulates around the world's oceans and between the five patches that currently exist. Consequently, matter and energy are exchanged between trash vortexes, jumping "from chaos to the beginnings of order in chaos" (Deleuze and Guattari 2014, 326), bringing a new order. Garbage vortexes are autonomous and spontaneous forms of creativity (Beaulieu 2016), a horrifying example of the autopoietic forces of life and nature that always manage to find new routes leading to emergent formations based on self-organization. This is similiar to "Joyce's neologism 'chaosmos' [which] expresses the fact that chaos and cosmos (disorder and order) are not opposites, but part of a larger continuum [...] There is a symbiosis between order and disorder and their boundaries are somewhat undefined" (Beaulieu

2016, 201). Prigogine and Stengers also discuss the relation between order and disorder, arguing that the law of entropy, which proposes that the world is irreversibly evolving from increasing order to increasing disorder, has been displaced by the realization that irreversible processes can actually organize matter, hence that order can and does emerge from chaos.

Beaulieu reflects on the idea that "Deleuze's chaosmological amazement and wonder has more to do with the non-familiarity and strangeness that is happening 'down here' than with the harmony 'up there'" (2016, 204). In cases such as these unpredictable oceanic plastic vortexes, the strangeness arouses wonder at the inherent transformative and creative capacities of the cosmos and a dynamic Earth in an "infinite and endless process of deterritorialization" (Beaulieu 2016, 207). It also awakens awe at all the new and strange forms of spontaneous self-organization that are constantly emerging, as the increasingly heterogeneous material nature of the world (natural and artificial) continues to thrive. The unpredictability, unfamiliarity, and strangeness of plastic vortexes reaffirm the fact that the "earth remains a source of unknown surprises" (Beaulieu 2016, 206). The virtual centripetal forces of the ocean currents forming spirals are evidence that the dynamics of earthly processes are cosmological in the sense that they are affected by impersonal, creative, and intensive forces of deterritorialization coming from an outside. Spiral formations can be found at different scales of magnification around the cosmos, from the very small to the very large.

In garbage vortexes, the chaosmos forms a material-discursive *agencement* characterized by a dynamic reciprocity and co-creation between "content" and "expression." Since "while something always exists materially, it is never materially available *in itself*" (Prokić 2021, 550), content is always linked to certain modes of expression. Moreover, garbage vortexes are a*gencements* where matter and discourse are linked "at a specific point in time and space," where the "entanglement of discourse and matter" is generated at different levels: as a *body without organs* and as *strata* (Prokić 2021, 547, 543). As a body without organs, the boundaries in these *agencements* between water's "free intensities or nomadic singularities" and microplastics' "mad or transitory particles" (Deleuze and Guattari 2014, 46) dissolve. Hence, the vortexes' *agencement* corresponds to "the mode of linkage that joints" (Prokić 2021, 546) water and plastic in a multiplicity, while maintaining as fundamentally unclear the distinction between an element and the whole, "where it begins and ends" (Prokić 2021, 546). Plastic is dragged, or plugged, right into the middle of ocean currents, surrounded by the dynamics of its creative forces, between rhythms and flows that move out of control in centripetal directions and absorb every entity they encounter.

Due to the extreme difficulty of separating microplastics from water, the "monstrosity" is given by the realization of the permanent integration of a non-biodegradable artificial agent into the "unformed, unstable matters" (Deleuze and Guattari 2014, 46) that permeate the ocean, the Earth, and the cosmos, and the scale of it. Plastic is "thrown into a milieu, becoming part of it, becoming the milieu" (Philippopoulos-Mihalopoulos 2016, 81). Plastic vortexes are characterized by a perpetual contradictory alliance, in between a conflict and a communion, between natural and artificial matters, which leads to the emergence of a new natural/artificial formation. The "space in the middle is a space of struggle [...] is a space of encounters with other bodies, a space in which a body affects and is affected by other bodies" (Philippopoulos-Mihalopoulos 2016, 82). Plastic vortexes are smooth spaces whose formation processes lead to the transformation of the "original" milieu and thus to the production of difference. In addition, as milieus, plastic vortexes are vibratory spaces, constituted by "periodic repetitions of [their] components" (Deleuze and Guattari 2014, 365), whose rhythms are autopoietic and self-organizing responses to chaos, and which are in between—that is, undifferentiated from—"that which is constructed and that which grows naturally, between mutations from the inorganic to the organic" (Deleuze and Guattari 2014, 364).

As *strata*, due to water's material forces behaving "like 'black holes' or occlusions striving to seize whatever comes within their reach" (Deleuze and Guattari 2014, 46), microplastics and water become inseparably captured together, composing an *agencement* in between territory and deterritorialization. Trash vortexes are not fully formed, stable, and consistent structures that hold these disparate and heterogeneous matters together because plastic spans the waters and covers large areas and even travels between patches. Regarding the *double articulation* of the strata, the *agencement* between water and microplastics does not fully achieve a second articulation—a consistency—remaining in between a form and a substance as "unstable particle-flows, metastable molecular or quasi-molecular units" (Deleuze and Guattari 2014, 47) or an "unformed" and "shapeless mass" (Deleuze 2009, 123) of heterogeneous matter, and does not fully achieve the construction of a molar compound or "functional, compact, stable structures (*forms*)" (Deleuze and Guattari 2014, 47), which could provide a stable and lasting order.

4. *Following a Plastic Vortex of Spinning Waste*: Flowing with Plastics

Following a Plastic Vortex of Spinning Waste[5] is an artmaking project-in-progress, an *agencement* that follows the problem of plastic contamination of the oceans and trash vortex formations. The main piece under construction is a crocheted plastic formation made from the re/collection of recycled plastic, whose trans/formation is produced through a variety of manual actions that combine "*metric*" striated and "*non-metric*" smooth-making strategies (DeLanda 2013). The trans/formation of the plastic into threads (spinning) is done through striated strategies consisting of consecutive actions: stretching, flattening–measuring–folding, flattening–measuring–folding (again), cutting–measuring–cutting, stretching, interlacing, and balling up (Figs 4.1 and 4.2).

The threads are knitted using a crochet needle and smooth-making strategies of *putting the variables into a state of continuous variation*. The aim is to seize the singularities of plastic and the knitting technique, a practice that, through repetitive and rhythmic movements, leads to the construction of aggregate surfaces whose singularities simultaneously materially present within themselves "the heterogeneity [and also the resemblance] of the diverse in a single [multiple] body" (Deleuze 2009, 308). Like plastic vortexes, crocheting is a practice of the milieu. It is located in between smooth and striated spaces, because it draws "a space

Fig. 4.1 Still from video of spinning process: stretching. © Macarena Rioseco

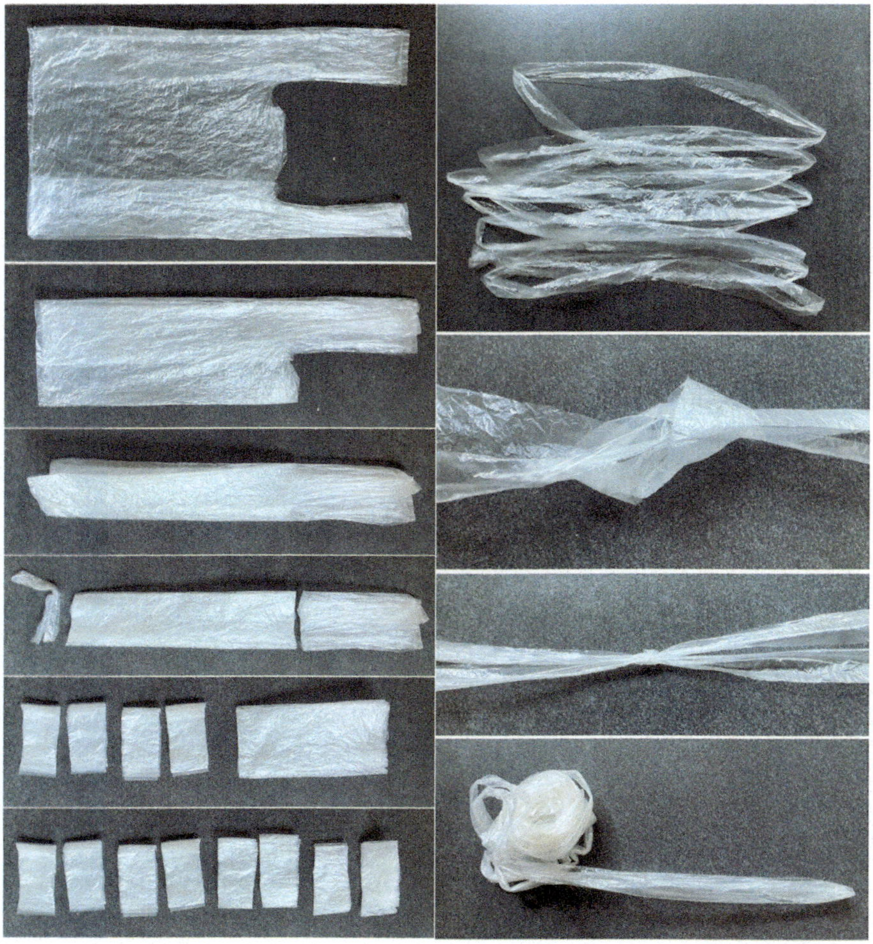

Fig. 4.2 Ten instances of striated strategy for spinning the plastic bags into threads. © Macarena Rioseco

that is prolongable in all directions—but still has a center" (Deleuze and Guattari 2014, 553). The shape of the piece under construction is being defined in the process, as it unfolds. It is the result of following the captured singularities of a soft and flexible material (plastic), the crocheting method, and the inner logics of the emerging structure, while also resonating with the singularities of the ocean and the cosmic rotating dynamics between material, forces, and flows. At the moment, it is a rounded, potentially flat, and circular formation with one stitch as its center (Fig. 4.3). Due to exponential duplication of the stitches (Fig. 4.4) at the beginning of the process, the perimeter of the formation

Fig. 4.3 Rounded, flat and circular formation with one stitch as its center.

Fig. 4.4 The formation folds and meanders due to exponential duplication of the stitches.

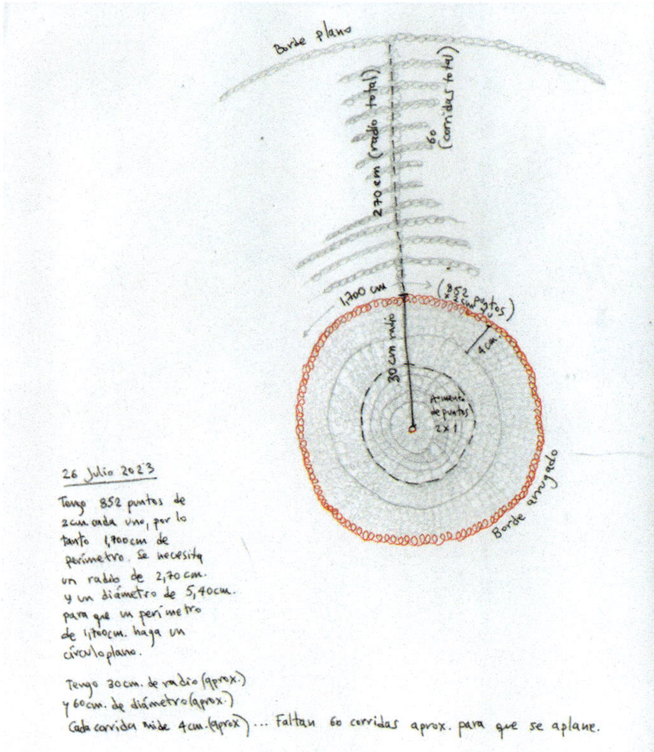

Fig. 4.5 Sketch of the potential future of the formation.

grew too fast in relation to the rate of growth of the radius: the length of the perimeter now is 1,700 cm (852 stitches) and of the radius 30 cm (Fig. 4.5). Consequently, the perimeter is too long for the current radius; for that reason the structure folds and meanders, and instead of being organized as a flat circular surface, it is arranged as a rounded coral-like volume (Fig. 4.6).

The lack of coordination of the growth rate between radius and perimeter brought disorder (chaos) to the unfolding process and the initially projected form, deviating in its course of development from its potential as a circle (a *form*) toward its potential to be affected by swirlingness (a *problematic figure*). Accordingly, variations of swirlingness are the singularities of this *agencement* seized by means of following its matters. Nevertheless, the projection is that, paralyzing the perimeter's rate of growth (by ceasing stitch duplication), while continuing the growth of the diameter (knitting layers around the perimeter), might bring some order back (Fig. 4.7).

Fig. 4.6 Rounded coral-like volume. © Macarena Rioseco

Fig. 4.7 Crochet knitting with plastics.

> For sublime deeds [...] one draws a circle, or better yet walks in a circle as in a children's dance, combining rhythmic vowels and consonants that correspond to the interior forces of creation as to the differentiated parts of an organism. A mistake in speed, rhythm, or harmony would be catastrophic because it would bring back the forces of chaos, destroying both creator and creation. (Deleuze and Guattari 2014, 362)

This knitting process-of-formation follows the cosmic material–forces interaction in the ocean, whose above-described water-current dynamics guide and resonate with the knitting process enacted (put into action). As a result, the developed practice is characterized by the enactment of smooth "qualified movements" (Deleuze and Guattari 2014, 432) that were encountered when searching for "individual differences and change" (Williams 2021, 494) by putting the variables of the *agencement* into a continuous state of variations.

During practice the gestures, movements, and rhythms of my hands—as operations of "deformations, transmutations [...] metamorphoses, generations and creations" (Deleuze and Guattari 2014, 422)—*follow* and are affected by cosmic forces in the oceans and further affect the encountered plastic materials (Figs 4.8 and 4.9). In other words, the wavy practice of crocheting resonates with the qualified invisible cosmic forces that populate the ocean as both give form to fictional plastic islands as vortexes of spinning plastic waste. Consequently, this formation is "composed of flows, folds, and weaves [...] of corporeal flows [...] that move together [...] and fold over themselves [...] in a woven knotwork [... its becoming is] immanent to the flow and flux of matter in motion" (Lucretius as discussed by Nail 2018, 11). Crocheting is the practice that, through a process of stratification, a methodical building up of plastic layers, allows this formation to exist, grow, and expand as a molar aggregate.

> Strata are Layers, Belts. They consist of giving form to matters, of imprisoning intensities or locking singularities into systems of resonance and redundancy, of producing upon the body of the earth molecules large and small and organizing them into molar aggregates. (Deleuze and Guattari 2014, 46)

This knitted plastic vortex is a material-discursive entanglement, characterized by the indivisibility of material and ideal reciprocal relations and co-creations. Through intra-actions between agencies, the overall project consists of a simultaneous and reciprocal "co-emergence of both meaning and matter" (Luraschi, Timeto and Yardimci 2023, II) in-the-making, that is, during the construction process itself. As a *compound*

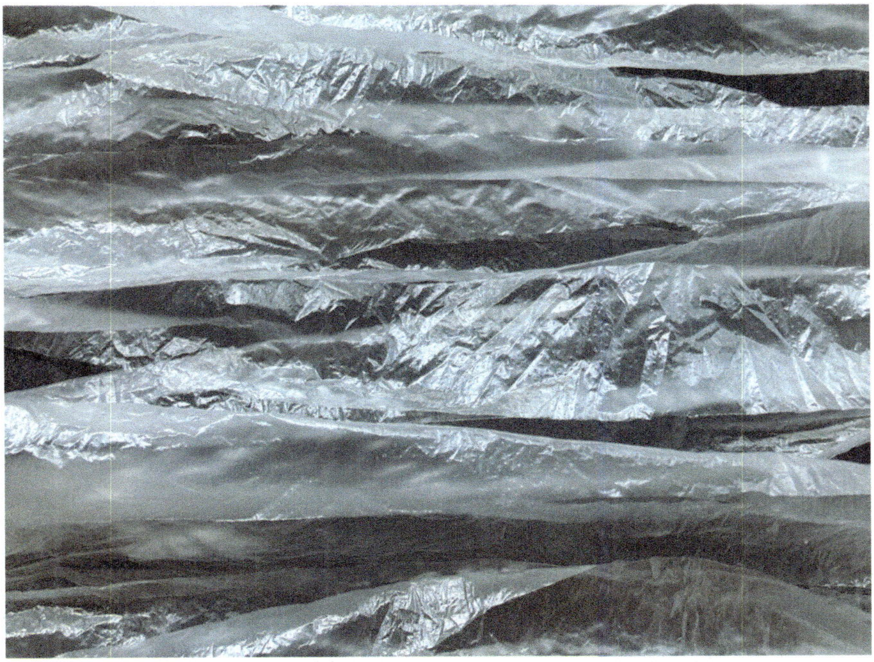

Fig. 4.8 Plastic bags. © Macarena Rioseco

Fig. 4.9 Still from video. © Macarena Rioseco

of sensations, it renders sensible a material-conceptual "expression of the virtual chaos and an artistic expression of the non-terrestrial forces" (Beaulieu 2016, 202) populating the Earth. Consequently, it is a material-discursive-affective entanglement that somehow brings to life the fictional idea of a floating island of trash by giving it a corporeality, whose wonderfully monstrous materiality is reminiscent of nature's chaosmotic dynamism and versatile capabilities for reinvention.

This formation will continue growing through the building up of plastic layers around its perimeter until it reaches a radius of approximately 270 cm. As calculated, this is the distance that will make the current perimeter (1,700 cm) flat and circular again, and for that, 60 knitted layers more are needed. The installation of the finished piece will seek to construct an immersive space evocative of plastic vortexes in the ocean. For that, the flat and circular plastic island will be installed horizontally as if it were floating at about 150 cm above the floor. This will divide the space into two: *surface-space* and *deep-space*. Visitors will be able to circulate around deep-space as if they were submerged in the water underneath a plastic island of trash (Fig. 4.10). Owing to

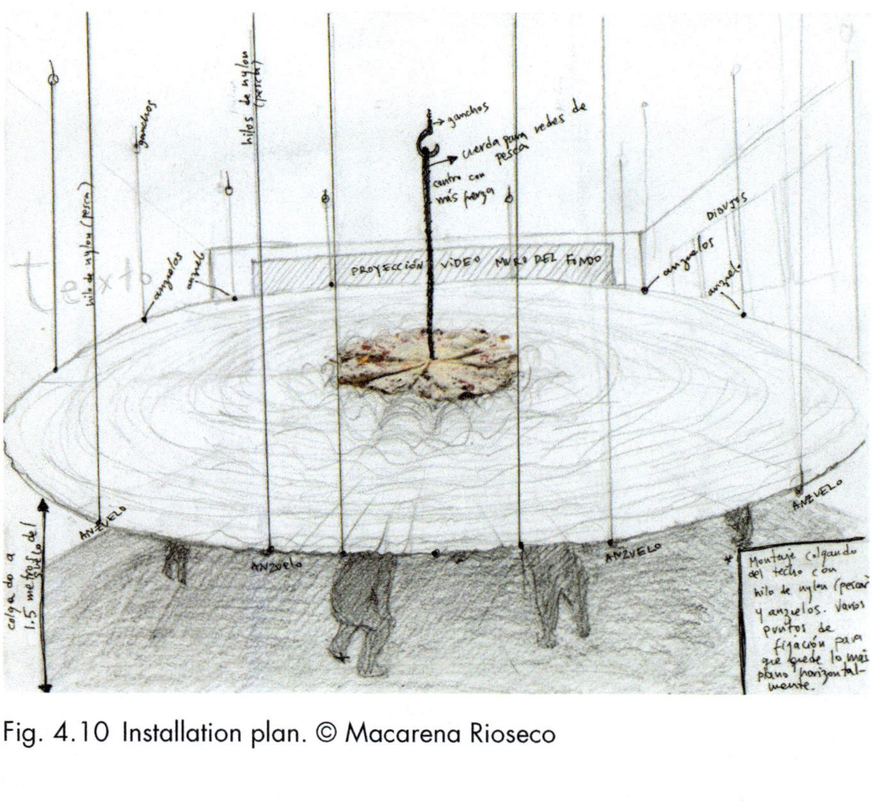

Fig. 4.10 Installation plan. © Macarena Rioseco

the height at which the piece will be hung, people will have to bend to go beneath the formation. This will create a claustrophobic and uncomfortable space that aims to provoke a sense of enclosure, which is the sensation that would be provoked by being underneath a plastic vortex in the middle of the ocean.

Conclusion

Following a Plastic Vortex of Spinning Waste is an artistic research project that has been developed as a dialogue between three *matters*: Deleuze and Guattari's cosmological thinking, environmental issues related to plastic contamination of the ocean, and the knitting of a plastic vortex. This chapter has explored the relations between order (cosmos) and disorder (chaos)—understood as continuum, a "chaosmos"—that emerge from the interrelations between these three matters and has analyzed the new kinds of order that emerge out of chaos. Through this project a synergy between practices, theories, and materials has simultaneously guided and shaped the evolution of one another, leading to the co-creation of a new order as a material-discursive-affective entanglement. Hence, the braiding between these three matters has allowed the development of a new perspective so as to see and to understand them all together and on their own.

The dynamic and immanent ontological view of creative processes in the universe offered by Deleuze and Guattari has relevant repercussions for thinking about creative processes at the human scale too. It provides a nonrepresentational framework, based on features such as following material forces and flows, which has proven to be insightful for thinking of these processes, and for being drawn by non-terrestrial chaotic forces that will lead the processes in unthinkable directions and generate difference. It also enables an understanding of the agency of nonhuman entities in the world, because their existences, affections, reactions, and transformations affect other entities and trigger multidirectional processes. Consequently, through this perspective, creativity, as an only-human potential, is decentered and fragmented to be shown as an undeterminable outcome emerging from the interrelations of a multiplicity of multidimensional agencies involved in an ongoing, ever-changing, and undetermined cosmic process as a negotiation between order and disorder.

Notes

1. This text forms part of the DIUMCE Project 05-2024-CIA: "*Following Recycled Plastic as Raw Material*", funded by the Universidad Metropolitana de Ciencias de la Educación.
2 "Potential" is the English word I am using for the term *puissance*, which Brian Massumi translates in *A Thousand Plateaus* (Deleuze and Guattari 2014) as "power." Massumi states that "Puissance refers to a range of potential. It has been defined by Deleuze as a 'capacity for existence,' 'a capacity to affect or be affected,' a capacity to multiply connections that may be realized by a given 'body' to varying degrees in different situations. It may be thought of as a scale of intensity or fullness of existence (or a degree on such a scale), analogous to the capacity of a number to be raised to a higher 'power'" (Deleuze and Guattari 2014, xvi).
3. In this text I use the term *agencement*, which is the original French word used by Deleuze and Guattari, to highlight the dynamic approach that this work endorses, expressed as interactions between multidimensional agencies that have emerged within and during the creative process of this project. Ian Buchanan argues that *agencement* is a word that has "'agency' at its core," and that describes "an ongoing process rather than a static situation" (2021, 27). At the same time, using *agencement* emphasizes the importance of "desire" as a productive force in this process, which is "the starting point" and "the basis" (Buchanan 2021, 60) that have made this project possible. I describe the role of desire as an affective resonance that mobilizes me and further "animates" (Buchanan 2021, 60) the *agencement*, that is, the multidimensional matters whose emerging agencies make possible the development of this creative process.
4. The Southern or Antarctic Ocean (encircling Antarctica) is the most southern portion of these three.
5. "Spinning" is used in this text in a double sense, in terms of "swirling" and of "drawing out and twisting" to convert fibers into yarn.

References

Åkervall, Lisa. 2021. "A Differential Theory of Cinematic Affect." *Deleuze and Guattari Studies* 15, no. 4: 571–92.

Ball, Philip. 2015. *H2O: A Biography of Water*. London: Orion (Kindle edition).

Barad, Karen. 2007. *Meeting the Universe Halfway: Quantum Physics and the Entanglement of Matter and Meaning*. Durham, NC: Duke University Press.

Beaulieu, Alain. 2016. "Introduction to Gilles Deleuze's Cosmological Sensibility." *Philosophy and Cosmology* 16: 199–211.

Berressem, Hanjo. 2021. "In the Light of Leibniz and Lucretius: An Encounter between Deleuze and New Materialism." *Deleuze and Guattari Studies* 15, no. 4: 497–522.

Buchanan, Ian. 2021. *Assemblage Theory and Method: An Introduction and Guide*. London: Bloomsbury.

DeLanda, Manuel. 2013 [2002]. *Intensive Science and Virtual Philosophy*. London: Bloomsbury Academic.

Deleuze, Gilles. 1993. *The Fold: Leibniz and the Baroque*. Translated by Tom Conley. Minneapolis: University of Minnesota Press.

Deleuze, Gilles. 2009 [1990]. *The Logic of Sense*. Edited by Constantin Boundas. Translated by Mark Lester and Charles Stivale. New York: Columbia University Press.

Deleuze, Gilles, and Félix Guattari. 2011. *What is Philosophy?*. Translated by Hugh Tomlinson and Graham Burchell. London: Verso.
Deleuze, Gilles, and Félix Guattari. 2014. *A Thousand Plateaus: Capitalism and Schizophrenia*. Translated by Brian Massumi. London: Bloomsbury.
Evers, John. 2023. "Great Pacific Garbage Patch." *National Geographic Society*. https://education.nationalgeographic.org/resource/great-pacific-garbage-patch/#undefined. Accessed September 8, 2025.
Gibson, James Jerome. 2015 [1986]. *The Ecological Approach to Visual Perception*. Mahwah, NJ: Lawrence Erlbaum Associates.
Gray, Chantelle, Delphi Carstens, Evelien Geerts, and Aragorn Eloff. 2021. "Deleuzoguattarian Thought, the New Materialisms, and (Be)wild(ering) Pedagogies: A Conversation between Chantelle Gray, Delphi Carstens, Evelien Geerts, and Aragorn Eloff." *Matter: Journal of New Materialist Research* 2, no 1: 200–23.
Ingold, Tim. 2013. *Making: Anthropology, Archaeology, Art and Architecture*. Abingdon: Routledge.
Jacob, François. 2022. *The Logic of Life: A History of Heredity*. Translated by Betty E. Spillmann. Princeton, NJ: Princeton University Press.
Luraschi, Silvia, Federica Timeto, and Sibel Yardimci. 2023. "Situating Research on Art, Technological Practices and Literature." *Matter: Journal of New Materialist Research* 7: i–vi.
Nail, Thomas. 2018. *Lucretius I: An Ontology of Motion*. Edinburgh: Edinburgh University Press.
Nail, Thomas. 2019. *Being and Motion*. Oxford: Oxford University Press.
Philippopoulos-Mihalopoulos, Andreas. 2016. "Milieu, Territory, Atmosphere: New Spaces of Knowledge." In *Knowledge-Creating Milieus in Europe*, edited by Augusto Cusinato and Andeas Philippopoulos-Mihalopoulos, 79–95. Heidelberg: Springer.
Prigogine, Ilya, and Isabelle Stengers. 1986. *Order out of Chaos: Man's New Dialogue with Chaos*. Glasgow: Fontana.
Prokić, Tanja. 2021. "From Constellations to Assemblages: Benjamin, Deleuze and the Question of Materialism." *Deleuze and Guattari Studies* 15, no. 4: 543–70.
Rioseco, Macarena. 2018. "Following Cotton Fabric, Oil Paint and Plastic Bags." In *Proceedings of the 9th SAR International Conference on Artistic Research*, edited by Geoff Cox, Hannah Drayson, Azadeh Fatehrad, Allister Gall, Laura Hopes, Anya Lewin, and Andrew Prior, 225–39. https://www.researchcatalogue.net/view/512748/512749. Accessed September 8, 2025.
Williams, James. 2003. *Gilles Deleuze's Difference and Repetition: A Critical Introduction and Guide*. Edinburgh: Edinburgh University Press.
Williams, James. 2021 "Matter and Sense in Gilles Deleuze's *Logic of Sense*: Against the 'Ism' in Speculative Realism." *Deleuze and Guattari Studies* 15, no. 4: 477–96.

PART II

INFINITE COSMOLOGIES: PERSPECTIVES, FOLDINGS, AND TIME

Autopoiesis: Actual Occasion emerged as an attempt to express my intuitive sense of temporal substance through the tactile weaving of mind and cosmos. Space dancers trip the light fantastic, emboldened by rope goggles of entanglement, harnessing the virtual analog of expressive technology. This painting, or aesthetic machine, addresses the tactile pleasure of temporal substance in episodic drops of experience. Time takes on the vitality of matter and we feel the meaning through intensities of prehended understanding. This region of performative becoming is where my paintings are situated, and life becomes what it is when virtual meets actual on the plane of immanence. My paintings are an attempt to act through the expressivity of the cosmos rather than the communication of an anthropocentric post-rational logic. An urge to make pictures has always been driven by a resistance to the verisimilitude of the art object that exists in a Euclidean grid of pregiven coordinates. Art and picture making are a transformational position of understanding the emergence of stuff in becoming. Paintings are self-identical and do not need to be emptied out of symbols to harness the vitality of matter—a matter in which vibrates the ubiquity of consciousness. My painting then is nonhuman autopoiesis entangling me with the cosmos to join a collective harmonious region of interrelated mutual altruism. By recognising the expressivity in cosmic matter we expand outside of the set patterns of the communication model to feel the joy of the capacity to create. My paintings seek to merge epistemology and ontology and to rediscover my indigenousness to nature.

—Mikey Georgeson

Mikey Georgeson, *Autopoiesis: Actual Occasion*, acrylic on unstretched canvas, 1.2 × 1.2 m (hung with eyelets on the plane of immanence)

Chapter 5

Points of View: Deleuze, Leibniz, and the New Cosmology

Debjyoti Sarkar

Gilles Deleuze claims that philosophy must be understood as creation. The creative gesture is obviously closer to the arts, but it also belongs to science. While the creation of functions belongs to the proper domain of science, it is in the arts that the infinite flow of colors and lines are given a form, as in a painting, or in music a melody is created from silences and noises, and it is in this condensation that the artist marks her name. Similarly, the task of philosophy is the creation of concepts from the thought-flow. It is in this creation that the concepts bear the name of the philosopher, like a signature, and, in using these concepts or relating to them, we become aligned to the philosopher—perhaps Kantian or Leibnizian. The concept is also like a scream, and there are philosophers who do not sing but scream. Such are the philosophers of pathos, as distinct from philosophers of logos. In this sense, the creation of concepts is already divided between two types of philosophers: those with a singular concept, like Descartes with his concept of the cogito as a specific idea of subjectivity in the form of a proposition; but there are others, exasperated philosophers like Leibniz, who cannot stop creating.

This is what interests Deleuze: Leibniz's endless creation of concepts, as if he does not know how/when to stop. Leibniz's influence on Deleuze is evident across his entire oeuvre, starting from the lecture "The Method of Dramatization," which would be a precursor or even a condensed version of ideas developed in his *Difference and Repetition*, where he, through Leibniz, claims that the Idea must be understood as distinct-obscure rather than clear and distinct. But it is with his last book, *The Fold: Leibniz and Baroque*, that Deleuze affirms the creative aspect of Leibniz by situating him in the Baroque world. It is Leibniz's screams that we hear in this Baroque world.

Leibniz's endless creation is symptomatic of his world, the Baroque, which marked a crisis of theological foundation. It is this world that

Leibniz, as we will see, tries to save—providing it with the philosophy it lacks. Deleuze's reading of Leibniz connects his philosophy to the Baroque world in all its manifestations—painting of light and shadows, architecture of floors that are high and low, music of harmony, mathematics of the calculus, and even biology of preformation. Deleuze not only derives the concept of the fold from Leibniz's work, where it is present only disparately, but also creates this singular concept of the fold. Because of this double task of reading and creation, we do not just find Leibniz in the pages of *The Fold*, but it is rather Leibniz-Deleuze who speaks.[1] This new concept of the fold, which is inserted in the history of philosophy, can now be discovered not only in the Baroque but also in periods before and after it. And hence, it also becomes a contemporary concept. But the specificity of the Baroque in Leibniz's work is that folds go on till infinity and are also hierarchically arranged according to forms of infinity on two floors, the pleats of matter occupying the lower floor and folds in the soul the upper, closed floor. This necessity of arrangement of forms of the infinite indicates that the problem of the infinite arrives as a crisis in the Baroque world, which now must be arranged or controlled.

One way of reading this crisis is to relate it to a new cosmology, which though inaugurated by Copernicus with the theory of a heliocentric cosmos as opposed to the Ptolemaic geocentric one, finds its radical assertion with the infinitization of the universe. However, it is not that this relation is a causal one, meaning that a new cosmology directly effects the thinking of a period; rather it should be understood in the commonality of how a certain period might think of the cosmos and of humans and their earthly relations. This shift in thinking about the universe can also be understood if we consider the change in thinking of a different object from art, like the object of theater. Deleuze notes, "For some time the world has been understood on a theatrical basis, as a dream, an illusion" (2013, 143). In the Christian Middle Ages, the metaphor *theatrum mundus*, the theater of the world, was mobilized to indicate the impermanence of the world or its illusionary nature. Theater became a problem for Christianity, and this mobilization of the metaphor worked in two ways: it allowed for a condemnation of theater, as a place of entertainment or falsity, but it also became a pedagogical tool, to preach that the world too is like a theater in its illusionary nature and must be sacrificed for the Christian truth, for a world to come.[2] But theater has also been understood as an object of systematization of hierarchies, of knowledge systems. The Renaissance perhaps shows such a desire.[3] This notion of theater understood as an illusion but also as a place of organization changes with the Baroque. Leibniz understands the Baroque as an Italian theater or even a Harlequin's costume, where everything is in transformation. This transformative nature will be affirmed by Leibniz, because through his philosophy it can be arranged.

This arrangement of the hierarchies, of assigning position to specific aspects, of which a closed theater of/in the world is indicative, is related to the Christian principle of *analogy*. This principle goes hand in hand with the hierarchical principle of the arrangement of the cosmos as present in Ptolemaic cosmology. This analogy considers a separation between God and humans, heaven and earth, supernatural and natural; this is how both medieval epistemology and its social order are arranged. In this sense, though Christianity does away with the notion of the eternal cosmos of the Greeks, it is unable to let go of Ptolemaic cosmology. The Copernican revolution poses a challenge to Christianity because a rethinking of the structure of the cosmos challenges the assumptions of Ptolemaic cosmology as the basis for any possible arrangement of knowledge. Jacob Taubes recognizes this challenge posed by the Copernican revolution to argue that after this scientific revolution, the basis of thinking based on the cosmos had to be rethought—from analogy one must move to dialectics (2009, 165–76). At the heart of such an understanding is that thinking can no longer be based on *any* cosmology; it can only be acosmic. But the Copernican model still puts forward a closed world with a different center.

Alexandre Koyré, in his book *From the Closed World to the Infinite Universe*, elucidates how philosophers and scientists of the sixteenth and seventeenth centuries started to think of an infinite universe. From Cardinal Nicholas of Cusa's metaphor of the "infinite sphere" to Giordano Bruno's affirmation of the existence of infinite worlds, the period marks a shift from the thinking of the macrocosm as reflected in the microcosm, an analogical relation, toward a belief in the infinite potential of the human intellect. Sjoerd van Tuinen notes that this world has not yet succumbed to the geometrization and mechanization of nature, and hence "the first principle for reality itself to become possible is neither God nor Nature, neither the Subject nor simply Being as differentiated from beings, but multiplicitous becoming" (2022, 6). This is the Baroque world, the world of Leibniz, which must be understood not as a theater with its stable identity, but as a world that becomes theatrical—hence the specificity of the Italian theater. Deleuze-Leibniz will indicate that this world is an infinite curve with no stable identity of the object. This new object which is always in relation to this infinite curve is what Deleuze calls "objectile," which is a singular event.

The notion of the subject also changes with the changing status of the object. The subject becomes a "superject," a term Deleuze borrows from Whitehead, such that it apprehends the variation of the object (Deleuze 2013, 21). The subject is a "point of view," and this introduces perspectivism into Leibniz's philosophy. But before discussing this specific idea of Baroque perspective, it is essential to understand Leibniz's creative gesture in formulating the notion of the subject.

Leibniz considers identical propositions which are either reciprocal or where the predicate is included in the notion of the subject. These propositions can also be called analytical. The principle of identity can then be put as: All analytical statements are true. Leibniz considers the reciprocity of this statement and proposes that "all true statements are analytical"; this is where we hear him "scream." In doing so he argues that anything said with truth about the subject must already be included in the notion of the subject. This is what he calls the principle of sufficient reason, which he considers the reciprocal of the principle of identity. The principle of identity is the principle of essences, justifying why there is something rather than nothing, while the principle of sufficient reason answers the question, why is there this rather than that? Leibniz shows that this subject must be unique, indicated by a name or concept. He reconciles this problem of substance and subject by introducing the principle of indiscernibility. "Caesar crossing the Rubicon" or "Adam sinning" must be already included in the notion of Caesar and Adam respectively. Such an *arrangement* of propositions solves the problem concerning the variety of propositions, be that of reactions of comparison or relation of activity. Hence, all forms of propositions become propositions of judgment.

All the examples that Leibniz uses are factual propositions that bring into consideration the problem of the world: "Caesar crossing the Rubicon," "Adam sinning," "Sextus raping Lucretia." If the reason for each predicate is present in the notion of the subject, what about the causes of predicates? Causes which are exterior are never sufficient; they go on indefinitely, as distinct from the interiority of reason. The problem of causes brings into question the problem of the world, and this is where the principle of continuity becomes essential. Leibniz proceeds to propose that each subject encompasses the totality of the world. But this does not mean that there is a universal subject, which would be a problem for the theological Leibniz. A subject encompasses the totality of the world from causes to the effects that go on infinitely, but Leibniz adds a restriction to it—he says that every notion of the subject encompasses the totality of the world but *expresses* it from a point of view. Deleuze says that in introducing this concept of point of view, Leibniz belongs to a tradition of perspectival philosophy, whose modern persona is Nietzsche. In Henry James's works, we find similar use of perspectivism; also in the geometry of Desargues before Leibniz. This notion is significant in a world which has lost its center. The following section elucidates this concept with specific reference to an infinite universe.

The Baroque Perspective

Taking Michel Serres' work on Leibniz, Deleuze explains why the point of view as a concept is both necessary and possible in an infinite universe that has lost its center. Deleuze attributes the beginning of such a perspectival philosophy to the Renaissance and to Cardinal Nicholas de Cusa. It is interesting to note that it is perhaps Nicholas de Cusa who first uses the metaphor of the infinite sphere with circumference nowhere and center everywhere for God. Karsten Harries argues that it is with this transference of the metaphor into the world that, for the first time, we have a notion of an infinite universe (2001, 42–63). It is also with de Cusa that we have a theory of perspective. In his *On Learned Ignorance*, he argues for a coincidence of opposites which is based on a perspectival understanding. Comparison between opposites like the small and the big is only possible in a finite understanding, but all these opposites coincide in the realm of the infinite. Similarly, all humans have different perspectives or points of view but what is necessary is the recognition that these are simply "perspectives." It is through this recognition or this "learned ignorance" that one moves beyond one's limited perspectival understanding. When one recognizes that one's point of view is one among multiple points of view, it is already a step toward moving beyond it. Though there is an infinite separation or distance from God, recognizing the respective perspectives is somehow to transcend these finite positions. In contrast to this, God's point of view is like the portrait de Cusa mentions in *On Learned Ignorance*—the portrait always looks back, independent of the position of the spectator. In taking these steps, de Cusa's thinking surely indicates an indefinite, if not infinite, world which is not based on the hierarchical cosmos. For him, the earth is no less illuminated than the sun and is like the heavens in its perfection.

De Cusa can be positioned at the beginning of the Renaissance, where perfection becomes a category. The use of perspective in paintings is advocated by Alberti in *On Painting*, a textbook that depicts the notion of beauty and perfection, but also teaches artists how to use the tool of perspective to make paintings that have *concinatas*, or a perfection based on proportions, colors, and figures. This creates what Alberti calls a *historia* or a fiction.[4] But this fictional world cannot exist in the Baroque. To understand why, it is necessary to contrast two thoughts of infinity in the Renaissance and the Baroque. While the thinking of an infinite space in the Renaissance creates the possibility of an idea of perfection in heaven and on earth, in the Baroque this infinity poses a challenge, indicating a disordered world which runs into chaos. This world cannot sustain the imagination of a painting as a window opening into the world, like that of the Renaissance, but is instead interiorized such that this interiority is marked by all sorts of grids and lines. This is also where one shifts from seeing to reading: "The painting-window is replaced by tabulation, the grid on which lines, numbers, and changing characters are inscribed (the objectile)" (Deleuze 2013, 30).[5] Deleuze shows how Leibniz, as a philosopher, must intervene to counteract this disordered world.

Using the concept of the fold, Deleuze reimagines Leibniz's conception of an interiorized subject. In this sense, Deleuze's move is anti-Cartesian, and he places Descartes and Leibniz in opposition on several occasions. If Descartes thinks through lines, it is Leibniz who thinks through curves, through folds. This specific opposition also plays a key role in understanding modernity as a "labyrinth." It can be said that, while Descartes recognizes that modernity is like a labyrinth, his attempt is to find a stable position, the cogito, which can escape this labyrinth. However, in doing this, the foundation that is the cogito separates itself from the world, making it a "world-picture."

Leibniz too recognizes the world as a labyrinth but provides a different solution to the problem. The concept of the fold problematizes the notion of the subject as a stable identity. It is not just that the world has a labyrinthine structure, but that there are labyrinths in the soul. The soul is a structure of absolute interiority which only communicates with the world because of its point of view. The point of view is a point of view on inflection, and these expressions differentiate the subjects. The world as an infinite curve and the points of view that constitute it are not to be considered in the sense of a gap or a void but a continuous differentiation. The inflections of this infinite curve, their variation or rather differentiation, constitute this point of view. Though the subject envelops the totality of the world, it does not understand it, and each subject or monad only reflects a little portion, "little perceptions," within it—making the world hallucinatory. In this sense Leibniz provides an equally hallucinatory concept for the hallucinatory world he belongs to.

The Leibnizian way is neither to fall into the depths of this illusion nor to emerge from it, but "rather [to realize] something in illusion itself, or of tying it to a spiritual presence that endows its spaces and fragments with a collective unity" (Deleuze 2013, 143). In this sense, the Baroque perspective is more of a transformative one.

This indicates that the point of view that Deleuze conceives is not a multiplicity of various points of view which would make truth relative—each according to one's point of view. Rather, the subject is constituted because of the variation of the points of view. In this sense, it is not a variation of truth, but a truth of variation.

The condition for imagining a philosophy of perspectivism is the existence of an infinite world where a given stable center cannot be found. While Descartes imagines a stable center, as if in response to this infinite universe, Leibniz invents the point of view. The new theory of conic sections, as Michel Serres also points out, supports Leibniz's desire for a philosophy of perspective. The point of view then becomes a power of "arranging cases," like the tip of the cone which arranges the finite point, infinite straight line, finite circle, infinite parabola, finite ellipse, and the infinite hyperbola. This power of arranging cases is also that of jurisprudence, and Leibniz, the thinker of the city, knows how this new center can arrange the variations. Deleuze also reminds us that Leibniz's philosophy must be considered as a philosophy which acts on different levels. Just like the construction of the Baroque house with its two floors, Leibniz's oeuvre is also based on a variation of levels. This distinction is also necessary to differentiate between soul and body, soul being at a higher level than the body.

There are three kinds of points: 1) the physical point, which is in variation but not exact, like a point-fold; 2) the point of inclusion or mathematical point, which is rigorous but not exact and becomes a focus or a site; this is the point of position and a point of view; and 3) the metaphysical point, projected onto the point of view as its modality. This metaphysical point is the point of inclusion. But all these points are inseparable and dependent on each other's existence. These three points are also types of singularities which have the form of explication–implication–complication. This logic of explication–complication is also present in de Cusa's thinking. Here, the universe is understood as a development or *explicatio* which is inadequate of God, whereas God is a *complicatio* which is a unity. Leibniz allows for the existence of individual substances, but the tip of the cone forms the unity to be understood not as universal but an individual unity. This is why he mobilizes the concept of the monad. Though it had existed since Plotinus, Leibniz's use, as Deleuze notes, allows him to "stabilize the concept. On the one hand, the mathematics of inflection allowed him to posit the enveloping series of multiples as convergent infinite series. On the other hand, the metaphysics of inclusion allowed him to posit enveloping unity as an irreducible individual unity" (2013, 25). This resolves the issue of thinking of the individual within an infinite universe without falling into the problem of relativism. The world as an infinite series is implicated in the soul, but it is the soul that also complicates the world—the point of view is on the variation where the infinite series converges:

> So, we discover a new idea of an individual. For some, the Nominalists, individuals would be the only existants, concepts being only carefully ordered words; for others, the Universalists, the concept has the power of being infinitely determinable, the individual referring only to accidental or extraconceptual determinations. But for Leibniz, at the same time only the individual exists, and it is by virtue of the power of the concept: monad or soul. Thus, this power of the concept (to become a subject) does not consist in determining a genre to infinity, but in condensing and in prolonging singularities. The latter are not generalities but events, or droplets of an event. (Deleuze 2013, 74)

With Leibniz, Deleuze also confronts two problems: the problem of representation that the Nominalist viewpoint projects and the subsumption of individuals under the "infinitely determinable" concept. This implies that we move from the percept to the concept, from something like seeing to reading. If the individual is the concept, then what we see about this individual is what is read in its concept. This means that the individual is interiorized, as if in a closed "theater of reading." Here Deleuze's reference to Mallarmé and his creation of a closed book is essential. The Mallarméan theater which is this monad is a theater of reading, where the singularities or inflections or even the events are read in the depths of the folds in the soul. From seeing to reading we move toward a signature, because this concept also carries a "name," and it is Leibniz who for the first time assigns concepts a name. There is a concept of Caesar which being in the world envelops all the events like "crossing the Rubicon" in its point of view, which is then read in its folds, leading to the signature or the name "Caesar."

What is implied in such an understanding of the concept is the necessity of the world. The monad relates to the world virtually. It is this world that is actualized in the monad. One can clearly understand the relation of the virtual and the actual which Deleuze brings into his reading of Leibniz. Against the Heideggerian argument of "being in the world," Deleuze posits "being for the world" as the condition of the monad's closure. For Heidegger, the monad does not require any windows because it is already outside, and hence in conformity with the world. Deleuze argues that God already creates the world, choosing the best from other infinite worlds where the monads exist. God not only creates "Adam the sinner" but the world in which Adam sins. The condition of the closure of the monad is because it envelops the world within it. Through its finite differential perceptions, it includes the infinite curve that the world is. In this sense, the world is also a production or a result. The closure of the monad allows the world to begin over again. The world is in the monad and hence it includes the series of this infinite curve, but the monad is also for the world, and hence does not contain clearly the reason of which it is the result. And this is the way of expression: "the soul is the expression of the world (actuality), but because the world is what the soul expresses (virtuality)" (Deleuze 2013, 28).

This condition of being for the world also marks Deleuze-Leibniz's separation from Nicholas de Cusa and from the specificity of the notion of the point of view. De Cusa understands point of view as a constituted category which must be given up through the recognition that it is just a point of view. This also implies that such an understanding does away with the notion of individuality. However, in Deleuze's reading, point of view constitutes the subject in the sense that it creates a unity though maintaining individuality. It is only this point of view which, as we will see, allows for comprehension.

Allegory and the Copula

Walter Benjamin's study of allegory in the Baroque provides a crucial understanding of Deleuze's thinking of the concept. Allegory and symbol are both forms of expression, but they must be opposed. While symbol stands for representation, allegory is an incomplete gesture which shows the failure of representation or rather its capture. In this sense allegory is a continuous gesture and is always incomplete. Allegory has the power of figuration, just like Bacon's paintings, which Deleuze says, "attempt to release the presences beneath representation, beyond representation" (2003, 51). These forms of expression have a different relation with respect to the object. In symbolism, the object is separated from its conditions, or the universe, and put in conformity not with the concept but rather with "an idea that develops this concept morally or esthetically" (Deleuze 2013, 144). This means that the object becomes a representation as long as it represents an idea, be it God, history, or nature, and finally stabilizes itself around this center.

Allegory, however, functions like a cone. Objects are not separated but flow from their frames so as to be connected to a series, and finally concentrated and interiorized in the concept. In this manner allegory also follows the structure, that we mentioned earlier, of seeing, reading and signature or the name. The new object, or the objectile, which flows becomes an event; in this sense, there is no distinction between natural and historical and, hence, they are neither attributes nor essences like those in symbols. Reading concerns the inscription of these events, where all propositions assume the structure of a predicate. This is where the issue of jurisprudence becomes crucial—becoming a predicate of any proposition allows it to be included in the domain of jurisprudence. Finally, the name or signature is what includes all these predicates; it is not a specific quality that the allegory defines but the quality of someone with a name.

Mogens Lærke (2010) argues that this version of the Leibnizian Baroque must be distinguished from the Benjaminian reading of the Baroque. Though allegorization is the central theme (Deleuze also mentions that for Leibniz it is the "allegory of the world" rather than the "symbol of the cosmos"), Benjamin's reading considers there to be a negativity at the heart of Baroque life. The infinite iteration which the allegory indicates is because of the lack of any authenticity, or a referent, and this iteration shows that the Baroque human can only be a mourning subject. This mourning subject lacks any subjective self-determination, and allegory shows the continuous deferral of authenticity.

The problem of freedom is at issue here. The Leibnizian version would be "delightful," in the sense that it ties the subjective determination of humankind with the objectivity of God. The lamentation then results only in the hatred of God, and to take the example of Sextus, he is only condemned because he does not recognize that he resides in the best possible world. It is then necessary to affirm this pure determinedness of God, rather than lamenting about one's subjective determination—that is, to recognize not just that this is the best world, but that God is the "good master." In this sense, delight is derived from finding the convergences from divergences—order derived from the chaotic world. It seems that what interests Deleuze in this Leibnizian version is precisely these two aspects: first, the presence of infinite within the finite and, secondly, this objective determination.

The first aspect concerns the principle of continuity, which is also the principle of existence. This is because it concerns being in the world and the series of causes and effects that go on infinitely. We are here in the domain of infinite analysis, infinite because of the infinite series of causes/effects and analysis because these causes/effects must be included in the concept of the subject. But individuals, as finite beings, are denied this power of infinite analysis, because it is only God who can carry out such a task. In choosing from infinite other worlds, God has already determined the best world for human beings. Yet Deleuze would disagree with such a reading of Leibniz. It is his claim that humans can take part in this infinite analysis. If the individual encompasses the infinite world without clearly understanding it, does that mean that humans as finite creatures do not comprehend this infinity, or, to put it differently, is it only God that can take part in this infinite analysis? Daniel Smith puts this succinctly:

> It's here that we begin to approach the originality of Deleuze's interpretation of Leibniz; for according to Deleuze, Leibniz indeed attempts to provide us finite humans with an artifice that is capable of undertaking a well-founded approximation of what happens in God's understanding, and this artifice is precisely the technique of the infinitesimal calculus or differential analysis. (2005, 138)

Thus, Deleuze tries to solve Leibniz's problem of a contradiction between the principle of indiscernibility and the principle of continuity, but he also departs from him by bringing in a new theory of singularities from mathematics. The law of indiscernibility argues that for each substance there will be a single concept, while the law of continuity ensures the existence of these substances because of an infinite series of causes: an individual has a concept, and the world is included in this concept. This means that the individual can be differentiated in its concept, but she also exists in a world which is undifferentiated. Deleuze's theory of singularities shows that the singular must be understood not as opposed to the regular, but as that which escapes this regularity. The notion of point of view thus corresponds to such singular points, existing within the ordinary in the infinite curve. In this sense the infinite curve must not be seen as constituted by these singular points, which are prolonged along its neighborhood.

The relation between humankind's subjective freedom and objective determinedness is the second aspect. Though Deleuze is not in favor of a theological principle, he agrees that subjectivity is not already given. So, as we will see in the next section, it cannot be said that the human player plays at his own will. It is rather that the play happens through him. For Leibniz, God makes the convergent series possible, and this becomes the condition for the point of view that constitutes the human being. In his creation of the world, the singularities are already given. The human task is then to affirm the singularities which he envelops and the world where he exists. The freedom of the monad is then tied to the objective principle of God's world, and it expresses the world that it inheres. In this way, the liberty of the monad is conjoined with the absolute determination of God. Deleuze will reject the principle of God and seek to determine the "impersonal and preindividual nomadic singularities that constitute the real transcendental field" (2019, 105). Thus he sides with Spinoza rather than Leibniz. But it is these two ideas from Leibniz that he retains. In this regard we agree with Hammond (2010), who claims that if Spinoza is the Christ of philosophy, for Deleuze, Leibniz is surely John the Baptist.

The Ideal Game, a Thinking of the Chaosmos

Though the Baroque, standing against the backdrop of the infinite universe, is a time of crisis marking the failure of classical reason, it should be noted that it is not *properly* modern. What does this mean? If modernism represents a falling into nothingness whose symptom is marked by nihilism, Baroque is both historically and conceptually the period prior to this. In the Baroque something is kept rather than nothing. Deleuze notes that prior to the neurotic phase of modernity, the Baroque shows a psychotic phase. From the theological reason where the principle is One to the modern condition with no principles, the Baroque posits a solution against this fall of the single principle. And for his proliferation of principles, endless creation of concepts, and, hence, screams, Leibniz remains the Baroque philosopher *par excellence*. He proposes that if there is no single principle governing the world one should multiply principles. "The Baroque solution is the following: we shall multiply principles—we can always slip a new one out from under our cuffs—and in this way we will change their use" (Deleuze 2013, 76). The new principles that Leibniz proposes are not like the theological principle which governs both essence and existence. Rather, Leibniz changes their use—he makes them reflective. The theological conception agreed that there is a unitary principle, and that objects must be evaluated based on what object responds to this given principle. This makes it a form of law that governs, and we noted earlier how the object of theater became a problem for such a law. In Leibniz's use, there is a reversal of this logic. There are hidden principles, but which principle corresponds to which object must be examined, and, in this sense, he makes these principles reflective. Therefore, for every object, principles are also invented.

This shifts Leibniz's domain from law to jurisprudence, which is evident in his conception of the propositional structure that he develops from the reversal of the principle of identity and which further reflects in the monad as enveloping the world. Hence Deleuze calls this "a honeymoon of singularity and the concept" (2013, 77). What follows from this is a new idea of the play: the play of principles or the play of singularities. The earlier idea of the play of the world or theater of the world exists because of the master player, God, as a spectator of this world. The modern sequence shows how with the fall of this divine principle, the figure of the child player becomes important. In both Nietzsche and Mallarmé, play is not based on any divine principle but on chance, which can become the only principle. Thus, the child player is not just the one not concerned about the rules of the game, but the one who breaks the board. This toppling of the board is also the world where divergence is affirmed. In Deleuze's vocabulary, the Leibnizian world is where the singularities converge. In this sense Leibniz's best is also not a selection among all the goods, but is a possibility only in the ruins of the good, a final attempt to save the good.

This notion of the world also provides a new understanding of the play. This play of principles refuses both chance or nothingness; in this sense, it becomes like a game of skill and of filling holes. Both these criteria are also essential to think of the ontology of the fold. A fold must be understood without any void, in the sense that the inflexion on the curve is a differential relation, a variable but never a void. This attempt also corresponds to Deleuze's mobilization of mathematical ideas from Riemannian geometry, a mathematics which does not think of holes or voids but rather a continuous variation.[6] This game is not

> guerrilla warfare but rather it is a nonbattle closer to guerrilla warfare than a war of extermination, more like go than chess or checkers: You don't catch your adversary in order to reduce him to absence, you encircle his presence to neutralize him, to make him incompossible, to impose divergence upon him. (Deleuze 2013, 77)

In this sense, Leibniz, who stands as God's lawyer, rebuilds the ruined world but knows that a return to the earlier divine principle is difficult. In his construction he allows for the liberty of humans, making it possible but not compossible with the world.

But our world is not the Baroque world, an infinite curve sustained by a harmony where the series converge, but, rather, a world of divergences.

In this regard, Deleuze recognizes this problem and departs from Leibniz. Leibniz provides Deleuze with a way to think of differences without representation and also indicates that the task of the philosopher must be the creation of concepts. But Leibniz's idea of God, the World, and the Self takes on a different meaning with Deleuze, and this is where Deleuze reads the contemporary. God is no longer a being who chooses the best, resulting in its existence along with the monads who occupy this compossible world with a pre-established harmony. Now God is replaced by a pure process which allows all virtualities to pass into existences as convergent and divergent series. In this new world, harmony is lost and divergences can be affirmed, becoming not a chaotic world but rather a "chaosmos." The new game of the world is based on chance and not a proliferation of principles. This changes the state of the self or monad too. It cannot be conceived as an absolute interiority but is made to be forced open as if by "a set of pliers," which are the divergent series (Deleuze 2013, 156). Baroque harmonics allowed for a selection of monads and a distinction based on the private and the public, dominant monads which are harmonic and those which are part of the masses, leading almost to the bottom as the place of the damned who are in discord with this world. This then allows for an elevation or raising of status, like the structure of the cone, or the pyramid, or the upper floor. Such a differentiation of hierarchy collapses, and monads are now nomadic. Thus Deleuze demands a nomadology instead of Leibniz's monadology.

Leibniz's solution to the fall of classical reason remains only temporary, and hence Deleuze's theory of singularities both takes from Leibniz but also departs from him to think about the contemporary situation. This is what also separates his idea of the game from the Leibnizian one. In "Tenth Series of the Ideal Game" from his *Logic of Sense*, Deleuze describes an ideal game which can be played only in thought. In the classical schema, the game is a part of human activity with certain rules which are already given, and the effect of chance is only present at certain points. This also means that it is the counterpart to human activities which are not games but those like morality and labor. Games, then, act as a caricature of these activities. This is like the Leibnizian conception of a game under police control—freedom is allowed only as long as it is guided by a rule, that of compossibility for Leibniz, under the principle of the best.[7] To think of another game which is the ideal game would be to consider a game without rules, in which with each move a new rule is invented. All throws or moves of that game will affirm chance such that chances are not distributed at specific points in the game, that is, they are not numerically counted. Each throw is a series in a time, which Deleuze, from the Stoics, will call Aion, and to this time the singularities correspond. But this point to which the singularities correspond is displaced continuously and becomes the aleatory point. Aion is like an infinite line, a labyrinthine line that Deleuze cites from Borges, that breaks the continuity and circularity of the Chronos which is the present. The throws are successive with each other, forming a constellation (or cosmos), but also simultaneous in relation to this point, which is the unique cast (or chaos), which is eternally displaced. This ideal game disturbs the economy of the world which can now not be convergent or harmonious; it disturbs its given configuration.

For this reason, the winner of this ideal game is not one who economizes chance by calculating each move. Rather, one wins only by affirming chance. However, this ideal game can only be played in thought: "For only thought finds it possible to affirm all chance and *to make chance into an object of affirmation*" (Deleuze 2019, 63). For this game, the world must be toppled, as with the child player. This game is possible only where divergent series exist along with convergent ones, as if "a periodic infusion of chaos into cosmos," as Deleuze cites from Borges's description of the Babylonian lottery (2019, 63). It is in this ideal game that Deleuze aligns himself with Nietzsche's profound notion of the point of view. In an infinite universe, Leibniz invents point of view as constituting the subject but places it under strict rules by rejecting divergences. Such a situation corresponds to the thinking of the cone, as we have seen, which allows the propositional forms and even allegory to function. This structure amounts to a form of jurisprudence. The toppling of the world indicates that such a structure is no longer possible, and now in a truly infinite universe, in its positive sense a world where divergences must be affirmed, what is only possible is a point of view on point of view, or a displacement of perspectives.

Notes

1. Deleuze says that a concept bears a signature, but it is interesting to note that the concept of the fold cannot be assigned the name of Leibniz, unlike other concepts such as the monad or compossibility. I take this combination of Leibniz-Deleuze from Badiou's conjunction of these two names in his review of Deleuze's book to indicate the double task of reading and creation.
2. Soumick De's work is of importance in this respect, because it shows not only how the metaphor of theater was deployed in the Christian Middle Ages, but also how this metaphor changed its meaning with Christian thinkers, notably Meister Eckhart, whose mystical orientation corresponds to a way of thinking this theater of the world without analogy, in being infinitely separated from God. In this

sense his thinking also corresponds to a new idea of cosmology but it finally leads to a closure of such radical affirmation (De 2022).

3. The Renaissance is witness to thinkers who used the structure of theater to organize knowledge. In his book *Universae Naturae Theatrum* (1580), Jean Bodin, the French jurist, uses theater not only as a metaphor but as an arrangement of positions where objects are fixed. This is what he calls *dispositio*. In another Italian thinker, Guilio Camillio, we find a similar desire to create an actual theater where the entire knowledge of the world can be arranged. Each position in the theater would house a certain form of knowledge, but there would also be movement between these positions. All of these issues or attempts are also connected to the mnemonic technique and the use of a certain "theatre of memory" (Rossi 2000).
4. It is Hubert Damisch's contention that with Alberti, perspective is made into a mere technique for the production of paintings, and it loses its original meaning present in Brunelleschi's invention of perspective, comparable to Thales' invention of geometry. In this sense Damisch's book, which borrows from Husserl's *Origin of Geometry*, puts forward the view that the origin of perspective must be seen as a demonstration of the thought of an infinite space, similar to how Thales separated geometry from empirical objects to make it into thought (see Damisch 1995).
5. This is why Deleuze also compares this specific interiority or this theater of reading, of lines and tabulations, to that of paintings by Rauchenberg. In this sense Rauchenberg represents a neo-Leibnizian attitude (Deleuze 2013, 30).
6. This is how Badiou distances himself from Deleuze. He argues that Deleuze's ontology remains organicist or of the Great Animal. The concept of the fold corresponds to such an ontology, and hence uses the mathematics of continuous curves like Riemannian geometry. Against this, Badiou aligns himself with Cantor's set theory, and hence the challenge to Deleuze's reading of Mallarmé as a thinker of the fold follows from this fundamental ontological decision (Badiou 2012, 243–7).
7. In his lecture on Leibniz, Deleuze cites "Funny Thought," where Leibniz imagines an academy which would allow for various types of games, but all under police control. The point of view then determines whether one thinks of it as institutional control or enjoys the game, even while recognizing the restriction itself (1998, 82).

References

Badiou, Alain. 2012. "Gilles Deleuze, *The Fold: Leibniz and the Baroque*." In *The Adventure of French Philosophy*, edited by Bruno Bosteels, 241–68. London: Verso.

Damisch, Hubert.1995. *Origin of Perspective*. Translated by John Goodman. Cambridge, MA: MIT Press.

De, Soumick. 2022. *Perversion Pedagogy and the Comic: The Problem of Theatre in the Christian Middle Ages*. New York: Routledge.

Deleuze, Gilles. 1998. "Vincennes Session of April 15, 1980, Leibniz Seminar." Translated by Charles J. Stivale. *Discourse* 20, no. 3: 77–97.

Deleuze, Gilles. 2003. *Francis Bacon: The Logic of Sensation*. Translated by Daniel W. Smith. New York: Continuum.

Deleuze, Gilles. 2004. "The Method of Dramatization." Translated by Michael Taormina. In *Desert Islands and Other Texts 1953–1974*, 94–116. Los Angeles: Semiotext(e).

Deleuze, Gilles. 2013. *The Fold: Leibniz and the Baroque*. Translated by Tom Conley. London: Bloomsbury.

Deleuze, Gilles. 2019. *The Logic of Sense*. Translated by Mark Lester and Charles Stivale. London: Bloomsbury.
Hammond, Matthew. 2010. "Capacity or Plasticity: So Just What is a Body?" In *Deleuze and the Fold: A Critical Reader*, edited by Sjoerd van Tuinen and Niamh McDonnell, 225–42. London: Palgrave Macmillan.
Harries, Karsten. 2001. *Infinity and Perspective*. Cambridge, MA: MIT Press.
Koyré, Alexandre. 1957. *From the Closed World to the Infinite Universe*. Baltimore, MD: Johns Hopkins University Press.
Lærke, Mogens. 2010. "Four Things Deleuze Learned from Leibniz." In *Deleuze and the Fold: A Critical Reader*, edited by Sjoerd van Tuinen and Niamh McDonnell, 25–45. London: Palgrave Macmillan.
McDonnell, Niamh. 2010. "Leibniz's Combinatorial Art of Synthesis and the Temporal Interval of the Fold." In *Deleuze and the Fold: A Critical Reader*, edited by Sjoerd van Tuinen and Niamh McDonnell, 65–88. London: Palgrave Macmillan.
Nicholas de Cusa. 1981. *On Learned Ignorance*. Minneapolis: A.J. Benning Press.
Rossi, Paolo. 2000. *Logic and the Art of Memory: The Quest for a Universal Language*. Translated by Stephen Clucas. London: Continuum.
Smith, Daniel W. 2005. "Deleuze on Leibniz: Difference, Continuity and the Calculus." In *Current Continental Theory and Modern Philosophy*, edited by Stephen. H. Daniel, 127–47. Chicago: Northwestern University Press.
Taubes, Jacob. 2009. *From Cult to Culture: Fragments toward a Critique of Historical Reason*. Stanford, CA: Stanford University Press.
van Tuinen, Sjoerd, 2022. *The Philosophy of Mannerism: From Aesthetics to Modal Metaphysics,* New York: Bloomsbury Academic.
van Tuinen, Sjoerd, and Niamh McDonnell, eds. 2010. *Deleuze and The Fold: A Critical Reader*. London: Palgrave Macmillan.

Chapter 6

The Lily of the Field Sings the Glory of the Heavens: The Cosmos of the Fold

Bryan Noonan

This chapter explores the cosmological implications of two relatively overlooked concepts in Deleuze's corpus—*the contemplative soul* and *the fold*—to paint a picture of Deleuze's cosmos of expression by explicating the strange resonance that he weaves between seemingly disparate thinkers like Leibniz, Riemann, Butler, and Whitehead. Often using the concepts of one philosopher interchangeably in his explanations of another,[1] our exegetical goals quickly become difficult, as we must cover a lot of ground to put these concepts into concert if we are to properly illustrate the madcap cosmological vision that Deleuze unfolds.

It goes without saying that the most important thinker in *The Fold* is Leibniz; therefore there is no better place to start. In his article on the interaction of Deleuze and Leibniz, Lærke provides a concise and astute summation that, for Deleuze's purposes, Leibniz's work contains four important *motifs*: "(1) a conception of objects as events, folds, and forces; (2) an expressionist and perspectivist account of subjectivity; (3) a conception of a differential unconscious; (4) a joyful affirmation of the infinite" (Lærke 2010, 25). I will add one important addendum to Lærke's account: for Deleuze, Leibniz is, anachronism aside, a thinker of *topological space*. Leibniz therefore is a precursor to Deleuze in thinking of space *intensively*, rather than merely *extensively*.

These five motifs provide us with a roadmap for what is admittedly a bit of a romp through a wide array of ideas in both Leibniz and Deleuze. The general strategy of this chapter will be akin to peeling an onion—with the passing of each propaedeutic layer, we will go deeper into the complexity of Deleuze's difficult concept of the fold and its implications for cosmology. We will begin with a broad vision of the expressive cosmos and then perform some exegesis on Leibniz's doctrine of metaphysical and physical points. As we shall see, these metaphysical points correspond to Deleuze's notion of *contemplative souls*, and the physical points will correspond to the material folds of the "lower floor" which are integrated and "contemplated" within a soul.

According to Deleuze, the cosmos is folded, or perhaps it would be even more accurate to say, it is a labyrinth, infinitely folded, folding, unfolding—a processual vision of the universe, which is not populated by elementary extensional points in space, but rather the anti-extensional folding of primitive forces. No matter how small bodies of our cosmos get, no matter what microscopic chemical strata we explore, we will always find that our world is infinitely "pierced with irregular passages" (Deleuze 1993, 5), always opening to more and more worlds swimming in the infinitely small.

Deleuze points to Michel Serres as noting that this Leibnizian cosmos is a model which has "lost notion of a center" and exchanged an infinity of points of view for this missing center. There is no center; wherever we delve, we continue to swim in the infinite: the cosmos "is the infinite curve that touches at an infinity of points an infinity of curves, the curve with a unique variable, the convergent series of all series" (Deleuze 1993, 23). There is not a universal, privileged point of view, but only an infinite number of perspectives—a decentered cosmos that stretches into the infinite, an infinite labyrinthine cosmos whose consequence is not the "relativity of truth," but "the truth of relativity" (Deleuze 1993, 23)—an infinity of points of view, an infinity of centers, a new harmony. This is a cosmos where the ontology of "*is*" finally gives way to an ontology of "*and* ... *and* ... *and* ... *or* ... *or* ... *or* ..."

As Deleuze notes, the Baroque captures this cosmic vision in its great works of art, in its representations of "water and its rivers, air and its clouds, earth and its caverns, and light and its fires, themselves infinite folds [...] jewelry that burns with folds of fire; vegetables, mushrooms, or sugared fruits caught in their earthy folds" (1993, 122–3). The world is a pond full of fish, but even when one examines the individual fish, each individual fish consists of only more ponds of fish, and so on to infinity. This is a vision of the cosmos in which the "totality of the universe resembles a 'pond of matter in which there exist different flows and waves'" (Deleuze 1993, 5).

The Metaphysical and Physical Points of the Cosmos

In the "New System," Leibniz lays out a profoundly novel schema for encapsulating the nature of the cosmos. In Leibniz's youth it became obvious to him that extended mass would not be enough to account for the dynamics of reality and that a notion of force was indeed necessary.

The classical model of passive matter, for Leibniz, was simply not enough to account for reality. The infinite multitude of the cosmos requires a true unity beyond mere mathematical points derived from extended masses. A proper philosophy of matter was required to show recourse to an indivisible, non-extended formal atom, since anything extended is infinitely divisible, and, therefore, incoherently posited as the primary unit of reality.

In his search for an alternative, Leibniz is forced to rehabilitate the notion of "substantial forms" (Leibniz 1989, 139), using the term *soul*. Out of these souls, there is an originary, primitive force that accounts for the vital energy of matter. These souls are not merely what animate humans but eternally "subsist" in reality. But we cannot confuse these souls with rational souls or minds; these are material souls, and not the "little gods" that rational souls are (Leibniz 1989, 141). When an organism perishes, Leibniz argues, it is simply reintegrated into the order of matter; its soul persists and becomes the genetic element of future forms. As Leibniz writes, "instead of the transmigration of souls, there is only a transformation of the same animal, according to whether its organs are differently enfolded and more or less developed" (1989, 141).

Leibniz says that things we believe to "begin and perish merely appear and disappear" (1989, 141). They are either covered over in a folding or uncovered in an unfolding. The machines of nature have an "infinite number of organs," which are indestructible, only changing in relation to the enfolding they undergo. The unity of these souls corresponds to what we call a "self"—but the unity of these material souls is due to a more profound immaterial soul which records them, organizes them. The mass of materiality which makes up these unities is akin to an "army or a herd, or a pond full of fish, or like a watch composed of springs and wheels" (Leibniz 1989, 141).

There are not atoms of matter, but rather atoms of substance, or "real unities absolutely destitute of parts." These unities are the "source of actions, the first absolute principles of the composition of things" (Leibniz 1989, 142). It is out of these souls that material things fashion themselves, organize themselves. They are *metaphysical points*, which "have something vital, a kind of perception, and mathematical points are the points of view from which they express the universe" (Leibniz 1989, 143). These souls inhere a point of view on the mathematical curve, include the curve's inflections into their souls, and then express the universe. These vitals points might not be the only things that are real, but they are the points out of which the real emerges—corporeal bodies are merely their expression.

Leibniz calls these points monads, borrowing "this name from the Neoplatonists who used it to designate a state of One, a unity that envelops a multiplicity, this multiplicity developing the One in the manner of a 'series'" (Deleuze 1993, 23). It envelops the singularities of the virtual inflections and then develops them into the unfoldings of the organism's activity. This *implication* and *explication* are only understood through the "universal Unity that 'complicates' them all and complicates all the Ones" (Deleuze 1993, 23).

However, Leibniz hit a snag here as he developed this doctrine: "after I established these things, I thought I was entering port; but when I began to meditate about the union of soul and body, I felt as if I were thrown again into the open sea" (1989, 142). How can the union between these metaphysical and physical points be explained? How can the body communicate and make something happen in the soul? How can we explain this harmony without recourse to a *deus ex machina*? Leibniz's answer is surprising, but also quite a beautiful vision of the cosmos: "we must say that God originally created the soul (and any other real unity) in such a way that everything must arise for it from its own *depths* [*fonds*], through perfect spontaneity relative to itself, and yet with a perfect conformity relative to external things" (1989, 143). Our internal sensations, or the sensations pertaining to our soul—not to be confused with our rational ones—are "internal perceptions in the soul itself" (Leibniz 1989, 143). These internal perceptions of the folds of matter *express the entirety of the cosmos*. Endowed with a vital force, the soul can produce, in relation to the capacity of its organs, the entirety of the cosmos in conformity with its perceptions of the material folds. Therefore, the extensive is the immanent product of the intensive's expressive movements: I read about Deleuze, I include the inflections of these lessons deep into the folds of my soul, and then I express these lessons, I actualize them into a conversation over a beer—a continual folding and unfolding.

But what are these inflections which are included in a contemplative soul? Simply put, they are *events*. The soul, and by extension the entire cosmos, is made up of the events which are explicated into the actualized bodies of the world. These events, according to Leibniz, are what make up a thing's concept; they are the predicates we use to describe a thing. This is obviously very strange, but a crucial point attesting to this cosmic vision. Normally when we think of predicates, we think of a thing's attributes, and attributes are merely the inherence of a general trope within the thing itself, and not the contingency of an event. This is the madcap Leibniz that the history of philosophy has tried to explain away or cover over in its majoritarian interpretation—a thinker rife with eerie and sublime notions which swell below the skin of his thought. As Deleuze puts it, "Leibniz pushes the presuppositions of classical philosophy as far as he can, down the paths of genius and delirium" (1980, Seminar V).

A New Cosmic Logic

Here, I propose that Leibniz's notion of "sufficient reason," the battle cry of Rationalism itself, is the very problem which propelled Leibniz into his "delirious conceptual creation" (Smith 2012, 46). The conclusions of this section will help illuminate some of the intricacies of Leibniz's expressionism, crucial to understanding *The Fold*. To begin, Leibniz states that all predication has a foundation in the nature of things (Deleuze 1993, 47). This denotes that anything truly predicated of a thing is inevitably encompassed or contained within the concept of the thing. Predications include not only the essence of the thing, where the principle of identity and the principle of sufficient reason converge, but also the entirety of the affections and events linked to the thing. Consider the proposition: "Caesar crossed the Rubicon." If this proposition holds true, Leibniz posits that the predicate "crossed the Rubicon" must be contained within the concept of Caesar. "Everything has a reason" signifies that all occurrences pertaining to something—its "differences" (Smith 2012, 45)—must be eternally encompassed or included in the individual notion of a thing. If we designate an "event" as something occurring to a thing, we may assert that sufficient reason comprehends the event as one of its predicates: the concept of the thing, or its notion. "Predicates or events," Leibniz posits (1989, 47).

This Leibnizian undertaking is one that had been eschewed by earlier philosophers such as Aristotle, who believed that at a certain point analysis must simply stop—the point of stopping being the lived experience of a particular. This is the notion of the "blocked concept," developed in *Difference and Repetition* (Deleuze 1994, 12). This is something that Leibniz seeks to overcome through his attempt at infinite analysis and can be exemplified by Moby Dick. Consider Moby Dick from the analytic viewpoint of categories: he is simply a whale recognized within the very general concept of "whale," which applies universally to all its ilk. One can also predicate more things of Moby Dick: he is a sperm whale, albino, and male. Yet we are still at the level of sanitized universals; we have failed to truly describe the singularity of Moby Dick. What can be done? Well, perhaps we can get even more specific, as Ahab does in the novel:

> "Corkscrew!" cried Ahab, "aye, Queequeg, the harpoons lie all twisted and wrenched in him; aye, Daggoo, his spout is a big one, like a whole shock of wheat, and white as a pile of our Nantucket wool after the great annual sheep-shearing; aye, Tashtego, and he fan-fails like a split jib in a squall. Death and devils! men, it is Moby Dick ye have seen—Moby Dick—Moby Dick!" (Melville 2003, 177)

Surely, we have gotten closer to identifying Moby Dick's complete concept, but it is here that categorical analysis is blocked; one can go no further to breach into the lived existence of Moby Dick. However, we still have not achieved our goal of escaping the domain of descriptive language and successfully describing his singularity. If we were to possess omniscience, a divine gaze extending infinitely, we could comprehend not only Moby Dick's current relations but also his past and future ones. We could see the entirety of the events which make up his life. Only this narrative of events, the white whale's weaving through the past and future, forms his "complete concept," excluding all others. From this divine standpoint, the notion of the blocked concept becomes relevant. Only the divine mind grasps the infinite concept clearly. While we, bound by finite intellects, capture only fragments of the universe, God comprehends the infinite concept. "The beauty of the universe could be learnt in each soul, could one unravel all its folds, but us humans with our finite minds can only comprehend small and isolated patches of the universe" (Hughes 2009, 36). Traditionally, the operation Leibniz is undertaking is anathema to classical formal logic.

Traditional logic distinguishes between the order of the concept, pertaining to generality, and the order of the individual, referring to singularity or particularity. Typically, philosophers have maintained that proper names are not concepts. At a certain juncture, the process of conceptual specification must conclude; one reaches the final species (*infima species*), amalgamating a plurality of individuals. Rather than adhering to the notion that concepts are unable to comprehend the individual, Leibniz seeks to elevate the concept to the level of the individual itself. In Leibniz's perspective, "Adam" and "Caesar" transcend mere proper names—they are *concepts*.

If everything attributed truthfully to a subject must be inherent in the concept of the subject, one is inevitably compelled to include not only the attributed thing but also the entirety of the world within the notion of the subject. Why? As Smith elucidates, this stems from "the principle of causality" (2012, 47). While the principle of causality asserts that everything has a cause, a series of causes and effects extending to infinity, the principle of sufficient reason demands a justification for causality itself.

Everything has a cause implies that A causes B, B by C, and so forth. Thus, the event "crossing the Rubicon" spawns a cascade of causes and effects, extending infinitely into the past and future. Asserting that this event is part of Caesar's notion necessitates including its myriad repercussions. This transcends mere inherence or inclusion; it embodies the fantastical Leibnizian concept of *expression*. The notion of the subject and its concept becomes a vessel, expressing the entirety of the cosmos. Individually, each soul/monad includes the whole infinite series, it "conveys the entire world" (Deleuze 1993, 27). However, despite expressing the entire universe, it only clearly expresses a small, clear portion—a portion whose bounds is the subdivision of one's body. However, it is crucial to note that the soul is still a "production" of the world.

Whitehead and the Cosmic Event

The cosmos is an *event*, and these events are, as we shall see, the *points of inflection*. But how do we get from this discussion in Leibniz to such a claim? As Deleuze points out, one of Leibniz's great followers, Alfred North Whitehead, is the one who continues Leibniz's initial plunge into the chasm of delirium and draws out the logical conclusions of his positing that events are predicates. As Keith Robinson helpfully notes, the four conditions of the Deleuze–Whitehead event are *extension*, *intension*, *prehension* and *ingression*: "Extension and intension are withdrawn or virtual conditions and prehension and ingression are processes of actualizing, individualizing or actual determination but the relation between the folds is given by a differential opening onto the creative" (Robinson 2010, 196). With Whitehead, "for the third time in the history of philosophy this cry reverberates: *Everything is event*" (Deleuze 1986–87, Seminar XII).

The anatomy of the event consists of "actual occasions," the fundamental units of a concrescence which combine with a specific number of preceding, already existing actual occasions. Actual occasions represent the ultimate paths of experience, each one alluding to other actual occasions. The concert this evening alludes to concerts that came before it. Those prior concerts allude to an occasion that was, and remains, the musical work itself—the subject of the concert. This work, in turn, alludes to other actual occasions representing the state of music in that period, other works, and so forth.

According to Whitehead, each actual occasion assimilates these prior occurrences and designates them as its given or data. Consequently, every actual occasion presupposes data originating from earlier actual occasions, forming a combination of data or preexisting actual occurrences. The birth of a new actual occasion signifies a novel occurrence, and the governing principle of this phenomenon is *creativity*. We transition from one concretion to another, with creativity manifesting as the continual generation of fresh actual occasions. Hence, actual occasions persistently become integrated as data in subsequent actual occasions. For example, my birth is an actual occasion, yet it involves incorporating my father and mother as data, and through them, an entire generational lineage.

In essence, the initial facet of the event involves the conjunction of actual occasions, where a new one arises within a world already populated by actual occasions, absorbing them as given data. The actual occasions are intimately tied to prehensions, as actual occasions are amalgamations of these cognitive grasps. Even though any actual occasion involves an active prehension of preceding events, emphasizing Whitehead's concept of prehension serves to clarify that no actual occasion is a passive outcome of previous events or existing occurrences. Instead, each creative actual occasion actively prehends preceding events, analogous to how tonight's concert assimilates past performances of the same composition. As Deleuze words it:

> A concert is being performed tonight. It is the event. Vibrations of sound disperse, periodic movements go through space with their harmonics or submultiples. The sounds have inner qualities of height, intensity, and timbre. The sources of the sounds, instrumental or vocal, are not content only to send the sounds out: each one perceives its own and perceives the others while perceiving its own. These are active perceptions that are expressed among each other, or else prehensions that are prehending one another: First the solitary piano grieved, like a bird abandoned by its mate; the violin heard its wail and responded to it like a neighboring tree. It was like the beginning of the world... (Deleuze 1993, 91)

This synonymy drawn between Leibniz and Whitehead is by no means a stretch, as paragraph 71 of *The Monadology* states:

> But we must not imagine, as some who have misunderstood my thought do, that each soul has a mass or portion of matter of its own, always proper to or allotted by it, and that it consequently possesses other lower living beings, forever destined to serve it. For all bodies are in a perpetual flux, like rivers, and parts enter into them and depart from them continually. (Leibniz 1989, 222)

Here, we see that there is a relation between the soul and the body and that this will be a crucial relation for understanding the nature of a folding cosmos, one expressed by the unfolding of an infinity of souls.

We have discussed the prehension of a litany of microperceptions and how the aggregation of these prehensions constructs a subject which can be thought of in terms of an event itself. It is this same process that Deleuze describes in his treatment of Foucault's notion of subject formation, a folding of the outside into an inside:

> An interiorization of the outside. It is not a doubling of the One, but a redoubling of the Other. It is not a reproduction of the Same, but a repetition of the Different. It resembles exactly the invagination of a tissue in embryology, or the act of doubling in sewing: twist, fold, stop, and so on. (Deleuze 1988, 98)

This operation is what Deleuze refers to as an act of contemplation which completes the formation of a quasi-subject. He posits that each organ, in prehending other prehensions, extends infinitely, designating "feeling" as the private manifestation of prehension. This imparts a serene confidence in persistence, portraying the individual as a composite of prehending subjects—a gathering of prehensions, all infused with a degree of self-enjoyment. A stalk of wheat contemplates its elements; it surveys them and creates a non-localized bond with them, creating a unity. We will investigate this claim in depth in the next section, peeling another layer of the onion to gain insight into the operation of a singular monad in its contemplative expression.

Contemplation

Deleuze's peculiar notion of contemplation is not born out of the doctrine of a single philosopher but rather seems to be an amalgamation of internal resonances between a series of thinkers: notably, Samuel Butler, Plotinus, David Hume, Raymond Ruyer, as well as Whitehead and Leibniz. In Leibniz, one merely needs to point to his concept of the *mens momentenea*, a momentary mind necessary for the being of a thing to persist, a sort of ontological memory. Ruyer is a figure whose immense influence on *The Fold* is admittedly understated by Deleuze, who only earmarked him once as "the latest of Leibniz's great disciples" (Deleuze 1993, 116). As Smith points out, there is an almost identical concept in Ruyer's metaphysics: *the absolute form*: "An absolute form is a domain or multiplicity in constant formation that has an irreducible unity—a 'being-together'—characterized by a non-dimensional or absolute survey of itself, which establishes *non-localizable bonds* between its constituent components" (2017, 123). Ruyer's technical formulation of this phenomena echoes the more poetic form Deleuze finds in Butler:

> All beings are habits, habituses, and these are also full of philosophical concepts, and the habitus was contemplation. And in some very lovely pages, he stated that wheat was contemplation of its own elements, elements from which it emerged, and through that, it was habitus, even full, says Butler, a very beautiful thing, of a "joyous and naïve confidence in the self." (Deleuze 1986–87, Seminar XIII)

These *molecular* microperceptions are genetic to our conscious *molar* perceptions—the snail vine perceives a manifold of microperceptions as it stretches itself out of the shadows of the fence toward the sun-soaked path out of a desire to bask in the sun's rays. Infinitesimal, intensive changes in air pressure, temperature, etc. mark the differential path toward its desired expression.

It is through the introduction of perception, or the point of view, that the object (or *objectile*) comes into play, "the implementation of perception establishes the folds in the soul, the folds whose monad is decorated on the inside" (Deleuze 1993, 112). The contraction of unconscious microperceptions gives rise to the objects which are made up of an infinity of molecular inflections: "Microperceptions or representatives of the world are these little folds that unravel in every direction, folds in folds, over folds, following folds" (Deleuze 1993, 98). Contraction, therefore, is intimately connected to contemplation. As Deleuze writes:

> The whole world is only a virtuality that currently exists only in the folds of the soul which convey it, the soul implementing inner pleats through which it endows itself with a representation of the enclosed world. We are moving from inflection to inclusion in a subject. As if from the virtual to the real, *inflection defining the fold, but inclusion defining the soul or the subject*, that is, what envelops the fold. (1993, 23)

Things are folded in order to be enveloped (Deleuze 1993, 16) and included within a subject: "folds are in the soul [...] they are pure virtualities, pure powers whose act consists in habitus or arrangements (folds) in the soul, and whose completed act consists of an inner action of the soul" (Deleuze 1993, 23).

Anywhere there is a life, at any metaphysical point on the plane of immanence, there is a soul which inheres a contractile power, a point of contraction and contemplation. The folding and unfolding of these singularities in the soul which envelops them is the act which expresses the world, such that

> the world is *actualized* in souls and *realized* in bodies. It is therefore folded over twice, first in the souls that actualize it, and again folded in the bodies

> that realize it, and each time according to a regime of laws that corresponds to the nature of souls or to the determination of bodies. (Deleuze 1993, 120)

This is the metaphysical point, the soul or the subject. It is what occupies the point of view on the points of inflection, it is what is projected in point of view.

Here we see the importance of expression, which is coterminous with explication, or unfolding. The fold and the unfold are the expressed and expression of the soul:

> There are rhythms of folds and unfolds that traverse the living being. That is what creates the flesh, the flesh that's irreducible to infinity, on the contrary, expressible, unfolding its own parts. The organism folds and pleats its own parts, unfolds and again pleats its own parts. It's very beautiful vision of the living being. (Deleuze 1986–87, Seminar I)

When the soul expresses itself, there is a sense of thrill, joy, and satisfaction emitted. As Deleuze writes, "wheat is thrilled with being wheat, but how is it thrilled to be wheat? Through contracting, and from the fact that it contracts the Earth and humidity from which it emerges, the philosophical version of lilies sing the glory of the heavens, plants sing the glory of God" (1986–87, Seminar XIV). This discussion directly echoes the above: "the plant sings of the glory of God, and while being filled all the more with itself it contemplates and intensely contracts the elements whence it proceeds. It feels in this prehension the self-enjoyment of its own becoming" (Deleuze 1993, 78). The cosmos, it can be said, is shot through with this joyousness; the cosmos sings with the glory of its own expression.

Against Atomism, the Manifold, and Baroque Mathematics

Now, we have a general vision of this Deleuzo-Leibnizian cosmos, but to increase our understanding we will go a level deeper and discuss the nature of the manifold. It should be noted that the manifold is a concept which was not available to Leibniz, but it is still crucial, despite the anachronism, to Deleuze's reading of the mathematics of the great Baroque philosopher. Also, we must note, the seemingly innocuous English translation of *multiplicité* into "multiplicity" has missed some of the nuance that exists in the original French. Chiefly, this translation misses the *pli* at the heart of *multiplicité*, a dimension of the term not missed in the alternative of mani*fold*. This subtle emphasis of "the fold" in "manifold" fleshes out the importance of folding to the entire Deleuzean corpus—as Deleuze himself points out, explication–implication–complication are the "triad of the fold" (1993, 23). Thus, the emphasis on the *pli* at the heart of these three terms should not be ignored. This is why the manifold is a crucial concept to keep in mind when making head or tail of this difficult text. As Plotnitsky puts it, the notion of the manifold "brings together geometry and topology and is crucial to all of Deleuze's philosophy" (2003, 101)

Herein the importance of topological thinking enters the fray, and the immense influence of Bernhard Riemann is most acutely felt. We get a cosmic vision in which everything is composed of smooth curves, folding these differential forces into our soul, and expressing the cosmos in our unfolding. The spaces constituting this cosmos are primitively *topological manifolds* made up of various neighborhoods, or points, "that can be treated as Euclidean, while the manifold as a whole cannot" (Plotnitsky 2003, 101). The distinction between the smooth and the striated forms of space maps onto, as Plotnitsky points out, the notions of metric and topological manifolds (2006, 190). Our point of view is restricted to our neighborhood, or point of view, but it is crucial never to lose sight of the existence of a corresponding manifold—the infinite curve that is the cosmos itself.

As Badiou notes, the fold renders the labyrinthine cosmos "irreducible to any elementary composition" (2017, 52). This is the case since such a composition presupposes a metric, extended, and Euclidean space into which the distribution can occur. However, as we have seen, the units of extension are always produced by a deeper regime of smooth, heterogeneous ideal events which unfold the actualized extended entity into a metric, homogeneous space. This is the essence of Leibniz's anti-Cartesianism—Descartes, of course, being a dogmatic thinker of extension, substance, essence. As Deleuze delicately points out, Leibniz's break from Cartesianism is no small matter; it is a rupture in thought itself (1993, 7). A movement beyond the geometry of Descartes to the curve of Leibniz's calculus—a movement out of the static distribution of being, toward a philosophy of dynamism, toward the topological, curved spaces of the manifold.

The link to Leibniz is certainly not out of bounds, since this notion is also found in Leibniz, who clearly states that "atoms of matter are contrary to reason" (1989, 142). To oppose the atom through the notion of the fold is to eschew an image of the cosmos as a primordial One containing and emanating a procession of beings in favor of a chaosmos conceived as an "infinite curve that touches, at an infinity of points an infinity of curves, the curve with a unique variable, the convergent series of all series" (Deleuze 1993, 23). These curves incur inflections, or becomings—what Deleuze in his early works calls "ideal Events" (1990, 52)—and these events are what properly populate the chaosmos.

The very inflection of these virtual points "opens onto fluctuation, it becomes fluctuation" (Deleuze 1993, 17). This is the definition of Baroque mathematics and the very reality which Leibniz's calculus is attempting to capture in its analysis—it is an attempt to capture the nature of *variation itself*. The nature of the infinite that Leibniz is so thoroughly invested in is the nature of chaos in terms of continual variation. The series converge, but they converge at the limit of an irrational number, which points to the porousness of even a simple line. Flows can flow through the holes, in and out, enacting folds within the unfolds, and unfolds within the folds. One may ask, why the line? Well, the line is contradistinction to the point; the line is a continuous becoming, the point being a merely concrete "thickening" or folding of the line.

The Differential Depths of the Cosmic Soul

Deleuze reads into Leibniz's repudiation of atomism something dramatic. Instead of conceiving the world as populated by a primordial, inert collection of points of matter, Leibniz proposes a cosmos inhabited by the propulsion of dynamic forces and their generation of events. These events are what cause the curve to inflect; this notion of the mathematical curve is the precise locus where the deep engagement between Leibniz and Deleuze on the notion of the differential relation begins. Instead of imagining the universe in terms of static, permanent points, the cosmos is instead dynamically imagined in terms of a differential equation $\left(\frac{dy}{dx}\right)$ which expresses the variations of the curve in its very curving, its very actualization in perception: "for example, the sound of the sea: at least two waves must be minutely perceived as nascent and heterogeneous enough to become part of a relation that can allow the perception of a third, one that 'excels' over the others and comes to consciousness" (Deleuze 1993, 101). As we have already reiterated once, the world is no longer a place of inert beings lying in wait but is populated only by variations of curves propelled by a primitive, intensive force of difference.

The points of inflection are the site of "cosmogenesis" (Deleuze 1993, 15), but the product of inflection is not a whole or a One. Rather it is a labyrinth, infinitely folded, whose folds are folded infinitely. A world of "ponds of fish" all the way down. A cosmos of the multiple, a cosmos that resists the orthogonal of Descartes and is instead engendered by a curve which "envelops an infinitely cavernous or porous world" (Deleuze 1993, 16). The Mandelbrot sequence of fractals and the existence of irrational numbers provide evidence that there is "an infinity of holes" in the figure of the line—the world is a labyrinth, one can ascend or descend their point of view on the cosmos, but they will only ever find more folds, more ponds of fish, more and more curves. "The continuous is a labyrinth that cannot be represented by a straight line. The straight line always has to be intermingled with curved lines" (Deleuze 1993, 18).

The points where these lines intermingle introduce a new point, the point of inclusion. What has been described so far is merely a cosmos of pure inflections, an infinite cosmos of mobile folds of matter—merely the folds which occur on the lower floor of the Baroque house. However, there is a point of view on these cosmogenic folds, and it is with the introduction of a "point of view" that we go from pure inflections of curves to their inclusion in a contemplative soul. Or, put slightly differently, a transition from the pure transmission of singularities to their inclusion into series.

The folds are always pushed to infinity but along two registers: the folds of matter and the folds in the soul. As Leibniz points out, the nature of a force is that *a differential of other forces*, and, therefore, we get a wholly anti-atomistic notion of spatiality, and we even get a genesis of the extensive out of the intensive souls, or *body without organs*.[2] This notion of the "lower floor" of the Baroque house (the folds of matter) corresponds to the notion of the depths in Deleuze's earlier works, and further, the folding of matter relates to the notion of the dynamic genesis in *The Logic of Sense*. As Deleuze writes, "On the floor below we find the material universe of bodies, like that of Commoners who are forever expressing movement, propagating waves, and acting upon one another" (1993, 101). There is simply the miasma of forces and their play, a chaos which needs a screen to intervene and organize it via an aleatory selection of certain singularities through an act of contraction/contemplation. "In the depths, there exists a contractile power, a body without organs, or a *contemplative soul*, which contracts the 'partial objects' which are the product of a differential of forces and enacts a 'dynamic genesis' which constructs a surface which surmounts" (Hughes 2011, 78) the lower floor unto the upper floor proper. As Deleuze writes, "events are produced in a sort of chaos, in a chaotic multiplicity, but only under the condition that a sort of screen intervenes" (1993, 76). Chaos is a cosmic "giddiness," an incognizable collection of possible perceptions, and it is only with the introduction of a screen, or the surface, that one can begin to make sense of it. How is this done? Through an act of "contemplation" (Deleuze 1994, 74). The fragmentary partial surfaces are conjuncted into a global subject/monad, whose microperceptions, affects, images, impulses, and surfaces all belong to and produce this hallucinated global subject. This is the move to the upper floor of the Baroque house, to the folds of the soul.

"To think is to fold" (Deleuze 1988, 118). The immaterial, impersonal material fragments of the depths, the mixture of material folds, are then transformed into incorporeal forms which belong only to thought, only to the soul itself. This corresponds to the "event," in which the material fragments and affections become incorporeal, de-actualized. The body dissolves, and the singularities are released like "spores" (Deleuze 1990, 213) before being reinvested into thought. This leads to a symbolization of these fragments, a reinvestment of them onto the metaphysical surface in the form of an Idea which allows them to resonate and communicate in the virtual. We have gone, thus far, from pure inflection to inclusion, from the dynamic genesis to its inclusion in the soul, in the body without organs in a point of view. "Inflection is a virtuality that currently exists only in the soul that envelops it. Thus, the soul is what has folds and is full of folds" (Deleuze 1993, 22).

Life in the Folds: A New Cosmic Harmony

The nature of the object is completely transformed by this new vision of the cosmos; the object becomes an *objectile*, the object becomes an event, a "temporal modulation" (Deleuze 1993, 19). The object is not the discrete reception of form to matter, but rather the continuous modulation of intensive forces. However, not only is the object rethought in this schema, but so is the subject, which is merely an "adventure" of the event (Deleuze 1986–87, Seminar XII). This is the basis of perspectivism, which does not amount to subjectivism. Rather, quite the opposite: the subject is what comes to the point of view, or what remains in the point of view. Every point of view is a point of view on variation; it is not the point of view which varies, but the point of view is rather an apprehension of the variation itself. It is not the truth which varies in perspectivism, but rather *perspectivism is the truth of variation*. As Deleuze elaborates, "points of view never meant that truth is relative to each person, but that there is a point of view starting from which chaos is organized, in which the secret is discovered" (1986–87, Seminar IV).

The true upshot though is that in Leibniz we do not go far enough—we are still subject to the old harmony of faculties, the good will of the thinker, and the banishment of incompossibilities by a God who ensures the good nature of thought. This "old harmony" is no doubt related to Leibniz's monadic perspectivism: a perspectivism where the entirety of the cosmos and its infinite points of view are directed upon the same object. As Deleuze says,

> Every perspective transforms into an alternative town, and these towns are connected solely by their separation, resonating through the divergence of their series, houses, and streets. Within the town, there is always another town. Each term serves as a means to traverse the entirety of another, following the entire distance. (1990, 174)

Nietzsche's perspectivism emerges as a far more profound art than Leibniz's point of view. In Nietzsche's worldview, "divergence is no longer a principle of exclusion, and disjunction no longer a means of separation. Incompossibility is now a means of communication" (Deleuze 1990, 174). The giddiness of Leibniz's preoccupation with the infinite is complete in Nietzsche's perspectivism, every cavernous perspective opening onto an infinity of caverns; the play of the cosmos now becomes the play that diverges (Deleuze 1993, 87). As Smith writes, the profundity of perspectivism in Nietzsche is the insight that "it is our drives that interpret the world, that are perspectival—and not our consciousness or perceptions" (2012, 179). These drives are the larval selves which swarm over the contemplative soul's manifold; intensities, all seeking to unfold and express themselves. The subject, not a unified monad, but an aggregate of contemplative souls.

What this chapter has hoped to eschew is the commonsensical notion of a cosmos-as-container populated by indivisible particles or atoms. Rather, it has unfolded an expressivist cosmos through the works of Deleuze and Leibniz, with some detours into Riemann, Ruyer, and Whitehead. So, what is the upshot of this cosmological vision? What does it mean for the contraction of cosmic elements to *enjoy* their own becoming as Saturn or a comet's tail? Let's end with a quick summary of this vision: we begin at the order of events, vibrations, inflections of the manifold, an infinite bed of chaos, the cavernous labyrinth of matter. These inflections are perceived, enveloped in the contemplative souls, the swarm of larval selves which contract these inflections into a habitus. There are no objects in the cosmos, but merely a vast sea of microperceptions, folding and unfolding anywhere and everywhere.

Notes

1. The best example of this is Deleuze's attribution of the manifold to Leibniz, who wrote a few centuries before Riemann coined the concept.
2. The primary difference between Leibniz's notion *soul* and Deleuze and Guattari's notion of the body without organs is rather straightforward. The Leibnizian soul is the monad, whereas the Deleuzo-Guattarian body without organs is the *nomad*, or perhaps it's better to say that it's *nomadic*.

References

Badiou, Alain. 2017. "Gilles Deleuze, *The Fold: Leibniz and the Baroque*." In *Gilles Deleuze and the Theater of Philosophy*, edited by Constantin Boundas and Dorothea Olkowski, 51–72. Abingdon: Routledge.

Deleuze, Gilles. 1980. "Leibniz: Philosophy and the Creation of Concepts." In *The Deleuze Seminars*. https://deleuze.cla.purdue.edu/seminar/leibniz-philosophy-and-the-creation-of-concepts/. Accessed September 10, 2025.

Deleuze, Gilles. 1986–87. "Leibniz and the Baroque." In *The Deleuze Seminars*. https://deleuze.cla.purdue.edu/seminar/leibniz-and-the-baroque/. Accessed September 10, 2025.

Deleuze, Gilles. 1988. *Foucault*. Translated by Sean Hand. Minneapolis: University of Minnesota Press.

Deleuze, Gilles. 1990. *The Logic of Sense*. Translated by Mark Lester and Charles Stivale. New York: Columbia University Press.

Deleuze, Gilles. 1993. *The Fold: Leibniz and the Baroque*. Translated by Tom Conley. London: Athlone.

Deleuze, Gilles. 1994. *Difference and Repetition*. Translated by Paul Patton. New York: Columbia University Press.

Deleuze, Gilles, and Félix Guattari. 1987. *A Thousand Plateaus: Capitalism and Schizophrenia*. Translated by Brian Massumi. Minneapolis: University of Minnesota Press.

Deleuze, Gilles, and Félix Guattari. 1994. *What is Philosophy?*. Translated by Hugh Tomlinson and Graham Burchell. New York: Columbia University Press.

Hughes, Joe. 2009. *Deleuze's Difference and Repetition: A Reader's Guide*. London: Bloomsbury.

Hughes, Joe. 2011. *Deleuze and the Genesis of Representation*. New York: Continuum.

Lærke, Mogens. 2010. "Four Things Deleuze Learned from Leibniz." In *Deleuze and the Fold: A Critical Reader*, edited by Sjoerd van Tuinen and Niamh McDonnell, 25–45. London: Palgrave Macmillan.

Leibniz, Gottfried Wilhelm. 1989. *Philosophical Essays*. Translated by Roger Ariew and Daniel Garber. Indianapolis: Hackett.

Melville, Herman. 2003 [1851]. *Moby Dick: Or, the Whale*. London: Penguin.

Nietzsche, Friedrich. 1997 [1881]. *Daybreak: Thoughts on the Prejudices of Morality*. Translated by Mauderie Clark, Brian Leiter, and Reginald J. Hollingdale. Cambridge: Cambridge University Press.

Plotnitsky, Arkady. 2003. "Algebras, Geometries and Topologies of the Fold: Deleuze, Derrida and Quasi-Mathematical Thinking (with Leibniz and Mallarmé)." In *Between Deleuze and Derrida*, edited by Paul Patton and John Protevi, 98–119. New York: Continuum.

Plotnitsky, Arkady. 2006. "Manifolds: On the Concept of Space in Riemann and Deleuze." In *Virtual Mathematics: The Logic of Difference*, edited by Simon Duffy, 187–208 Manchester: Clinamen.

Robinson, Keith. 2010. "Towards a Political Ontology of the Fold: Deleuze, Heidegger, Whitehead and the 'Fourfold' Event." In *Deleuze and the Fold: A Critical Reader*, edited by Sjoerd van Tuinen and Niamh McDonnell, 184–202. London: Palgrave Macmillan.
Ruyer, Raymond. 2016. *Neofinalism*. Translated by Alyosha Edlebi. Minneapolis: University of Minnesota Press.
Smith, Daniel W. 2012. *Essays on Deleuze*, Edinburgh: Edinburgh University Press.
Smith, Daniel W. 2017. "Raymond Ruyer and the Metaphysics of Absolute Forms." *Parrhesia* 27: 116–28.

Chapter 7

The Fold and String Theory

Ewa Szumilewicz

Deleuze's Fold

This chapter introduces the concept of Deleuze's *fold* to elucidate the sensitivity of interconnectivity and the notion of perspectivism as it is seen for Deleuze. I will introduce string theory along with its interconnectivity, together with a brief description of quantum mechanics (QM) and general relativity (GTR), and then discuss the Maldacena paradox of string theory. I will relate Deleuze's perspectivism to the Maldacena paradox and deliberate on the multidimensional notion of time, which is a valid and possible implication of string theory and simultaneously is one that results from Deleuze's philosophy. Finally, I will relate string theory to Deleuze's fold, and, more widely, to his philosophical sensitivity as such. I draw some final conclusions concerning Deleuze's perspectivism and his notions of time (as there is always more than one interpretation, always more than one dimension) that might serve as a tool to interpret string theory.

Deleuze's book *The Fold: Leibniz and the Baroque* explores the concept of the fold as a way of understanding the universe and the relationship between the virtual and the actual, that is to say the fold "in the world" and the one "in the soul." String theory, on the other hand, is a theoretical framework which suggests that the fundamental constituents of the universe are one-dimensional "strings" of pure energy rather than particles (Susskind 2013). In *The Fold* Deleuze describes curves similarly to the way strings in string theory are perceived. Folds are perceived as string-like and interwoven, rather than one-dimensional (the reduction of variable to a single and unique variability) and curved:

> It is because the Fold is always between two folds, and because the between-two-folds seems to move about everywhere: Is it between inorganic bodies

> and organisms, between organisms and animal souls, between animal souls and reasonable souls, between bodies and souls in general? (Deleuze 1993, 13)

Thus, Deleuze's intuition departs from a granular depiction of the world toward interconnectivity. What distinguishes a certain object from another one is not a thing as such (as there is none as such) but a certain point of view, through variation, which Deleuze calls *point of view*. He insists that it has little to do with relativism but exactly the other way around: the perspectivism shows *the truth of change*.

> To the degree it represents variation or inflection, it can be called *point of view*. Such is the basis of perspectivism, which does not mean a dependence in respect to a pregiven or defined subject; to the contrary, a subject will be what comes to the point of view or rather what remains in the point of view. [...] It is not a variation of truth according to the subject but the condition in which the truth of a variation appears to the subject. (1993, 19–20)

In *The Fold* Deleuze discusses a new way of understanding objects: the *objectile*. This new understanding is based on the idea of a series of curves that are described by a single and unique variability, which he refers to as the fold. Instead of being defined by an essential form, objects are now defined by a pure functionality, which is framed by parameters that are inseparable from a series of possible declensions or from a surface of variable, self-describing curvature.

> There exist thus a series of curves that not only imply constant parameters for each and every curve, but the reduction of variables to a "single and unique variability" of the touching or tangent curve: the fold. [...] This new object we can call *objectile*. [...] The new status of the object no longer refers its condition to a spatial mold—in other words, to a relation of form–matter—but to a temporal modulation that implies as much the beginnings of a continuous variation of matter as a continuous development of form. (Deleuze 1993, 19–20)

Deleuze argues that this new understanding of objects is based on a temporal modulation that implies a continuous variation of matter and a continuous development of form. In other words, the object is no longer defined by a spatial form, but by a continuous and perpetually variable framework. This manneristic object becomes an event, not an essentialized object. This point of view represents variation or inflection and is the basis of perspectivism. Perspectivism is a procedure that goes back to Nietzsche's intuition:

> "Point of view" does not signify a theoretical judgement; as for "procedure," it is life itself. [...] With Nietzsche, on the contrary, the point of view

> is opened onto a divergence which it affirms: another town corresponds to each point of view, each point of view is another town, the towns are linked only by their distance and resonate only through the divergence of their series, their houses, and their streets. [...] Nietzsche's perspective—his perspectivism—is a much more profound art than Leibniz's point of view; for divergence is no longer a principle of exclusion, and disjunction no longer a means of separation. (Deleuze 1990, 173–4)

This perspective is a peculiar point of view in which the affirmed discrepancies, or more precisely the nodal points of discrepancies, determine the direction of divergence in the next, successive discrepancies. Deleuze roots his reasoning in Nietzschean divergent thinking, and postulates that perspectivism should be understood as a condition of a certain truth of change appearing to the subject. Hence, it is not only a tool of thinking, without concern for the true character of the shape of thoughts, but rather a true convergence, the truth of change. "If the status of the object is profoundly changed, so also is that of the subject," Deleuze postulates (1993, 19). The operation can be metaphorically described as a fractal construction of the image of thought, where the result of such an operation is the emerging image as such.

Perspectivism sensitizes us to "affirmative divergences." Thus, returning to Deleuze's metaphor of a town, the question of these distances between towns would at first sight appear to be unreasonable (if referred solely to the area of cold science). It would therefore have to be assumed that the question of what is between the objects in the town is irrelevant. It wouldn't matter whether there was a lawn, a forest, or, for example, streets. In turn, if this problem were to be transferred into areas of, for example, mathematics, then it would be a question about properties and attributes of the space in which elements of a given set are located. From a mathematical point of view, this question is generally overlooked. Examples of three elements will be subject to the same analysis whether they are in a vacuum, water, or elsewhere. However, Deleuze points out that the procedure refers to the sphere of life rather than to axioms as such. I will requote the fragment that is quoted above: "'Point of view' does not signify a theoretical judgement; as for 'procedure,' it is life itself" (Deleuze 1990, 174).

Perspectivism brings attention to the issue of background. Removing elements from space is not just removing elements as such, but is also a situation in which you discover the space that surrounds those elements. Forgetting about buildings is at the same time an opportunity to look at the forests, streets, and lawns in a sensitive way—the forests that fade away every day and disappear under the pressure of the houses.

In summary, Deleuze discusses a new way of understanding objects that is based on a temporal modulation and a series of curves. This new understanding changes the status of both the object and the subject and is the basis of perspectivism. And I find it comparable to the string hypothesis, namely the Maldacena paradox that will be elucidated in the next paragraphs.

String Theory—An Introduction

The theory of relativity combines time and space in a four-dimensional continuum. Time and space do not exist in an absolute way but are related to each other. In the light of quantum theory, elementary particles can evolve in time either forwards or backwards[1]—which is not possible with general relativity. The General Theory of Relativity describes the gravitational field in the macrostructure of the world; quantum mechanics describes the relations that occur between elementary particles. Physical processes taking place in the background of the spacetime continuum are not independent of it, because they can and do shape the spacetime continuum itself.

> The relativism of the general theory of relativity is expressed in the statement that concepts and time lose their absolute meaning in GTR; the absolute sense has in a sense the space-time itself. Physical processes taking place in the background of the space-time continuum are not independent of it, because they can and do shape the space-time continuum itself. Their influence on space-time manifests itself through its curvature. Curved does not become time and space separately, but space-time—a fundamental object of relativistic theory, to which the laws of physics refer. (Szydłowski and Tambor 2016, 235)

By contrast, quantum theory takes account of discontinuity. The quantum description of the electron gives only the area in which the density with the highest probability of finding the electron occurs.[2] Heisenberg's principle resonates well with Deleuze's intuitions concerning the paradoxical status of the I–Other relation, where one can only be constituted through the annihilation of the other. Deleuze acutely articulated this co-dependency in his essay "Michel Tournier and the World without Others" (1990, Appendix). There Deleuze says:

> The mistake of theories of knowledge is that they postulate the contemporaneity of subject and object, whereas one is constituted only through the annihilation of the other. [...] The Other assures the distinction of consciousness and its object as a temporal distinction. The first effect of its

> presence concerned space and the distribution of categories; but the second effect, which is perhaps more profound, concerns time and the distribution of its dimension—what comes before and what comes after in time. (1990, 311)

Although both GTR and QM enable a full understanding of all processes occurring in nature, their synthesis is not possible: after combining their equations of quantum-mechanical probability, some processes have infinite values in the results. Also, the difference between the language of GTR and the language of QM can be expressed in a similar way. The difference is not a structural one, but it is a difference between the planes on which the operation takes place: either the immanent, internal plane (QM) or that of the external perspective (GTR). In other words, we can see the difference as that between the ontological plane (QM, Deleuze) and the "symbolic" plane, where this "symbolism" would be understood as a relation taking place "in the world." GTR and QM require different positions of the observer. In GTR the observer is a part of the phenomenon and an "element of the game." In QM the observer is an external frame of reference for the phenomenon that takes place "beyond"; the observer occupies "the impossible" place, that is to say—no place. The inherent incompossibility of the two mutually exclusive planes provides the philosophical meaning of the two theories' incompatibility (Szumilewicz 2021, 49–85).

On the macroscopic level, GTR explains the movement of the stars and galaxies by means of gravity. So far, the two theories (GTR, QM) and four forces (gravity, electromagnetism, strong and weak nuclear forces) cannot be explained within a single coherent system. On the verge of the two, GTR and QM, paradoxes[3] appear—singularities, black holes, pendulum movement etc. We have now two mutually exclusive theories, both "working" on a supposed scale as either micro (QM) or macro (GTR).

There is one beautiful attempt to unify GTR and QM: string theory, which assumes that everything is constituted of tiny, vibrating strings of energy. The unification of all particles and forces can be attained through the assumption that a vibrating string is the most basic element of nature and a factor forming all matter. The analysis of string theory, which remains unverifiable, can reveal a whole new perspective of philosophical inquiry into fundamental ontological issues: the beginning of the universe, the number of its dimensions, or the very structure of time and space (Dawid 2013). String theory is a typical example of a theoretical model of reality, which, due to the limited research apparatus of modern physics, will probably remain experimentally unverified for years to come.[4] String theory is a bold extrapolation from previously known theories of fundamental interactions. In string theory, particles come into being as vibrations of the string of energy. In all interactions (including the quantum theory of gravity), the existence of a finite string size avoids the infinities. String theory also removes the paradox of divisible infinity, since the string is the element that can no longer be divided into a smaller one.

A Theory of Everything?

A Theory of Everything (TOE) is the Holy Grail of physicists. TOE is a hypothetical framework which would explain all the laws of physics and would link together all aspects of the universe. There is an emotional element in unification, for it gives hope for a deeper understanding of the universe. Nowadays, we have two physical theories: GTR and QM. Both theories work on supposed scales but are mutually exclusive. Merging GTR and QM is seen as an attempt toward TOE.

String theory is one of the candidates for TOE, as it merges both GTR and QM. Still, it is worth noticing that for a theory to be seen as TOE it must not only merge GTR and QM but also fully explain all aspects of the universe. This quest is seen as impossible for some scientists (such as Stephen Hawking). Also, Gödel's incompleteness theorem[5] is a strong argument against any TOE ever being discovered. Gödel's argument states that in any axiomatic system that is seen as complete there are true statements that cannot be proved by the axioms of the system; in consequence every theory is inevitably incomplete and for that reason cannot be seen as TOE. Similarly to Gödel, Deleuze criticizes science that dreams of unity, which dream he calls a mere opinion. "Opinion offers us a science that dreams of unity, of unifying its laws, and that still searches today for a community of the four forces. Nevertheless, the dream of capturing of a bit of chaos is more insistent [...]" (Deleuze 1994: 206). Likewise, Deleuze views the scientific unification of the four forces (the strong force, the weak force, the electromagnetic force, and the gravitational force) as a dream. But what is more, the perfect unity of science results in rigidity, completeness, lacking an element of chaos. This lack of unpredictability, the lack of incompleteness (to use Gödel's taxonomy), is Deleuze's strongest objection against TOE, and it is what brings his argument close to Gödel's.

The biggest weakness of string theory is its unprovability, which is what also makes it outdated for some scientists,[6] as it has been almost forty years since string theory was introduced and still it is unproven and not likely to be proved due to our limited apparatus. Whether string theory is considered obsolete or not, however, it is valuable for philosophy in its formal aspect.[7] The major strength of string theory is its simplicity and beauty.

The Maldacena Paradox of String Theory (T-duality)

According to string theory, we can assume that our universe with a radius of R is equivalent to a universe of radius 1/R. Both universes are physically indistinguishable because when the circular dimension shrinks to the Planck scale, string theory undermines any such attempt by reversing geometrical relations. This is the Maldacena paradox, which I will elucidate. After my elucidation the analogy with Deleuze's perspectivism will be vivid.

With further discovery and analysis of string theory, our standpoint on dimensions of time or space will probably change. String theory requires at least ten dimensions (Witten 1995a; 1995b). Let us now assume that we exist in a ten-dimensional space. We use nine spatial dimensions plus time. Let us consider one of the nine dimensions of space and extract a circle of radius R. Traveling along the circle we can return to the point of origin, namely the beginning. A string is different because, besides traveling around the circle, a string can also fold in on itself. The windings of a string can be numerous. Or to use Deleuze's taxonomy, the number of times the string "folds" can be great. As such, due to the string folding in on itself, it changes a great length scale into a small length scale.

QM describes the behaviors and properties of elementary particles. This theory holds true when gravity is so weak that it can be neglected. In other words, quantum theory works when we assume that gravity does not exist. GTR works when we assume that the laws of classical mechanics govern our universe and that quantum mechanics is not necessary for our description of reality. String theory, alongside, among others, loop quantum gravity theory, is a candidate for a theory of quantum gravity. In string theory, particles arise as excitations of strings. If an ideal quantum theory existed, then the particle that would carry the force of gravity would have zero mass and two spin states. This theoretical particle is the graviton. In string theory, the graviton is a closed string. The considered gravitational force involves the exchange of closed strings, which trace paths resembling tubes in spacetime. The summation of gravitational forces does not lead to infinite values.

String theory is a finite theory of quantum gravity. Not only was the graviton discovered through string theory, but adding the graviton to quantum field theory manually is also conceivable, though calculations become useless in that case. This happens because particle interactions occur at individual points in spacetime, at zero distance between mutually interacting particles. For gravitons, calculations lose meaning at zero distance. According to string theory, strings collide over a small but finite distance. Calculations then make sense.[8]

String theories are divided into strings that are closed or open loops, and those whose particle spectrum includes fermions. To incorporate fermions into string theories, a special type of symmetry called supersymmetry must apply. Supersymmetry means that for every boson (a particle that transmits force), there is a corresponding fermion (a particle that makes up matter). This supersymmetry assigns forces to particles that transmit forces and matter-building particles. For the currently known particles, their supersymmetric partners have not been observed.[9] Physicists believe this is because supersymmetric particles are too massive to be detected by accelerators. Proof of high-energy supersymmetry would imply that string theory is a good mathematical model of nature, valid even on the smallest scales. Thus, if string theory is to be a theory of quantum gravity, the average size of a string should be the size of the Planck length. Unfortunately, this means that strings are too small to be observed using the research apparatus of modern physics.

When considering fermions, the need for supersymmetry arises. The supersymmetric version of string theory is called superstring theory. There are six distinct string theories: Type I, Type IIA, Type IIB, Heterotic-O, Heterotic-E, and Bosonic theory. In string theory, physical models are loops on the scale of the Planck length. Therefore, for the physical model of the string, we have two different mathematical descriptions; two symmetric images (mirror reflections). This is the mirror symmetry (Rickles 2013). However, we can theoretically describe each image of the mirror reflection equally well, but the characteristics of the physical processes resulting from such an operation for each space are different. For such simplified solutions, a change of topology is possible without side effects. It turns out that under such conditions, physical processes are possible. This is the correspondence between Ads/CFT (anti-de Sitter/conformal field theory), known as the Maldacena conjecture or the Maldacena paradox (Maldacena 1999).[10] String theory develops through mirrored manifolds that allow physical processes to occur when changing the topology of space. Today, it is known that all six string theories are interconnected. Each of the six theories is a specific case of the general theory called M-theory. These theories are interrelated through transformations called dualities. If two theories are interrelated by duality transformation, it means that one theory can be transformed in a certain way to become the other theory.

For the two separate, particularly chosen physical properties of folded space (different sizes, different shapes) it is possible to receive identical results—identical images. In relation to point particles, this is an unattainable result. In other words, according to string theory we can assume that our universe with a radius of R is at the same time equivalent to a universe of radius 1/R.

Due to limited research apparatus, it is impossible to conduct an experiment that would ultimately prove whether we live in a universe of 10^10 light years (R universe) or in a microscopic universe of 10^-85 meters (1/R universe). If our universe consists of light strings, then it is large and expanding. If it consists of heavy strings, however, it is very small and shrinking. Both possibilities are equally valid.

The two universes are physically undistinguishable because, when the circular dimension shrinks to the Planck scale, string theory undermines such an attempt by reversing the geometrical relations. The greater the stress of a string, the smaller its vibrations. The smaller the radius of a string, the greater its vibrational energy. Great values of the radius of the string result from high energy of the winding (folding) and small energy of the vibrations. The winding number of a string and its energy of vibrations are inversely proportional. The greater the winding number, the smaller the vibrational energy of a string, and the greater the energy vibrations, the smaller is the winding number. Knowing that each large radius (high winding energy or, so to say, high folding energy) has a correspondingly small radius (high energy of vibrations), it is not difficult to observe that masses and charges of particles in an R-radius universe are identical to the masses and charges of particles in a 1/R radius universe. Both universes (R and 1/R) are physically equivalent, because it is the masses and charges of particles that determine the shape and the laws of physics. This is T-duality or the Maldacena paradox. T-duality in string theory resembles Deleuze's perspectivism in the sense that both approaches are based on interrelativity. It is a mutual relation of the elements. Elements are not being studied as separate but as interconnected, where the change of one influences the change of the other.

String theory also solves the problem of a scale smaller than the Planck scale. Such a scale does not exist, for the universe nearing the Planck scale begins to expand again. The collapse is replaced by a rebound. In string theory, there exist two equivalent concepts of distance that connect large scales with small scales. Because string theory contains a minimal, finite distance (the Planck distance), the boundary shifts from infinity to the smallest scale, but one that it is possible to calculate. However, string theory has also raised new doubts (Smolin 2006). How would we conduct the analysis of the notion of space, once it is assumed that universe R is equivalent to universe 1/R: the Maldacena paradox? In the light of string theory, it might be apt to question the objectivity of both time and space.[11]

Time and Dimensions

String theory is at least ten-dimensional. If there were more than one time dimension, how could we then characterize the flow of time? In accordance with common interpretations of string theory, the universe has nine spatial dimensions and one dimension of time. Three spatial dimensions (and the time dimension) are large and extended and the remaining ones are small and curled up (folded). Why is it assumed, then, that six dimensions are curled up? Why is it assumed that out of the ten dimensions only one is a dimension of time, while the others are spatial? Would any new philosophical inclinations be raised by the hypothetical existence of extra time dimensions? Could we ever have reliable scientific tools with which to analyze the differences between, for example, psychological time, biological time, imaginary and cyclical time?[12]

Deleuze's philosophy provides tools to characterize more than one time dimension/more than one interpretation of time. For Deleuze, the starting point is always more than one interpretation of time. In *The Logic of Sense* there are two dimensions of time: Aion and Chronos. In *Difference and Repetition* the syntheses of time constitute the living present and the past and the future as dimensions. To bring Deleuze's taxonomy close to a physical one, I would say that Deleuze's notion of time is always multidimensional as it always has more than one interpretation. Deleuze's time is at the start intrinsically "cracked" or "in parts," "the one" having more than one interpretation. In *The Logic of Sense*, time is Aion and Chronos:

> *There are two times, one of which is composed only of interlocking presents; the other is constantly decomposed into elongated pasts and futures.* There are two times, one of which is always definite, active or passive; the other is eternally Infinitive and eternally neutral. (Deleuze 1990, 62)

Aion, the place of the past-future depth, paradoxically does not contain "this" past-future direction. Chronos is a present that measures, like a

Swiss precision watch, past-future forms, cutting them into thin, single instances of "now" on the surface. Looking at the structure of Aion, one can see that this *depth* time is more a set of possibilities than states existing as such. Aion is a collection of all kinds of possible configurations, which are actualized only with the gesture of the Chronos's "cut."

As it has been sketched, according to Deleuze time is both Aion and Chronos. Aion is, using Deleuze's terminology, eternally infinitive and eternally neutral. Aion is an empty form of time and, as such, is boundless. Aion does not depend on matter, so it is incorporeal. It is the time of events-effects: it spreads in the past and the future. It seems that everything that Aion is, Chronos is not. Thus, Chronos is defined as active or passive. As it is not an empty form of time, Chronos is always living presence. Matter limits and fulfills Chronos. Deleuzian sense emerges on the verge of Aion and Chronos synthesis. With the cut of Chronos, the event of the Aion is articulated to the shape of a word.

> At most, there was between them a shift of orientation: in the case of Aion, the becoming-mad of the depths was climbing to the surface, the simulacra in turn were becoming phantasmata, the deep break was showing as a crack in the surface. (Deleuze 1990, 165)

The extreme fragility of the structure grasped should not be forgotten. The last sentence from *The Logic of Sense*, from the Twelfth Series "On the Paradox," is as follows: "This presentation of a total deployment at the surface is necessarily affected, at each of these points, by an extreme and persistent fragility" (Deleuze 1990, 81). According to Deleuze, the emerging of sense is secondary to time. The tension between Aion and Chronos results in the birth of a word. The time of the depths (Aion) can be seen a set of all possible configurations rather than the configurations as such. Aion as the set of all possibilities can be actualized only with Chronos's cut.

Deleuze's intuition of the notion of time in *Difference and Repetition* functions in a similar way, and creates a similar mechanism. Three syntheses of time can be seen as contradictory and simultaneously inclusive constructs. The third—the active synthesis of time as empty form—has morphological convergence with Aion; the first synthesis—habitus—constitutes time as always living presence, and Chronos comes to mind here. The three syntheses are rooted in the context of polemics with Freud; the second synthesis—of memory—with its dependency is compared to the way the pleasure principle functions, whereas the third is related to Thanatos—the death drive. In short, it is paradoxical and contradictory but also a symbiotic construct. Deleuze puts it as follows:

> The synthesis of time constitutes the present in time. It is not that the present is a dimension of time: the present alone exists. Rather, synthesis constitutes time as a living present, and the past and the future as dimensions of this present. This synthesis is none the less intratemporal, which means that this present passes. (1994, 76)

Here, what matters most is that we again have two interpretations—two dimensions of time, as it was with Aion and Chronos in *The Logic of Sense*. In *Difference and Repetition*, the present is not a dimension of time as the present alone exists. It is the past and the future that are the dimensions of time.

Deleuze postulates the contemporality of subject and object as dependable coexistence, as time is inscribed within them. Deleuze's tradition perceives time as the internal disposition of the subject in relation to an Other. Thus, when no Other is, neither is past, nor is there a One to which to refer. Deleuze's relation I–Other is mediated in time, to the degree that the relation is reverse and correspondent: the more "I," the less "You," and the other way around. It is a co-dependency within the frames of time. To recall again: "The mistake of theories of knowledge is that they postulate the contemporaneity of subject and object, whereas one is constituted only through the annihilation of the other" (Deleuze 1990, 311).

Deleuze understands time as an individual's disposition. His intuition regarding the notion of time is similar throughout his philosophy, at least in terms of the construct (time) being rather the individual's disposition than the result of extrinsic relations. For Deleuze, the starting point always consists of more than one interpretation of the notion of time. Deleuze's time is always from the start "cracked" or "in parts," "the one" having more than one interpretation, having more than one dimension.

String Theory and Deleuze's Fold—Doubts and Conclusions

As sketched in the section above, the notion of multidimensional time in the context of string theory could be well depicted by Deleuze's intuitions concerning the notion of time. As to the granularity of the reality versus some kind of interconnectivity, both Deleuze and string theory take a similar approach: that it is something completely in opposition to, for example, the dualistic view, like the Cartesian body/mind or subject/object distinction, but rather represents the continuum of events as such. Deleuze even calls it the "event," whereas in string theory, there are one-dimensional strings of energy.

The "fold" represents a non-hierarchical way of understanding the relationship between the virtual and the actual, in which everything is interconnected and interwoven. Similarly, string theory challenges traditional binary distinctions by proposing a model of the universe in which everything is composed of tiny, vibrating strings of energy. We might also consider how the idea of vibrating strings in string theory might be seen as a metaphor for the folding and unfolding of the universe in Deleuze's view. Both theories (the fold and string theory) challenge traditional dualistic views of the world and suggest a more fluid and interconnected understanding of reality. Both string theory and the fold address the relationship between the virtual and the actual. Deleuze's concept of the fold suggests that the virtual is not separate from the actual, but rather is folded into it. In other words, the virtual is not a separate realm of existence, but instead is immanent in the actual. Both Deleuze's concept of the fold and string theory challenge traditional binary distinctions between subject and object, by blurring the lines between them. Deleuze sees the *objectile* rather than the object and "the point of view" rather than the subject (Deleuze 1993, 19).

Still, when we talk about Deleuze, the frame of reference might be drawn as positing two folds: one being "the fold of the soul" and one being the fold which is "in the world." Thus, we can refer one *fold* to another *fold*. This view in Deleuze's landscape is called perspectivism. From the referral, depending on the point of view chosen, we see different landscapes, and can obtain different results. Although in the world of string theories there is only one—as it were—"frame of reference," since everything is "worldly," we still have at least ten dimensions.

String theory is a way of describing the universe and all the forces and all the particles by one single vibrating string of energy. But the very act of "describing" must be done "by someone." In other words, the notion of the subject, or the perspective from which we are describing the theory, is not "string-like." Therefore, we are describing all the kinds of string vibrations. It might be that the very process of describing is similar, speaking metaphorically, to wave particle dualism. It might be that, like the double split experiment, the very act of observing changes the nature of what is being observed. This is the moment of the theory that might not be parallel with Deleuze's philosophy, or, one could say, the limit of the parallels between the two. String theory does not depict the dynamic of the structure well, as the notion of the subject is not inscribed in the landscape of string theory. Still, the sensitivity of continuity and changeability in both approaches is similar.

What makes Deleuze special is his sensitivity to happening and becoming. I am thinking here of a certain fragility and the related dynamics of structure, which the philosopher reflects on in many metaphors, such as the metaphor of the body without organs (where it is worth remembering the premise of the existence of the body as such). This metaphor reflects on the dramatic dynamics of resistance to structural deadness. There is also the classic example in *A Thousand Plateaus* of a wasp and an orchid which, through feeding and fertilizing, create a symbiotic, emergent cell (Deleuze and Guattari 1987, 31). The wasp and the orchid permeate each other; they become, and they are becoming. As a marginal note, it is worth remembering the premise that orchids and wasps, as such, must have existed before this was established.

Nonetheless, the notion of the subject or the position of the observer, or the frame of reference, is the one lacking description in string theory. Deleuze's perspectivism could be the inspiration for string theory. Deleuze's perspectivism radically changes the status of the subject and object. He shows the world of interconnectivity, of the string-like landscape (the landscape of the fold), and gives answers to questions of the "subject" as such. The subject is the *objectile*; it is a kind of fluid continuum which is still a fine form within it, a temporal modulation, whereas the object is what he calls point of view.

In summary, I would say that Deleuze's philosophy could be an inspiration for string theory because it is a philosophical framework that considers the notion of time in more than one dimension. Furthermore, there is a similar sensitivity in both string theory and Deleuze's approach to grasping the continuity rather than granularity of the world. Also, similar solutions are made in terms of the Maldacena paradox and Deleuze's object–subject relation in the sense that entities are rather being interrelated than examined as separate. Processes and events are interrelated both for Deleuze and in string theory. Recalling the heart of the Maldacena paradox: the greater the stress of a string, the smaller its vibrations; the smaller the radius of a string, the greater its vibrational energy. Deleuze postulates "If the status of the object is profoundly changed, so also is that of the subject" (1993: 19). Both approaches (Deleuze's and string theory), uttered in different languages, the one of science and the other of philosophy, sound similar in terms of the interrelation of entities being scrutinized. However, within the framework of string theory, there is an absence of description regarding the concept of the subject, or the observer's position, or the frame of reference. String theory might draw inspiration from Deleuze's perspectivism in addressing this gap.

Notes

1. See further on retrocasuality: https://plato.stanford.edu/entries/qm-retrocausality/, along with the first successful experiments: https://www.nature.com/articles/s41598-019-40765-6; and on causality as a symmetrical relation, see https://arxiv.org/pdf/0706.1347.pdf (all accessed September 10, 2025).
2. Heisenberg showed that the uncertainty of measurement of the position multiplied by the uncertainty of measurement of the product of speed and particle mass is always greater than the Planck constant.
3. In this context it is interesting to read an article by Michał Eckstein and Paweł Horodecki, "The Experiment Paradox in Physics." Modern physics is founded on two mainstays: mathematical modeling and empirical verification. These two assumptions form a prerequisite for the objectivity of scientific discourse. Here it is shown, however, that they are contradictory, leading to the "experiment paradox." The authors reveal that any experiment performed on a physical system is—of necessity—invasive and thus establishes inevitable limits to the accuracy of any mathematical model. They track its manifestations in both classical and quantum physics and show how it is overcome "in practice" via the concept of environment. The authors argue that the unraveled paradox induces a new type of "ontic" underdetermination, which has deep consequences for the methodological foundations of physics (Eckstein and Horodecki 2022).
4. Still, there have been indications of indirect evidence surfacing regarding the correctness of string theory. One such indication is research into neutrino oscillation, which was meant to reveal how one flavor of neutrinos changes into another. As a result of these experiments a new type of neutrino was observed, called the sterile neutrino. See Biermann and Kusenko (2006).
5. Gödel's incompleteness theorem ~∀n[Πx proves Pw(w)] = Pk(w) implies the following property: any formal system containing the axioms of natural number arithmetic is either internally inconsistent or incomplete; then there exist (within that system) sentences whose truth cannot be derived from the axioms and theorems of the considered system.
6. Major supporters of string theory are, among others, Thomas Banks, Robert Brandenberger, Peter Goddard, Michael Green, Brian Green, David Gross, Steven Gubser, Gerardus Hooft, Igor Klebanov, Juan Maldacena, Emil Martinec, Lubos Motl, Yoichiro Nambu, Holger Nielsen, Joseph Polchinski, Alexander Polyakov, Joel Schelk, John Schwarzm, Nathan Seiberg, Andrew Strominger, Leonard Suskind, Comrun Vafa, Gabriele Veneziano, and Ed Witten. Critics of the theory are Richard Feynman, Daniel Friedan, Lee Smolin, Scheldon Glashow, and Peter Woit.
7. I am grateful for discussion to Professor Paweł Horodecki.
8. For the reader wanting to know more about the basis of string theory, I recommend the lectures of Dr David Tong that are aimed at beginning graduate students. They were given at the University of Cambridge, Department of Applied Mathematics and Theoretical Physics, Centre for Mathematical Sciences, 2009. The script is available at: https://www.damtp.cam.ac.uk/user/tong/string/string.pdf (accessed September 10, 2025).
9. https://home.cern/science/physics/supersymmetry (accessed September 10, 2025).
10. The AdS/CFT conjecture or correspondence (proposed by Juan Maldacena in 1997) says that physics with gravity inside an anti-de Sitter space is exactly equivalent to a conformal field theory (no gravity) living on the boundary of that space.

11. In the context of Maldacena's conjecture, it is worth reading a thought experiment concerning doubling the universe within one night: the nocturnal doubling. Schlesinger's thought experiment explores the concept of whether everything can double in size overnight. The experiment presents a hypothetical scenario in which, upon waking up, individuals find that all objects, including themselves, have doubled in size. The thought experiment encourages readers to question the stability of their own identities and the consistency of the physical world. It serves as a tool for exploring philosophical notions of change, perception, and the nature of the physical world (Schlesinger 1964).
12. In the context of space and time dimensions, it is worth reading an article which introduces an extension to the concept of special relativity within a 1+3-dimensional spacetime. This extension serves to accommodate the presence of superluminal inertial observers. It reveals that this extension inherently invalidates the established dynamics that govern mechanical point-like particles. Consequently, the adoption of a field-theoretical framework becomes a necessary course of action. The findings establish a direct and intriguing connection between this expanded special relativity and the foundational principles of field theory. This underscores that the emergence of field theory is a direct outcome of the modifications made to special relativity to account for superluminal observers, leading to a paradigm shift in how we comprehend the relationship between these theoretical frameworks. In short, it proves that it is possible to construct a theory for superluminal initial observers within the frames of GTR, where there are three dimensions of time and one dimension of space. See Dragan et al. (2023).

References

Bierman, Peter L., and Alexander Kusenko. 2006. "Relic keV Sterile Neutrinos and Reionization." *Physical Review Letters* 96, no. 9. https://arxiv.org/pdf/astro-ph/0601004. Accessed September 10, 2025.

Dawid, Richard. 2013. *String Theory and the Scientific Method*. New York: Cambridge University Press.

Deleuze, Gilles. 1990. *The Logic of Sense*. Translated by Mark Lester and Charles Stivale. London: Athlone.

Deleuze, Gilles. 1993. *The Fold: Leibniz and the Baroque*. Translated by Tom Conley. London: Athlone.

Deleuze, Gilles. 1994. *Difference and Repetition*. Translated by Paul Patton. New York: Columbia University Press.

Deleuze, Gilles. 2001. *Pure Immanence: Essays on A Life*. Translated by Anne Boyman. New York: Zone Books.

Deleuze, Gilles, and Félix Guattari. 1987. *A Thousand Plateaus: Capitalism and Schizophrenia*. Translated by Brian Massumi. Minneapolis: University of Minnesota Press.

Deleuze, Gilles, and Félix Guattari. 1994. *What is Philosophy?*. Translated by Hugh Tomlinson and Graham Burchell. New York: Columbia University Press.

Dragan, Andrzej, Kacper Dębski, Szymon Charzyński, Krzysztof Turzyński, and Artur Ekert. 2023. "Relativity of Superluminal Observers in 1 + 3 Spacetime." *Classical and Quantum Gravity* 40. https://iopscience.iop.org/article/10.1088/1361-6382/acad60. Accessed September 10, 2025.

Eckstein, Michał, and Paweł Horodecki. 2022. "The Experiment Paradox in Physics." *Foundations of Science* 27, no. 1: 1–15.

Heidegger, Martin. 1996. *Being and Time*. Translated by Joan Stambaugh. New York: State University of New York Press.

Heisenberg, Werner. 1958. *Physics and Philosophy*. New York: Prometheus Books.
Maldacena, Juan. 1999. "The Large-N Limit of Superconformal Field Theories and Supergravity." *International Journal of Theoretical Physics* 38: 1113–33. https://doi.org/10.1023/A:1026654312961.
Popper, Karl. 2002. *The Logic of Scientific Discovery*. London: Routledge.
Rickles, Dean. 2013. "Mirror Symmetry and Other Miracles in Superstring Theory." *Foundations of Physics* 43, no. 1: 54–80.
Schlesinger, George. 1964. "It is False that Overnight Everything Has Doubled in Size." *Philosophical Studies: An International Journal for Philosophy in the Analytic Tradition* 15, no. 5: 65–71.
Smolin, Lee. 2006. *The Trouble with Physics*. Boston: Houghton Mifflin Harcourt.
Susskind, Lenny. 2013. "String Theory." *Foundations of Physics* 43: 174–81, https://doi.org/10.1007/s10701-011-9620-x.
Szumilewicz, Ewa. 2021. *On the Paradox of Cognition*. Berlin: Peter Lang.
Szydłowski, Marek, and Paweł Tambor. 2016. "Relacjonizm i substancjalizm w kontekście teorii grawitacji i kosmologii." *Przegląd Filozoficzny—Nowa Seria* 25, no. 4: 224–44.
Witten, Edward. 1995a. "On S-Duality in Abelian Gauge Theory." *Selecta Mathematica* 1: 383–410. https://doi.org/10.1007/BF01671570.
Witten, Edward. 1995b. "String Theory in Various Dimensions." *Nuclear Physics B* 443: 85–126.

Chapter 8

Aion/Chronos and Living Time

Russell Duvernoy

In a well-known passage, Rudolf Carnap relates that Einstein was "seriously worried" by "the problem of the Now":

> He explained that the experience of the Now means something special for man, something essentially different from the past and the future, but that this important difference does not and cannot occur within physics. That this experience cannot be grasped by science seemed to him a matter of painful but inevitable resignation. (Schilpp 1963, 37)[1]

While Carnap insists that this psychological "peculiarity" will eventually be explained, Einstein does not share his confidence: "Einstein thought that these scientific descriptions *cannot possibly satisfy our human needs*; that there is something essential about the Now which is just *outside of the realm of science*" (Schilpp 1963, 38, emphases added).

Although Carnap quickly moves to deflate this worry, insisting that "we both agreed that this was not a question of a defect for which science could be blamed" (Schilpp 1963, 38), this chapter begins with the provocation of Einstein's worry as impetus for considering Deleuze's repurposed Aion/Chronos distinction in *The Logic of Sense*. This is not intended to establish an ontological alliance. Despite their mutual love for Spinoza, Deleuze and Einstein could hardly be mistaken for intellectual allies.[2] However, by showing an entanglement between cosmological images of time and existential orientations, such a juxtaposition supports the view that *The Logic of Sense* is concerned with ethics in a non-traditional manner.[3] Rather than normative decision procedures, Deleuze shows how sense-making *about* time changes existential attitudes and care for the time we are *in*, demonstrating the "profound link between the logic of sense and ethics, morals, or morality" (1990, 31).

1. Problematizing the Now

What does Einstein mean by the problem of the Now? Carnap implies that the problem is primarily lack of fit between experience and ontology. Phenomenologically, the Now is experienced as locus from which the lived order of time, oriented from a past toward a future, emerges. However, in the classical image of physics, because time is a symmetrical variable, there can be no particular uniqueness to any single instant.[4] Einstein's painful resignation acknowledges that a physical reality governed by timeless deterministic laws cannot account for the seeming particularity of the experiential now. This worry need not stem from any implication of relativity. However, if we do take account of Einstein's special theory of relativity, then the worry adds a new dimension. The phenomenological lack of fit becomes also cosmological.[5] An experience of the Now, which in Einstein's work is bound up with simultaneity, is always relative to the observer's position in spacetime. Special relativity thus additionally denies a single, cosmic Now that can be identified in any meaningful way.

Both the phenomenological and the cosmological dimensions are relevant for Deleuze's repurposed use of the "Stoic" Aion/Chronos distinction.[6] To explore this, we must first note the gap between Deleuze's commitment to philosophical thought not taking any received distinctions for granted, and Carnap's contrasting acceptance of a distinction between the objective and subjective as foundational and constitutive of science. Without striking into many well-trodden worries around this assumption, we simply observe Deleuze's suggestion in *Difference and Repetition* that "it is a question of someone [...] with the necessary modesty not managing to know what everybody knows" (1994, 130). Philosophy cannot presume to know, as a beginning, the difference between epistemic and lived questions, nor metaphysical from experiential. Though these distinctions might remain emergently important, they cannot be *a priori* premises.

This makes for an acute challenge in *The Logic of Sense*. In investigating transcendental conditions for sense without presuming a given transcendental subject or ego, philosophical thought dwells at an ever-emerging edge between lived and thought, experienced and ideated, felt and conceptualized. Such an edge of sense, on which Deleuze's work consistently hangs, unsettles a complacent reliance on a stable distinction between descriptive and generative thinking. Thus, Deleuze's engagement with Stoic thought is not conducted in the descriptive mode of the exegetical scholar or historian but *is* motivated by a shared commitment to the generative importance of philosophical thought for life.[7] Despite the hypercomplexity of Deleuze's metaphysical studies, such studies are always attuned to lived imperatives and effects, even as these resist categorical prescriptions and are often expressed in non-intuitive and non-traditional language. As Sellars observes, for both Deleuze and the Stoics, "the aim of philosophy [...] is a transformation of one's mode of existence or way of life" (2006, 159). For Deleuze, for metaphysical reasons, such transformation is not simply a descriptive matter of representing a presumed objectively separate reality. Rather, his "spiritual exercises" with their aim of "intensifying experience" follow a metaphysical understanding of reality as an open and ongoing nexus of relational processes within which what we call the subject is emergent and participatory (Sholtz 2021, 56). This challenges any complete separation between relational processes that we typically align with the subject and material and ideational flows that exceed while constituting the subject. With the Stoics, but for different detailed reasons, questions of how to live are inseparable from cosmological and metaphysical concerns.

From this meta-philosophical orientation, the Chronos/Aion distinction as two ways of "grasping time" (Deleuze 1990, 5; in French: *saisir le temps*) impacts all dimensions of the problem of the Now. From Einstein's linking to lived experience, Deleuze extends the scope, since some form of now becomes a transcendental condition for the possibility of *any sense at all.* This is not only an epistemic issue; Deleuze's distinction leads to a pragmatics of time, showing how time's variable images have lived ethical and sociopolitical effects.

2. Triple-Barreled Time

To clarify this claim, we must observe three different emphases that are often under-distinguished in discussions about time:

1. metaphysics or physics of time;[8]
2. phenomenological experience of time;
3. sociocultural images, norms, and narrations of time.

Alhough a full account of relations between these modalities is beyond the scope of this chapter, the distinction helps set up the entanglements I wish to more closely consider.[9] We have seen how Einstein's problem of the Now expresses lack of fit between the first and second emphasis (physics and experience), without being extended to the third in any meaningful way. Indeed, Isabelle Stengers has shown how the early twentieth-century image of physics defines its concern with modality 1 through the exclusion of any constraint that 2 or 3 could place on it.[10] Sociocultural images, norms, and narrations of time remain the realm of humanists and social scientists. There are some good reasons for this: innovative work in metaphysics or physics is often at odds with orthodox common sense; moreover, the explicitly political contestation of such work, as for example when Philip Lenard and Johannes Stark sought to discredit Einstein's relativity theory on racist antisemitic grounds, provides justification for distinguishing between emphases 3 and 1 (Kumar 2008, 132–3).[11] The winds of politicized rhetoric shift frequently; serious inquiry into time cannot take its tack through these winds.

Nevertheless, observing a distinction does not mean enforcing total compartmentalization. Such compartmentalization, as Stengers argues, is often in service of a hierarchical ordering that presumes 1 as sole ground of objective reality, with 2 and 3 demoted as ontologically lesser. If Deleuze's concerns with transcendental conditions of sense-making, or better, *eventings* of sense, primarily modulates between metaphysics and experiential phenomenology, the dynamism of this modulation challenges the coherence of an exclusively fixed referent from which to establish a logical or ontological order of operations between the three modalities. This does not just affect the relation between metaphysical understanding and phenomenological experience, but also the lived consequences of differing images of time.

Let me give a few examples to concretize what I mean. In recent articles, Potawatomi philosopher Kyle Powys Whyte has critiqued how a dominant image of time feeds into an "epistemology of crisis" that perpetuates settler-colonial power structures and continued exploitation of Indigenous peoples and lands in the name of climate crisis and emergency (Whyte 2020; 2021).[12] This epistemology functions with an image of time that Whyte expresses as a "ticking time clock [...] track[ing] climate change according to [qualitatively identical] linear units of time" (2021, 41) presented as neutral, objective, and independent of events. In the context of climate change, the image underwrites a sense of approaching deadline and emergency that licences unreflective action according to established norms and power structures.[13] Whyte's point is not to deny the seriousness of climate change, but rather to show how this image of time underwrites a particular kind of response to it. By contrast, Whyte offers an Anishinaabe-informed notion of "time as kinship," in which time's ordering features are emergent *through* practices and the maintenance of relationships between and across human and nonhuman kin. Caring relationships *manifest* time, and as such, the priority is "making the time" necessary to care for these relations. Response to climate change would not be measured in terms of objective deadlines, but rather as call for the work of healing, rebuilding, and caring for relationships.

A second example from a different context: In his 2013 *Time Reborn*, theoretical physicist Lee Smolin argues against the view of a timeless physical universe in favor of a process-relational universe with an *irreversible and open* sense of time. Smolin identifies as a "false idea" the notion "that what is bound in time is an illusion, and what is timeless is real" (2013, 256). He observes how this idea, grounded in images of physics traceable to Plato and Parmenides, tacitly structures other intellectual inquiry: "the greatest harm done by the metaphysical view that reality is timeless is through its influence on economics" (2013, 258). Presuming a timeless reality leads to markets being conceptualized as "path-independent" systems with a single equilibrium state. Smolin suggests that such systems should instead be seen as "path-dependent": details of their pasts matter for how they manifest into the future, and there is no single timeless equilibrium *telos* ordering this development.[14]

In both cases a (theoretical) image of time has extra-theoretical social, material, and affective effects. In the case of Whyte's critique, an image of time as a neutral, linear metric undergirds climate responses that perpetuate colonial power structures and attitudes. In the case of Smolin, an entrenched metaphysical assumption frames a purported non-metaphysical empirical science (economics), with numerous material consequences. This is not to say that such images and their effects *determine* time ontologically. It would seem that *all* images of time run up against a temporal reality that exceeds and conditions them. This is no doubt a motivation of the physicist or metaphysician, to get beyond the image of time in some way, to get at what time "really is." We feel in our bones that time exceeds theory, an affective certainty memorably expressed by poet C. D. Wright: "Nothing in the world beats time" (1998, 86). Even as differing *images* have lived consequences, we cannot just conceptualize time however we want.

Einstein's problem acquires an added complexity. Lived experience of the Now is inflected and conditioned by received images of time. The Now can be the latest instant in a linear series; a locus of caring attention; a center of network of relations; a moment of non-predetermined choice effecting potentials to come. With this complexity in mind, we next consider how Deleuze's Aion/Chronos distinction might contribute to a more reflective pragmatics of time.

3. Readings of Time: Deleuze's Aion and Chronos

In *The Logic of Sense*, Deleuze introduces the difference between Aion and Chronos at the juncture between the corporeal and incorporeal and between the cosmic whole and its limited "parts."[15] Chronos expresses a corporeal present that in principle extends to the cosmic whole; Aion, by contrast, is an ideal incorporeal instant.

Chronos is the time of the corporeal. From the perspective of bodies, there is only the present (Deleuze 1990, 162). This present is both "limited" (to a corporeal present) and "infinite" (in its inclusion of the entire physical cosmos). This limited corporeal present, however, should not be presumed as equivalent to any Now conceived as a discrete simultaneous instant or durationless, mathematical point. The present of Chronos is durational and its "measure" is not fixed: "there is always a more vast present which absorbs the past and the future" (1990, 162). What counts as present is thus perspectivally variable: "God experiences as present that which for me is future or past, since I live inside more limited presents" (1990, 162). And yet, even at the maximal scale when Chronos corresponds to "the greatest present, the divine present," this present is "not unlimited" temporally, since "it pertains to the present to delimit, and to be the limit or measure of the action of bodies, even if we are confronted by the greatest of bodies or the unity of all causes (Cosmos)" (1990, 163). This scalar variability of Chronos is by degrees of extensive inclusion rather than any incommensurate perspectivism. Chronos is necessarily unified even as it is qualitatively heterogeneous and thick with duration.

This leads to Chronos's "infinite" extension. Because only the present exists in Chronos, "it absorbs and contracts in itself the past and the future [such that] from contraction to contraction, with even greater depth, *it reaches the limits of the entire Universe and becomes a living cosmic present*" (Deleuze 1990, 61, emphasis added). This infinite extension is also described through reference to circularity and cyclicality: "the time of the present is *always a limited but infinite* time; infinite because cyclical..." (1990, 61, emphasis added). Chronos, to summarize, is always "the physical and cyclical [time] of the variable living present" (1990, 61).

How should we understand Deleuze's reference to the living present? It is intuitive to assume that because Chronos is corporeal, it is the time *of* living or bodily life. Deleuze seems to intend this in referring to "the living present in bodies which act and are acted upon" (1990, 5). But given his suggestion that "Chronos wants to die" (1990, 164) and Chronos's alignment with a cosmic whole, it is also possible to read this as a reference to a living cosmos and, perhaps, time itself as, in some sense, alive. There are some antecedent suggestions of such an understanding in the Stoics, though given Deleuze's practice of philosophical repurposing and collage, the question of whether or not this is a faithful reading of the Stoics is not the primary concern here.[16] In addition to the influence of Émile Bréhier and Victor Goldschmidt on Deleuze's reading, Chronos as living present also emphasizes Stoic cosmological elements as precursors of a Nietzschean eternal return, what Ryan Johnson refers to as a "sinuous strand of this twisted genealogy, beginning with Heraclitus' 'ever-living fire' (πῦρ ἀείζωον)" (Johnson 2020, 214).[17] Moving the eternal from a Platonic position outside of bodies to instead conceive the material cosmos as cyclically alive resonates with what Michael Bennet calls "cosmobiological" readings of Stoicism, with the cosmos as "a kind of cosmic organism [...] acting on itself" (2017, 46).[18] However, if there are some Stoic roots in Deleuze's emphasis of a living present, his account also de-emphasizes the importance of a rational principle or logos organizing this cosmic whole.[19]

The question of life (and correlatively death) is thus linked with an image of time. This linkage has primarily been noted in terms of *how to* live, but negotiations of how to live depend as well on what we conceive *as* living.[20] This is why we must take care with reductive "materialist" readings of Deleuze that presume too neat a line between living and nonliving, organic and inorganic.[21] This also allows for a finer point on difficulties that arise considering Deleuze's contrasting of Aion with Chronos. In contrast to the cosmic living present, Aion aligns with the ideal, the incorporeal, an instant always "infinitely divisible into past and future" (Deleuze 1990, 5).[22] Sense or meaning, though it may be *about* bodies, is itself incorporeal. Bodies by themselves just are what they are. Sense emerges in an ideal instant which is, paradoxically, never experienced as self-stably present and identical, but rather always divides this present into future and past: "Instead of a present which absorbs the past and future, a future and past divide the present at every instant and subdivide it ad infinitum into past and future, in both directions at once" (1990, 164). Thus, when Deleuze refers to Aion as "pure empty form of time," he expresses its lack of any necessary determinate content and, conversely, its formal necessity as condition of sense of *any* possible content. This is why he calls this "pure empty form" that is "always already passed and eternally yet to come" "the eternal truth of time" (1990, 165). This eternal truth might be called structural, rather than the living truth of the cosmic present in Chronos, which is a matter of the corporeal mixture of bodies in relation.[23]

In contrast to the *limited infinite* of Chronos, Aion is *unlimited finite*: finite insofar as conceived as ideal instant, but also unlimited in its mobility between relative past and future: "Aion is unlimited, the way that future and past are unlimited, and finite like the instant" (Deleuze 1990, 165). The mobility of its demarcation is relative to different emergent sense-events, making it in principle "limitless in either direction" (1990, 165). This limitlessness conditions the contrast between Chronos's circularity and Aion as "pure straight line" (1990, 62). In the time of Aion, "the present is nothing" because it is only posited in relation to an, in principle, unlimited future and past. This unlimited line means that neither future nor past can repeat. But this perspective is formal only—which is to say empty of content and "independent of all matter" (1990, 62).

Vectors of exegetical and existential decision converge on how to interpret this independence from matter: the living and dead, the ideal and material, the contingent and necessary, the determined and free. Aion as "locus of incorporeal events" and condition of all sense is existentially indispensable insofar as life expresses sense or meaning (Deleuze 1990, 165). But Aion's mobility also implies a contingency: different sense-events are possible.[24] Some sense-making events (Aion) are themselves *about* time, even as they are *within* time (Chronos). Because Aion is "unlimited," there is always a more and different sense that can be made, such that we cannot presume a single determinate linear time series extending across all sense-events.[25] Because the instant is ideal as mobile selection, different selections of sense might order future and past differently. How do such alternative orderings rearrange Chronos's living present? Aion, as eternal truth of time, is truth without content: neither determined unilaterally by Chronos nor capable of exhausting the cosmic present of Chronos definitively. To say that different emergent sense-events are possible is to observe an immanent and participatory perspectivism. How could we even conceive of a final description of a cosmic universe that, as description, is not itself emergent within time and space from a particular perspective?

Insofar as readings presuppose a reader, from what vantage does Deleuze describe Chronos and Aion as "two simultaneous readings [*lectures*] of time" (1990, 5)? We can ask this question about Deleuze and about ourselves: From where and when do we think? Material bodies reading material books: ink letters, geometrical shapes, paper pages, reconfigured trees. A Columbia paperback *Logic of Sense* printed in the

USA in 1990 on "permanent and durable acid-free paper" will not, for all that, persist throughout the "greatest [or] divine present" of Chronos, and yet now, in this moment, it is part of the present cosmic mixture of bodies passively and actively in relation. *When* are its ideas? If Aion and Chronos are two simultaneous readings of time, we know from Einstein that simultaneity is relative to an observer situated in spacetime. The readings, as sense-making distinctions, arise through instants of Aion: they are paradoxically within the living Chronos present and outside of that present.

4. Living Metaphysics: Why Does Chronos Want to Die?

This paradox leads Johnson to employ the topographical image of the Moebius strip or "twisted surface," with Aion and Chronos as two sides of a single surface, one extensive and one intensive (2020, 203–4). This is also why Arkady Plotnitsky stresses the importance of topology rather than geometry in Deleuze's thinking: a topological conception of time allows for an understanding of both the continuity of Chronos and the discrete points of Aion without flattening them into a single geometric line (see Plotnitsky 2006, 190–3). Aion's variability is conceived as arising from different place-moments in Chronos without collapsing all objective ordering of time. Spacetime is thus thought through topological continuities rather than geometrical magnitudes and metrics.

However, as the concluding passage of the last section stated, the relation of Chronos and Aion as two "readings" of time is not only conceptual. It is also lived. Aion, as we have seen, is linked to the experience of sense. But the embodiment of this experience complicates any exclusive disjunction. In focusing full present attention, we attend to Chronos: here now physical presence. But Aion keeps butting in—a ceaseless (mental) emergence of propositions, questions, thoughts, ideas, concerns. And while the sense-making of this distinction is grammatically expressed in spatial language, with Deleuze referring to a perspective "inside Chronos" (1990, 162) or Aion as "stretche[d] out in a straight line," Aion is uninhabitable (1990, 165–6). That is, while Chronos is necessarily spatial in its corporeality, Aion is a "being of reason," an ideal instant which is not the corporeal here now, but rather *atopon*, without place (1990, 164–5). It is a placeless Now. This intensifies the living paradox, presenting a double bind whereby meaning-making is always tempted toward alienation from corporeality.

One response to paradox is to relegate it to appearance by establishing an order of priority between its incompatible prongs, elevating one at the expense of the other: either Chronos determining Aion (materialism) or Aion determining Chronos (idealism, constructivism). Such a relegation would allow a sorting of legitimate from illegitimate orderings of time. But Deleuze specifically blocks this: "The greatness of Stoic thought is to show at once *the necessity* of these two readings *and* their reciprocal exclusion" (1990, 61, emphasis added). What is excluded is any unilateral causal priority: Aion is not determined by Chronos but nor does Chronos require Aion.[26] And yet both readings remain necessary for passages from matter to meaning and vice versa, as well as passages from physics to ethics that occupy the Stoics. Such biconditional passages inform each other while resisting efforts to ground one in the other.

What can it mean to live in paradox? Deleuze describes a perspective toggling: "*sometimes* it will be said that only the present exists' (Chronos) ... *Sometimes, on the other hand,* it will be said that only the past and future subsist" (1990, 62–3, emphases added). Is there a procedure for determining how this toggling takes place? Can we enact it willfully?

James Williams uses "aspect" in characterizing Chronos and Aion: "they are whole or entire on their own grounds *under one aspect,* yet completed by the other time, *under another aspect*" (2011, 150, emphasis added). By conjuring the duck-rabbit image associated with Wittgenstein, the language of aspect offers an analogy for thinking the existential situation. It is possible to see the duck as only a duck, or the rabbit as only a rabbit, and yet we are missing something if we see only duck or only rabbit. To acknowledge a seemingly self-consistent perspective (duck, rabbit, Chronos, Aion) as incomplete is to gesture toward an additional level or whole. Once I become aware of the duck-rabbit entanglement, I can move back and forth between them. I might even invent a new term, say a dabbit, to express what I see. But in the case of temporal experience, I am inside the duck-rabbit image—I am part of the entanglement. I am mixed up in it.

Mixture is the prevailing condition of living the Chronos/Aion relation. The corporeality of Chronos is "the time of mixtures and blendings" whereby "the present measures out the action of bodies and causes" (Deleuze 1990, 162). The effort to normatively isolate the relative priority of Chronos and Aion is thus a denial of the present mixture. A symptom of separation, it is a neurotic insistence that it must be either a duck or a rabbit but not both. Or we can attempt to install the ground in "simply what the lines are," prior to ducks and rabbits, which are after all conceptual names. This desire is also a symptom of separation. If Chronos is the time of the corporeal present, if lines of the image exist corporeally independent of their conceptual rendering, the effects of Aion are part of this mixture at an affective level: "the future and past are rather what is left of passion in a body" (1990, 162–3). They are not only thought in Aion, but also *felt* in the corporeal mixture of the present. *How* we see the duck-rabbit, even as "just what the lines are," is layered in past sedimented habits and experiences that impact the corporeal present.[27]

In what Deleuze calls the "aleatory point" or "paradoxical instant" (1990, 166), we cannot know for certain how we see the dabbit before we see it, but once seen, subsequent perception-events are layered with lingering effects in the mixture of the present. Given the volatility of the aleatory point, there is a desire to establish some directional measure, some way of ensuring that it proceeds in accordance with a rational or determined order. Deleuze accordingly locates a polarity between "Good Chronos" and "Bad Chronos," where the former offers an apparent harmony of the cosmological "unity of corporeal causes," and the latter the danger in presuming this harmony can be taken for granted as proceeding with regularity *ad infinitum* (1990, 163–4). Instead, a volatility of change is harbored in Chronos's depths, where "the pure and measureless becoming of qualities threatens the order of qualified bodies from within" (1990, 164). Such "Bad Chronos" is bad from the perspective of single measure or static identity. Against this perspective, Deleuze refers to a "becoming-mad" of Chronos that "expresses the revenge taken by the future and the past" on the present *in* the present (1990, 164). This is how Chronos "wants to die," a death drive indicating how its living present is haunted by that which it both cannot be and cannot prefigure or determine in a final measure: future and past (1990, 164).[28] Try as we might, we cannot know what the dabbit is before the aleatory event of sense.

Why does Deleuze personify Chronos and ascribe it a death drive? Motivated by worries about anthropomorphism (where to personify is to humanize) or the assumed good sense of a metaphysical separation between time and life, we could choose to see this as a "literary" gesture or poetic flight "only." But I would suggest that Deleuze follows what a living present can really mean while remaining on the edge of a disjunction, affirming in this way "divergence and disjunction" (1990, 172). In doing so, he performs the simultaneity of Chronos and Aion: sometimes it is a duck; sometimes it is a rabbit. Both claims express Aion instants of sense. But we can also gesture toward a felt privileging of Chronos as living cosmic present that *just is what it is* far beyond our restricted and partial claims about it. And yet, Deleuze suggests, even in this present there is the virtual haunting of its ephemerality and instability.

5. Becoming Worthy of Time?

Emphasizing paradox as the living condition of sense leaves us before an aleatory point. How can we understand effects of this image of time in relation to Deleuze's injunction to "not be unworthy of what happens to us" by "becom[ing] the quasi-cause of what is produced within us" (1990, 148–9)?[29] Might the sense-making affirmed in these incorporeal propositions *about* time help us learn how to "affirm destiny *and* deny necessity" (1990, 169) as a fundamentally ethical challenge? What is called for is an orientation that rejects necessity when it denies choice and responsibility, while also acknowledging that we cannot just make up whatever reality we like. We are conditioned by and responsible to pasts that amount to a kind of destiny.

Becoming worthy of Aion/Chronos means asking what the image generates. If we understand its entangled dependencies as a relation of "reciprocal capture," then we can ask how it "create[s] value" through the establishment of requirements and obligations (Stengers 2010, 35–40).[30] Although, transcendentally, Deleuze's account is descriptive, it destabilizes any hard boundary between the descriptive and the generative. Claiming the vantage of an external observer as philosophical judge, we might be tempted to see Aion's variability as licencing an unbounded, anything goes "relativism" that "requires nothing of us" (Stengers 2010, 106). But placing ourselves within the Aion/Chronos description, Aion is never an Aion *as such*, but always a specific Aion of an aleatory sense-event. Its relation to corporeal Chronic action is placed under question. But nor should we presume only one possible action determined by a given corporeal mixture of Chronos. This would also remove obligation from the relation by construing it as complete determinism.

Both moves fail to satisfy the dynamism of the Aion/Chronos relation and affective intuitions about polarities of living time. We cannot go back to undo what's done. We live in continuing mixtures of the Chronic corporeal present, but also feel that time remains liable to alternatives, full of the contingencies of forking paths and subsequent rearrangements of mixtures. To worry or hope, cherish or regret: if infinitives of these verbs abide virtually, each event of cherishing or worrying actualizes within a cosmic present of Chronos. How we live an image of time is not a choice that stands outside the conditions of Chronos. The problem of the Now calls for a pragmatics of time images—of knowing what image of time is appropriate for the needs of the present. We feel the slipping of time, the continual experience of loss, while also the granting of graceful reprieve and time that heals all wounds. In Einstein's case, this healing flows from understanding time's flow as illusory; he seeks to become worthy of what happens in consoling the family of his deceased friend Michele Besso through a perspective that emphasizes Chronos's reversibility: "For people like us who believe in physics, the distinction between past, present, and future is only a stubbornly persistent illusion" (cited in Rovelli 2018, 109).

But is such an attitude *always* consoling or appropriate? For the Stoics, the "sage" learns to become worthy of whatever mixtures condition reality in this cosmic present—to see from the perspective of the cosmos is to affirm rather than resent one's destiny—always a matter of the "unity and link of physical causes amongst themselves" (Deleuze 1990, 169). But even the Stoics recognize the instability of this perspective oscillating between two orders, the physical causal and quasi-causal: "Is there not already in the Stoics this dual attitude of confidence and mistrust?" (Deleuze 1990, 163). The mistrust arises in the fault line between denying necessity and affirming destiny. It can always be otherwise and yet this is how it is. This dual attitude is not primarily epistemic in a propositional sense, but rather follows the difficulty of discerning, from *within* the Chronos/Aion double aspect, how to create the value that best expresses this cosmic whole. This is a question of acting, not knowing (Bennett 2017, 113). If the sage becomes worthy of destiny, this is not through solace in a necessity that annuls one's reality as actor. Though "the bodies" and "mixtures" of Chronos *might* "possess enough justice and perfection [...] to avail a principle of an immanent measure," this is restricted to the "level of the cosmic Zeus" (Deleuze 1990, 163). Discerning those sense-events that enable resonance with this cosmic level from those of the "poisonous" or "monstrous" mixtures that increase the perturbations of *ressentiment* is a perennially jurisprudential task.

Ultimately, this discernment means experiencing incorporeal sense-events, despite their infinite divisibility into past and future, as affirming the infinity of the cosmic whole. All sense-events of Aion resolve into a single event: "nothing other than the Event subsists, the Event alone, *Eventum tantum* for all contraries, which communicates with itself through its own distance and resonates across all its disjuncts" (Deleuze 1990, 176). This formal point is balanced against the experiential partiality of relative sense-events. Deleuze's quasi-pragmatism of meaning runs into a participatory mysticism:

> It is at this mobile and precise point, where all events gather together in one, that transmutation happens: this is the point at which death turns against death; where dying is the negation of death, and the impersonality of dying no longer indicates only the moment when I disappear outside of myself, but rather the moment when death loses itself in itself, and also the figure which the most singular life takes on in order to substitute itself for me. (1990, 153)

If Deleuze's shift to a participatory mysticism is presented as a necessary implication of a transcendental account of time *as such*, then its ability to create experiential requirements that resist fatalistic apathy seems limited. For this gathering into transmutation to resist capture by eschatological logics, especially in a present context of proliferating eschatological images in secular-scientific hybrids of climate catastrophe and geopolitical warfare, alternative images and narratives of time are critical. This stresses a pragmatics of images of time, how different images might have different lived effects. For Deleuze, this role in *The Logic of Sense* is played by a latent hybrid Stoic/Nietzschean eternal return. The singularity of the transmutation is perpetually renewed and perpetually virtually present, always Now, in the Chronic present with no end to come. But does the Chronos/Aion distinction really force an action to participate in renewal? Does its affirmation ask enough of us? Or is Deleuze's reliance on transmutation liable to a fatal quietism, one that fails to force a worthy becoming?

6. Coda

Because of the variability of the sense-making instant, the Chronos/Aion distinction calls us to balance any definitive single image of a time order against a plurality of other possible orders. The lack of equivalence between these orders, their heterogeneity and difference, would generate new sense-events of meanings indeterminable before their event. If Deleuze's affirming of this condition is expressed in a spirit of playful contingency throughout *The Logic of Sense*, this spirit and this plurality are nonetheless deadly serious.

"Is this the good or bad mixture?" Deleuze asks, and then "This question finds an answer only when the two terms end up being indifferent [...] when the status of virtue (or health) has to be sought elsewhere, in another direction [or] element" (1990, 163). From what direction or element could we seek such indifferent health or virtue? Is it a matter of living something called time that is given and independent of *how* it is lived? Time is a challenge, then, because it "waits for no one"—because Aion's becoming destabilizes all mistaken desires for permanence. Or, conversely, is time an expression of a living cosmos? This calls for understanding how we might learn to care *with and for* time, through sense-making practices as expressions of this care in the haunting of life with death.

Notes

1. Carnap's conversations with Einstein occurred at the Institute of Advanced Study in Princeton during Carnap's residence there from 1952 to 1954 (Schilpp 1963, 36–9). See also Prigogine and Stengers (1984, 214).
2. Their differing affective economies can be seen in the contrast between Einstein's insistence that God "does not play dice" and Deleuze's affirmation of the aleatory "dice throw" as condition of an ideal game.
3. Johnson (2020), Sellars (2006), and Sholtz (2021), among others, all read *The Logic of Sense* as concerned with ethics in a non-traditional sense.
4. Prigogine and Stengers use "reversibility" to refer to time's symmetry in classical views (1984, 60–1, 71). For a sympathetic discussion of this view from a contemporary perspective, see Rovelli (2018, 19–36). Though he will ultimately be critical, Smolin (2013, 54–75) also discusses why Einstein's relativity strengthens the case for time as a symmetrical variable.
5. For a non-technical discussion of this point, see Rovelli (2018, 41–5); Smolin (2013, 58–61).
6. I place Stoic in scare quotes to acknowledge Sellars' work in complicating assumptions that Deleuze's usage is exegetically direct. As Sellars shows, the evidence in extant Stoic texts of a "dual theory of time" is spotty at best (2007, 178).
7. Sellars (2006), Sholtz (2021), and Johnson (2020) all observe this affinity.
8. For the purposes of this heuristic, it is not necessary to distinguish between physics and metaphysics. Both, with different methodologies, consider the real ontology of time.
9. Comprehensive investigation would be transdisciplinary, following a rough mapping of these distinctions to academic disciplines, the first as the province of philosophy and physics, the second philosophy and psychology, and the third social sciences and humanities.
10. The paradigmatic figure for Stengers' claim here is Max Planck, who maintained the "physicist's vocation" as "the construction of a position of judgment that gives the 'physical world' the power to transcend all other realities" (Stengers 2010, 32).
11. These two men, both Nobel laureates whose experimental work contributed to some of Einstein's innovations, founded the Working Group of German Scientists for the Preservation of Pure Science, whose prime aim was "to denounce Einstein and relativity" as "Jewish physics" (Kumar 2008, 132).
12. The rhetoric of climate change as unprecedented threat rationalizes the displacement of Indigenous peoples in the name of land grab emergency measures. Crisis is used as justification with the false claim that "we" "don't have a choice" (Whyte 2020).
13. Whyte (2021) employs the analogy of the difference between speed chess and untimed chess in terms of innovating or experimenting with alternative strategies.
14. Smolin situates his text as part of what he calls the "relational revolution," an interdisciplinary shift away from atomistic substances toward relational

process. A key element of this "revolution" is that time is real; hence, the future is open. Such "relationalism" has numerous antecedents in the history of philosophy, though his account tends to emphasize Leibniz (2013, xxvii–xxxi).

15. We should take care to bracket commonsense mereological assumptions about this part–whole relation, as the ensuing discussion will demonstrate.
16. There exists valuable work tracking the particularities of Deleuze's idiosyncratic reading (Sellars 2007; Bennett 2017; Johnson 2020). For the present argument, it is more important to consider the conceptual implications of *how* Deleuze employs this distinction, rather than clarifying its historical provenance.
17. Deleuze's account of Aion/Chronos in particular follows Goldschmidt's contrasting treatments of the infinite and limited in *Le système stoïcien et l'idée de temps* (1953, 30–40). From its first personification in Hesiod, Johnson claims the Pythagoreans made Chronos into an "extra-cosmic medium" that the "living, breathing creature" of the cosmos "inhales." This encounter between the limited living cosmos and the unlimited extra-cosmos generates number as the length of a cosmological breath (Johnson 2020, 214).
18. Sambursky is an example of such a reading, where the Stoics see "the cosmos as an entity regarded [...] as a living [...] being endowed with *psyche* and *nous*" (1959, 107).
19. Bennett describes the two cosmological principles in Stoicism as an "ungenerated active principle" associated with logos, rational order, and God, and a passive principle associated with matter (2017, 45–6).
20. As Johnson notes, for the Stoics, "time is the incorporeal proper to ethics" (2020, 203).
21. Though I will not pursue this in detail, such questions are germane to the cosmological context of Whyte's image of *time as kinship*, as well as Smolin's speculative proposal of a "cosmological natural selection" in which time is real and physical laws accordingly emerge and evolve over time (Smolin 2013, 124–9).
22. This characterization closely follows Goldschmidt (1953, 37–44).
23. Because this living truth of the cosmic present does not require a simultaneous instantaneous now, it remains potentially consistent with Einsteinian relativity, even as the necessity of Aion for sense resonates with Einstein's "problem of the Now."
24. Deleuze is influenced by Bréhier, for whom the incorporeal order of sense-events (*le plan des faits*) that "plays on the surface of being [*se jouent à la surface de l'être*]" presents "a multiplicity of incorporeal beings without connection or end [*une multiplicité sans lien et sans fin d'êtres incorporels*]" (1962, 13).
25. Presuming a stable subjective/objective pivot hinders closer consideration of the effects of Deleuze's reading. While it has the implication that there is not one single linear time series, this doesn't mean that time loses all potential objective ordering, but only that we cannot presume that this objective ordering is equivalent to any single linear rendering of it. This point seems consistent with Einsteinian relativity, provided we do not presume that Deleuze's account is simply recoiling into what Einstein calls "the subjective concept of time" that "refers to the arrangement of the experiences of the individual" (2001, 143).
26. As has been noted, Deleuze's account of Aion's "quasi-causality" departs from orthodox readings of the Stoics, since he does not subordinate the incorporeal event to bodies necessarily (Bennett 2017, 106–7; Sholtz 2021, 58–9).
27. See Clark (2013) for how this point is borne out micro-phenomenologically by contemporary neuroscience.
28. Deleuze's account explicitly diverges from the neo-rationalism of Stoic cosmology, if we presume this rationalism results in an epistemic ability to *know* the cosmos's active principle (Bennett 2017, 113–14; Sholtz 2021, 54–5).

29. This injunction describes a Nietzschean *amor fati* where the event is not just accepted, it is affirmed (Sellars 2006). Sholtz observes that this also resembles a fundamental Stoic attitude: not seeking to "have events happen as you wish but wish them to happen as they do happen" (2021, 54).
30. Stengers expands on what is typically translated as "double capture" in *A Thousand Plateaus*. What is important for our purposes is that reciprocal capture, in different ways, involves the "co-invention" of identities through reference to the other (Stengers 2010, 36).

References

Bennett, Michael J. 2017. *Deleuze and Ancient Greek Physics: The Image of Nature*. New York: Bloomsbury Academic.
Bréhier, Émile. 1962. *La Théorie des Incorporels dans l'ancien Stoïcisme*. Paris: Vrin.
Clark, Andy. 2013. "Whatever Next? Predictive Brains, Situated Agents, and the Future of Cognitive Science." *Behavioral and Brain Sciences* 36, no. 3: 181–204.
Deleuze, Gilles. 1990. *The Logic of Sense*. Translated by Mark Lester and Charles Stivale. New York: Columbia University Press.
Deleuze, Gilles. 1994. *Difference and Repetition*. Translated by Paul Patton. New York: Columbia University Press.
Einstein, Albert. 2001 [1916]. *Relativity: The Special and the General Theory*. New York: Routledge.
Goldschmidt, Victor. 1953. *Le système stoïcien et l'idée de temps*. Paris: Vrin.
Johnson, Ryan J. 2020. *Deleuze, A Stoic*. Edinburgh: Edinburgh University Press.
Kumar, Manjit. 2008. *Quantum: Einstein, Bohr, and the Great Debate about the Nature of Reality*. Thriplow: Icon Press.
Plotnitsky, Arkady. 2006. "Manifolds: On the Concept of Space in Riemann and Deleuze." In *Virtual Mathematics: The Logic of Difference*, edited by Simon Duffy, 187–208. Manchester: Clinamen.
Prigogine, Ilya, and Isabelle Stengers. 1984. *Order Out of Chaos: Man's New Dialogue with Nature*. Boulder, CO: Shambhala Publications.
Rovelli, Carlo. 2018. *The Order of Time*. Translated by Erica Segre and Simon Carnell. New York: Riverhead Books.
Sambursky, Samuel. 1959. *Physics of the Stoics*. London: Routledge and Kegan Paul.
Schilpp, Paul A. 1963. *The Philosophy of Rudolf Carnap*. La Salle, IL: Open Court.
Sellars, John. 2006. "An Ethics of the Event: Deleuze's Stoicism." *Angelaki: Journal of the Theoretical Humanities* 11, no. 3: 157–71.
Sellars, John. 2007. "Aion and Chronos: Deleuze and the Stoic Theory of Time." *Collapse* 3: 177–205.
Sholtz, Janae. 2021. "Deleuzian Exercises and the Inversion of Stoicism." In *French and Italian Stoicisms*, edited by Kurt Lampe and Janae Sholtz, 53–74. New York: Bloomsbury Academic.
Smolin, Lee. 2013. *Time Reborn: From the Crisis in Physics to the Future of the Universe*. Toronto: Knopf Canada.
Stengers, Isabelle. 2010 [2003]. *Cosmopolitics I*. Translated by Robert Bononno. Minneapolis: University of Minnesota Press.
Whyte, Kyle P. 2020. "Against Crisis Epistemology." In *Handbook of Critical Indigenous Studies*, edited by Brendan Hokowhitu, Aileen Moreton-Robinson, and Linda Tuhiwai-Smith, 1–15. Abingdon: Routledge.
Whyte, Kyle P. 2021. "Time as Kinship." In *The Cambridge Companion to Environmental Humanities*, edited by Jeffrey Cohen and Stephanie Foote, 29–55. Cambridge: Cambridge University Press.

Williams, James. 2008. *Gilles Deleuze's Logic of Sense: A Critical Introduction and Guide*. Edinburgh: Edinburgh University Press.
Williams, James. 2011. *Gilles Deleuze's Philosophy of Time: A Critical Introduction and Guide*. Edinburgh: Edinburgh University Press.
Wright, Charles D. 1998. *Deepstep Come Shining*. Port Townsend, WA: Copper Canyon Press.

PART III

COSMOLOGIES OF THE QUANTUM

The general theme behind *asstrOnOmical affected autOmata* is the linking of the human body's intimate openings to the huge unhuman alienness of deep-black cosmological space. This was conceived in a spirit of Deleuzian erotic and animistic connectivity and cosmic spirituality. My hope is that a contemplative viewer might imaginatively place him or herself within a Deleuzian/Dionysian flux of cosmological grandeur. The painting poetically suggests something personal to each viewer along the lines of what I would call an Intimate Astronomy. Influences on the painting were Gilles Deleuze and Félix Guattari's book *A Thousand Plateaus*; the transcendental black metal music of Liturgy, an American metal band from Brooklyn led and philosophically theorized by Hunter Hunt-Hendrix (now Haela Hunt-Hendrix); the Speculative Realism philosophical movement (especially Manuel DeLanda's book *Philosophy and Simulation*); Rahsaan Roland Kirk's avant-garde saxophone music; and general automata theory.

I have made art using computer technology since 1986 and have been contaminating my paintings with home-made computer viruses since 1991. In 2001 I combined my initial digital virus project (which was a reflection on my personal experiences of risk and loss during the AIDS epidemic) with the principles of artificial life (a sub-division of AI). In other words, I have been creating art systems that parallel the behavioral characteristics of living systems based on the model of the host–parasite. For this painting I used my cellular automaton-based computer virus to create connections between the intimacy of human life forms and empty non-life space. For me, the viral metaphor is particularly strong, as in biology viruses are considered neither dead nor alive. That is why I choose to use my a-life virus here to paint the pink velvety parts of the human figure; knitting filaments and assembling them to gradually reveal something warm and tender and very human about our place in the universe.

—Joseph Nechvatal

Joseph Nechvatal, *asstrOnOmical affected autOmata*, computer-robotic-assisted acrylic painting on velvet, 112 × 168 cm, 2011; collection of the Brooklyn Museum, New York City © Joseph Nechvatal

Chapter 9

Deleuze and Decoherence: Approaching the Measurement Problem through Intensity

Michael J. Ardoline

The central issue in the metaphysics of quantum mechanics is the measurement problem. To answer the measurement problem is to give an account of what (or at least what appears to be) the determination of a quantum system in superposition into a measured (thereby, determinate) value. There seems to be a straightforward application of Deleuze's metaphysics where the unmeasured system is virtual and is actualized through measurement.[1] This interpretation would put Deleuze in the category of modal interpretations of quantum mechanics, an approach that has not received wide support.[2] While on the right track, this reading of quantum mechanics in Deleuzian metaphysical terms is missing two important pieces: the physics of decoherence and the metaphysics of intensity. Rather than propose a Deleuzian answer to the measurement problem here, I want to approach this issue somewhat slowly and in detail for two reasons. First, Deleuze's account of modality is often reduced to the actual and the virtual, rather than the actual, intensive, and virtual.[3] Similarly, the processes of individuation and actualization are often confused or treated as one process, even by Deleuze scholars, despite Deleuze's protestations to the contrary in *Difference and Repetition* (1994, 247). Second, because the field of the foundations of quantum mechanics is a well-established field of the philosophy of science where dozens of interpretations and solutions to the measurement problem have been proposed, it would be rash to assume that Deleuzian modality offers a unique and yet unthought-of solution, though this is an open possibility. In other words, Deleuze's modal metaphysics may be cut to fit with an existing interpretation of quantum mechanics. Therefore, we are best served by first establishing the problem to be solved, translating it into Deleuzian concepts, and then asking if these new conceptual resources clarify anything in the problem, exclude any class of interpretation of quantum mechanics, and provide support for any others.

Due to the number of interpretations of quantum mechanics, such a project is larger than can be undertaken in one chapter. By exploring the connection between decoherence and intensity, I will argue that we see a more robust Deleuzian interpretation of quantum mechanics, and that this interpretation opens new avenues for critiques of other proposed solutions to the measurement problem, even if it is not yet developed into a solution. I will focus on the latter project here with a Deleuzian critique of the Many Worlds interpretation. This chapter will lay out a short introduction to decoherence as it pertains to the measurement problem. It will then give a reading of decoherence and measurement in terms of Deleuze's metaphysics of intensity and the virtual/actual divide. The goal of this chapter is to put these physical and metaphysical concepts into communication so as to establish that the modal concepts of actuality, intensity, and virtuality are useful in making sense of decoherence as a physical phenomenon. While it will not amount to a Deleuzian answer to the measurement problem, it is the first steps toward one.

The Measure of the Problem

The measurement problem is the fundamental question that most interpretations of quantum mechanics are proposed to solve. It is by now well known, but in brief: quantum mechanics is, mathematically speaking, a probabilistic theory. Systems are described as a vector in Hilbert space and their change over time (often called their evolution) is described by Schrödinger's wave equation. The wave equation describes an amplitude. Squaring this amplitude gives a probability density that describes the probability of potentially measurable values (this procedure is known as the Born rule). In experiments, however, we cannot measure a probability density directly; a detector does not read out that a particle has 33.3% likelihood to be spin-up and 66.6% to be spin-down. Instead, the measurement apparatus produces a particular, determinate value, a measurement. In cases where the predicted events being measured have a probability that is not zero or 1, the same experimental set-up will measure different results over repeated trials. However, the frequency of these results will, when calculated, correspond to the probability density as defined by the Born rule. So, if our apparatus has two possible measurements, say A and B, and A has probability of 75% and B has a probability of 25%, running the experiment enough times, we will see event A happen three times as often as event B. This movement from a system in an undetermined superposition to a determinate, measurable state is often known as wave function collapse; however, it is best at this point to avoid this terminology as it privileges some interpretations over others (for example, Copenhagen interpretations over Many Worlds). Rather, I will stick with as neutral as possible a description of what the experimenter encounters, and so will refer mainly to mathematical formalism, calculations, experimental operations, and measured results.

Because we often take probability as a sign of uncertainty, incompleteness of knowledge, or a lack of full understanding, it is worth remarking here that quantum mechanics is the most experimentally successful theory humans have ever devised. The measurements always end up matching the probability density. In other words, the probabilistic nature of the theory does not make it any less rigorous or exact. This is one reason to take the measurement problem as an ontological rather than an epistemic problem or a sign of an incomplete model of reality (though there are many who disagree, such as Einstein's famous refusal to accept quantum mechanics as a complete theory of fundamental physical reality).

So where is the problem? The problem exists for human beings trying to make sense of what is being predicted, what is being observed, and the apparent gap between them. The mathematical formalism of the theory does not predict specific outcomes, only probabilities. The measurement outcomes, however, are always specific, determined observables. We do not see in one measurement events A and B simultaneously as a superposition with 75% and 25% likelihood, respectively. Moreover, it is not even clear what that would mean. When we run the double-slit experiment and see an interference pattern, we are not looking at a probability density; we are looking at the registration of determined events that, when taken as a whole, conform to the probability density. Stranger still is that, if the mathematical description is all there is and quantum mechanics is the true description of fundamental reality, then we as experimenters should equally be expected to be in a superposition as entangled observers of these phenomena. Yet we instead measure and experience these outcomes as specific, determinate values. The measurement problem is then the question of how to make sense of this apparent gap between theoretical prediction and observation despite the incredible accuracy of the predictions.

We might also quote Tim Maudlin's (1995) summary of this version of the problem as such.[4] Take the three following statements as true of quantum mechanics:

1. The wave function is a complete description of the system.
2. The wave function evolves deterministically.
3. Measurement of the system gives a determinate result.

These are three reasonable beliefs about quantum mechanics, but they are inconsistent. The mathematical formalism provides only probabilistic predictions for outcomes. The process of measurement returns determinate (that is, non-probabilistic) results, which when the frequency of determinate results is taken, nonetheless confirm the probabilistic predictions with stunning accuracy. However, there is nothing in the mathematical description that hints at, let alone describes, the determinate nature of these results. Therefore, because the wave function does not predict a determinate measurement, propositions 1 and 2 are inconsistent with 3. Either there is more to the description of the system than the wave function (though it seems irrelevant to prediction, which would be odd), or the wave function evolution and measurement results are not determinate (which again would be odd, because we do observe determinate measurement results). Hence, the problem.

A solution to the measurement problem would either 1) give a reason why we are epistemically limited in such a way that we cannot know how superpositions become measured values; 2) give an account of the processes that lead from an undetermined state like that of a superposition to a determined measurement value (the wave function is complete but its prediction that the outcome of the experiment is a superposition is wrong); 3) accept that the wave function is an incomplete description of reality and add something to the ontology or physics; or 4) explain why there is no measurement problem, or rather that the measurement problem is simply an ill-posed problem following from our holding onto a theoretical posit that has no meaning in quantum mechanics (Albert 2015, 162). Different interpretations have tried each of these possibilities. For 1, we can think of epistemic Copenhagen interpretations where the measurement changes the system such that the state before the measurement cannot be recovered. For 2, there are objective collapse theories wherein the wave function is claimed to, upon interaction with something else, take on a determinate value. Option 3 includes theories such as Bohmian mechanics, which add configuration space to the formalism and particles with locations to the ontology. Bohmian mechanics thereby makes the theory completely deterministic (probabilities are then only a result of the incomplete theory). For 4, we have the Many Worlds interpretation, in which it is claimed that all possible events happen but they do so in separated or branching systems (worlds) and that observers at a world only "see" one of those outcomes. Type 1 solutions embrace our limitations as knowers. Type 2 solutions take the process of determination to be real. Type 3 take the wave function to be an incomplete description of reality. Type 4 solutions deny the reality of determination. Type 3 and 4 solutions can embrace either indeterminism or determinism, but what they reject is that what is not determined ever becomes determined, or vice versa (for example, because of the branching of worlds in the Many Worlds interpretation, the determination of a single outcome of a supposed probabilistic process is an illusion).

And so this is the measurement problem and several of its proposed solutions as they are most popularly known. However, a refinement of quantum theory has added further details to this issue, one that several interpretations, especially Many Worlds, claim bolsters their position. This refinement is known as quantum decoherence.

The Decoherence of the Physicists

As we are moving between the formal theory as such and various interpretations of quantum mechanics, we should be clear on the status of decoherence. Decoherence is accepted physics, so it is open only to the challenges that the physical theory itself is, not the challenges that the interpretations are (unlike, say the claim that the mathematical formalism implies the existence of objective wave function collapse or of many branching, causally isolating worlds). As my expected audience is not physicists, nor even primarily philosophers of physics, I will give a non-technical overview of decoherence.[5] Decoherence is a description of what takes place in any situation where a quantum system is not isolated from its environment. First, a note on the name. Decoherence refers to interactions in which a quantum particle comes to act more classically. What "decoheres" is the *quantum* nature of the particle. This is a somewhat confusing name, because, by decohering, the particle actually becomes more stabilized, more determinate, and more consistent with its environment. In short, it behaves more like a classical particle (referred to as quasi-classical state). So, if one imagines a particle falling apart or breaking down when we say that the particle decoheres, one has the image backwards. Instead, one should imagine a particle becoming less like a superposition of simultaneous different properties and more like a classical particle with determinate properties.

A quantum particle on its own is described as a probability cloud (or at least it is if we add nothing to the formalism). It is not in a specific place but is rather a distributed likelihood of the locations it might be. It is a superposition of internal differences of position, momentum, and spin. Or at least this is its quantum nature if it is not interacting with anything. We get this view of particles from Schrödinger's wave equation. However, Schrödinger's equation only applies to closed systems. Most of the popularly known experiments, like the double-slit experiment, are constructed to approximate this closed system to control nature's chaos and test whether we can detect what the mathematical formalism predicts. But quantum mechanical systems, like all systems, can only ever be said to be closed as an ideal approximation. All systems are open systems (barring the universe itself, if it can be said to be a single system). Since any quantum system is open, it has some interaction with the environment outside itself. This means that particles are potentially interacting with their environment. Fascinatingly, the chaos of these interactions between quantum particles *qua* probability cloud and whatever other scattered particles make up its environment actually make the particle behave more determinately. You can think of this as the particle constantly being partially "measured" by its interactions with its environment, though we will see why this is not quite right. This is quantum decoherence. It is the process by which an otherwise indeterminate particle is determined into a quasi-classical state through the interactions with its environment.

The measurement problem is simplified by decoherence, not solved by it. Decoherence explains how it is that quantum systems go from superpositions characterized by non-classical states whose mathematical description requires complex numbers, to a set of classical states that are characterized by a probability distribution requiring only positive real numbers. In further functional terms, the process inputs a superposition and outputs a classical state space. It is important to note that this is not yet a measurement. Real alternative possibilities still exist, or at least are described by the mathematical formalism, at this stage. Probability becomes a determined outcome only through measurement. I should say that this is somewhat controversial: decoherence, like the measurement problem, does not offer its own interpretation; we will see a slight alternative below when we look at Wallace's defense of the Many Worlds interpretations through decoherence.

We then have three stages that we can separate by the mathematical entities that they require: the Hilbert space and wave function description which requires complex numbers, the state space with a probability density description of a decohered system, and the determinate values of a measurement requiring real numbers. These three stages are linked by two processes: decoherence and measurement. Here we have enough detail to establish an isomorphism with Deleuze's modal concepts.

Deleuzian Modality and Quantum Mechanics

By the numbers alone, we have a very promising start for a Deleuzian interpretation of quantum mechanics. In our description of quantum mechanics through decoherence we found three important modes or states of systems: quantum (Hilbert space), quasi-classical (probability density), and classical (determinate, measured result). We have also isolated two processes for moving between these modes: decoherence (from quantum to quasi-classical) and measurement (from quasi-classical to determinate). Deleuze's metaphysics of modality is also based on three modes (intensity, actuality, and virtuality) and two processes (individuation and actualization). To establish a basis for a Deleuzian interpretation of quantum mechanics, I am going to argue that we have more than a convenient coincidence of the numbers of modes and processes; we have a structural isomorphism. I will do so by arguing that each mode and process in the description of quantum mechanics corresponds not just functionally, but in essential features, to those in Deleuze's account of modality.

Let's start with intensities.[6] Intensities are non-metric, self-differing, and, by extension, cannot meaningfully be said to be identical with each other. In fact, on Deleuze's account, any establishment of similarity, equality, or identity requires a cancellation or "covering over" of some differences. "It is nevertheless true that intensity is explicated only in being cancelled in this differenciated system that it creates" (Deleuze 1994, 255). Intensities are metaphysically prior to identity and similarity. They can be explicated only by changing in kind (that is, into something determinate) through interactions. Because of this, they can be conceived of only as pure becomings.

Here intensities share an important and unique feature with entangled quantum particles: the violation of the principle of the identity of indiscernibles (PII). In cases of entanglement between particles of the same type, seemingly distinct particles cannot be distinguished by their properties (even relational properties). This failure of the identity of indiscernibles and self-differing nature of quantum particles is even taken to require new logics in order to handle it. For example, da Costa and Krause's Schrödinger Logics (1994) have been developed to handle the unique logical requirements of theories in which some objects cannot be said to be self-identical.[7] These are logics developed for the sake of being adequate to this strange lack of identity in some cases in quantum mechanics. Formally, this is achieved by having two types of objects in this logic. For one type, identity statements are not well formed; that is, for these objects, statements like x = x or x = y are barred as nonsense. In other words, these objects cannot be said to be equivalent to either themselves or any other object. The violation of the PII is a controversial topic in the philosophy of physics, but one with strong support (see, for example, French and Redhead 1988). While I will not rehearse the arguments in this debate here, I will point out that one of the common refrains against this position is that treating quantum particles as violating the PII is conceptually unclear (see Friebe 2014, 90). One advantage of the isomorphism being argued for is that, if it holds, then what seems unclear given a more traditional metaphysics should become clear when making use of Deleuzian concepts which are created explicitly in response to the problem (among others) of making sense of situations where identity for individuals fails.

Once we have intensity in place, the rest of the isomorphism is relatively easy, and our path is determined by the metaphysics itself. From the intensive alone, the process of individuation is possible. "The essential process of intensive quantities is individuation. Intensity is individuating, and intensive quantities are individuating factors," and "individuation precedes differentiation [actualization] in principle" (Deleuze 1994, 246, 247). In other words, in Deleuze's metaphysics, the only process that can be carried out from intensities alone is individuation. So, since we began with intensities (and we were right to, since intensity is the characterization of difference itself), we cannot yet make use of the virtual, the actual, or the process of actualization, because each of these depends on intensity and individuation. Individuation begins when "differences in intensity are distributed here and there in the form of excitations" (Deleuze 1994, 96). Through interaction across difference, intensities come to form relations. These relations co-determine their relata. This determining means that some of their differences are cancelled, they begin to co-vary together and form the conditions for a system with an identity. "The act of individuation consists [...] in integrating the elements of the disparateness into a state of coupling which ensures its internal resonance" (Deleuze 1994, 246). The result of this process is the production of an object, a thing with an identity. Because that identity does not precede the coming-together of these intensities, there is no prior structure or hidden variables that explain what gets produced.[8] We can conclude that the process leading from intensity to the object is isomorphic with the process leading from the wave function to the probability distribution. That is, individuation is isomorphic with decoherence. Again, we see this in their structural similarities. Both are processes whereby something that cannot be made sense of in extensive terms alone is transformed into something with the *potential* to give rise to a determinate extensivity. Furthermore, it does so through the interaction of elements which themselves do not have an identity and are in themselves indeterminate. Indeterminate here is different from undetermined. The probability is undetermined before it is measured, what a quality will look like exactly is undetermined until it is actualized, but an intensity and a complex probability (that is, a probability with an imaginary component) are indeterminate as they cannot yet be specified as what they are like. This is why the Born rule is necessary.

Now, with the production of objects with identities secured, we can move forward. For Deleuze, each object then has two aspects: virtual and actual (1994, 279–80). The virtual is the structural component that defines the structure and capacities of the object. The actual is its determinate, current manifesting properties, its qualities and extensions (Deleuze 1994, 279). What is potential and what is determinate can change through the process of actualization. The virtual then defines a range of potential actualities. The virtual is not this range itself and should not be confused with it, because that would be to understand the virtual as merely a collection of possible actualities. To do so would fall foul of Deleuze's critique of misunderstanding the virtual as possibility. The actual is the determinate qualities as they follow from the various interactions that the object has with other things.

Furthering the isomorphism, the virtual lines up with the probability density produced by applying the Born rule to the wave function. This is both conceptually, because they each define a range of potential outcomes of the system, and mathematically, as one is literally a mathematical state space and the other, the virtual, Deleuze describes in terms of state spaces (see DeLanda 2002 for an in-depth account). Furthermore, the virtual is non-localizable and internally defined spatially (Deleuze 1994, 183), as are quantum systems. Next, the actual, as a determined outcome of the object's interaction with something else, is straightforwardly structurally similar to the measurement outcome of an experiment. In fact, on Deleuze's account, the actual is the qualities and extensions of an object. It is precisely what can be measured directly (or measured without changing in kind). If these nodes are correct, then the processes between them are also isomorphic, that is, actualization and measurement. We have then completed the isomorphism.[9]

If this isomorphism is correct, then Deleuze's metaphysics would imply a solution to the measurement problem of the second type mentioned above: an explanation of how what is ontologically indeterminate becomes determined. If so, then it also provides reasons why two other types of explanation cannot succeed. For the first type, epistemic interpretations would fail because the process of determination would be a real process, not an illusory artifact of a physical limit to our knowledge. The fourth type would fail in different ways given the different interpretations, but in each, it would be for some reason of ontological insufficiency. Let us look at an example for this in the Many Worlds interpretation. Proponents of Many Worlds take decoherence as support by reading into it a solution to the measurement problem. Let us look at why first; then I will show, using a Deleuze-inspired critique, why it fails.

The Decoherence of the Philosophers

Many Worlds, or the Everettian interpretation, is currently a favored interpretation among philosophers of physics. This favoring by philosophers is because of the belief that Many Worlds does not require adding anything to the mathematical formalism (we will note a challenge to this claim below). De Witt famously remarked that "*The mathematical formalism of the quantum theory is capable of yielding its own interpretation*" because that formalism "forces us to believe in the reality of all the simultaneous worlds represented in the superposition [...] in each of which the measurement has yielded a different outcome" (DeWitt and Graham 1973, 160–1, original emphasis). The Many Worlds interpretation is opposed to theories which require extra theoretical or mathematical additions, such as spontaneous wave function collapse in GRW, configuration space in Bohmian mechanics, and so on. In line with this, and expanding on earlier versions of the interpretation, Wallace argues that the worlds of the Many Worlds interpretation are found in the formalism of quantum decoherence itself (2010; 2012).[10] They are dynamically isolated structures that result from various branchings, that is, they are the different outcomes of quantum events (Wallace 2010). As particles interact with their environment, the different outcomes of those interactions become further entangled with their environment. This means that these branches spread and become causally isolated, which decoherence does provide the formalism for. On this interpretation, these different branches, these different worlds, are real. For Many Worlds proponents, the worlds are equally real, wave function collapse doesn't exist, and we only see an apparently classical world because we are in one dynamically isolated branch of a much larger structure of decoherent histories.

This is shockingly similar to Deleuze's account of incompossible worlds that result from divergent series: "the incompossibility of worlds is decided in the vicinity of those singularities which give rise to divergent series between themselves" (1994, 263). However, there is a central difference, which can be seen when Wallace invokes a favorite term of Deleuze's, chaos, in saying "Where there is chaos, there is branching" (Wallace 2010, 88). This invocation of chaos gives the hint of the Deleuzian response, because for Deleuze chaos is not actual; it is virtual, as are incompossible worlds. If so, then such branchings would only ever be structures and potentialities, not determinate actualities. Wallace, in arguing for the reality of these branches, equates the real with the actual. His argument is simple but powerful. We are justified in believing in the descriptions of reality described by our best theories (a premise I agree with). Next, he points out that quantum mechanics is the most successful scientific theory human beings have ever produced (a fact it would be silly to argue against). And finally, quantum mechanics describes isolated, quasi-classical worlds branching off of each other at each quantum event. Therefore, these branches are real, and the Many Worlds interpretation is not only true, but it is the natural, inherent interpretation of quantum mechanics' own mathematical formalism. However, this depends on an assumed set of common concepts in order to see one interpretation as more "natural" than the other.

The Deleuzian response is that there is more than one way to be real. There are three: to be real is to be either virtual, intensive, or actual.[11] If so, then Wallace's argument only establishes that these branches are either virtual, intensive, or actual. It does not establish that all branches exist in the same way. Wallace discusses different ways in which a thing might be real, including quasi-particles, cases where a collection of particles can be described as acting as if one, collective particle. Wallace then claims that branches are real in this way, basing this claim on Dennett's account of real patterns, but makes no distinction of ways in which something is real; that is, he makes no modal distinctions (Wallace 2010, 59–60). It is exactly these sorts of cases in which a Deleuzian would invoke the virtual; quasi-particles are not actually one big particle, but they function virtually as if they were. Furthermore, they do so in some sense because of their "pattern" or structure, and "the reality of the virtual is structure" (Deleuze 1994, 209).

I will give three reasons why I think we should treat these branches as virtual rather than actual. The first is mathematical. Quantum mechanics is a theory that gives probabilistic outcomes. If Many Worlds is correct, then those outcomes are not actually a measure of chance or probability; instead, they are a measure of some feature of these branches (sometimes referred to metaphorically as the thickness of the branches, higher probability being more "thick"). We simply do not observe such a feature. It is not even a measure of the percentages of worlds in which events happen, because the number of branches is not well defined without an arbitrary scale choice (Wallace 2010, 67). This is known as the preferred basis problem: essentially, depending on which basis one chooses for the system in question, it will generate different worlds (for complete details, see Hemmo and Shenker 2022). Which worlds, then, are we supposed to take as real? An absolute basis, one not up to choice in how one describes the system, would solve the problem, but there is no such basis to be found. If worlds are an artifact of our choice in setting out the formalism, then Wallace's argument that the formalism compels us to accept them as real loses its force.

The second reason is metaphysical. If we accept Wallace's view, then not only is probability not real, but potentiality is not either. Everything that can happen, does happen. It just all happens in different branches. For potentiality to make sense, it requires that some things that could happen do not actually come to happen. However, without potentiality, it is difficult to understand just what it is that a dynamical system, such as quantum mechanics, is describing. That is, if potentiality loses sense, then so does dynamism. It is then unclear what central parts of the physical theory, such as wave function evolution, mean.

The third reason is that Many Worlds equates individuation with actualization; that is, once a world is individuated, it is immediately actualized along with all other possibilities. That is, it is both a denial of actualization as a separate process (yet an affirmation of actual worlds with determinate properties) and a confusion of the virtual with the actual (because it affirms structure as determinate in the way the actual is, not as virtual). While this at first might not appear as interesting or convincing if one is not already a Deleuzian, there are independent metaphysical reasons for accepting that individuation and actualization must be different processes, and the Deleuzian critique of the confusion of virtuality, possibility, and actuality is well-trodden ground. We cannot retread this ground here, but in short, these independent reasons are to make sense of ontological novelty and potentiality, and to be able to distinguish these two cases (based on the fact that classical accounts of possibility cannot make sense of either). Again, confusing the two processes of individuation and actualization is, according to Deleuze, a mistake as bad as confusing the virtual with the possible. This would then be a problem for the Many Worlds theorist, because their description of reality is incomplete. If decoherence and worlds are emergent, then worlds are not the whole story. There are processes "underneath" worlds, that is, there are events that are not branching events, and there are purely quantum phenomena that are not quasi-classical. These would be whenever the interaction between a quantum system and its environment is loosely enough coupled that the system does not decohere. This fact supports the isomorphism I argued for above, because purely quantum phenomena undergird decoherence in the same way that intensities individuate into objects with virtual and actual aspects.

Such a confusion is central to the Many Worlds interpretation. That is, there is a single description of the universe wherein, at a non-fundamental level, increasingly causally isolating branches develop (worlds), all of which are equally real and actual.

> There is no mechanism by which one of those states is preferred (is "actual," if you like), and all others are mere possibilities: at the fundamental level, all are components in a single, unitary evolving state, and no one is preferred over the other. But at the emergent level, each term evolves independently of the others. (Wallace 2012, 4591)

This is how measurement is taken out of the Everettian picture. In Deleuzian terms, the Many Worlds interpretation requires a denial of actualization as a real process and the rejection of virtuality in favor of the multiplication of actualities. All physically possible states are taken to be equally actual. This is best seen in how Many Worlds reinterprets the probability measure of these states. Since all worlds are equally actual, the probability assigned to each state cannot mean anything like the likelihood of its obtaining; all described states already do or will obtain. Instead, the probability of an event is taken to measure the "thickness" of the branch or world in which that event takes place. What this thickness actually means, and how it is to be reconciled with what we generally take probability to mean, is unclear and possibly incoherent.[12] If these metaphysical arguments hold, then following our isomorphism, measurement (or rather, a process whereby undetermined probabilities become determined) is then ineliminable from an adequate interpretation of quantum mechanics.

Coherence

Wallace describes an approach much like the one we have taken in this chapter: "Decoherence could, however, play a role in a solution to the measurement problem that leaves the equations of quantum theory alone and treats the objective macroscopic reality we see around us as itself an emergent phenomenon" (Wallace 2012, 4592). However,

Wallace claims that this is equivalent to the Many Worlds interpretation, where the trade-off being made is that one has to accept all branches or "worlds" defined by the underlying quantum mechanics as equally real. This we have rejected for the Deleuzian reasons that 1) equally real does not imply equally actual; 2) Many Worlds is exactly the illusion one would expect if they projected from the virtual to the actual without taking into account a process of actualization; and 3) metaphysical arguments for actualization provide independent reasons for accepting measurement as a real part of the theory, not an extraneous artifact of human practice.[13] That said, this is not yet a full solution to the measurement problem, because the details of this particular process of actualization have not been worked out. What has been shown is that the interpretation of these portions of quantum mechanics in terms of Deleuze's metaphysics makes sense of these phenomena in a way that makes other interpretations, here Many Worlds, less compelling.

What is most of interest here, I think, is that Deleuze is not explicitly drawing from quantum mechanics when he is formulating his theory of modality and individuation. While he and Guattari do indeed make connections between the two in *What is Philosophy?*, this account is first worked out in detail in *Difference and Repetition*. There, the account is proposed as a solution to at least three major conceptual problems: 1) the flaws found in classical accounts of possibility, 2) developing a metaphysics capable of accounting for ontological novelty, and 3) explaining how the fundamental priority of difference can give rise to stability and identities. If Deleuze were merely ontologizing the sciences, these isomorphisms would be uninteresting; they would be baked in from the start of the project.[14] But if Deleuze is in fact solving philosophical problems, and this gives rise to a metaphysics adequate to making sense of otherwise hard to interpret quantum phenomena, then we have established something interesting. In other words, in the difference between the sciences and Deleuze's metaphysics we find a productive encounter. If the isomorphism I have argued for holds, then this makes several new connections for the philosophy of physics and the relevance of quantum mechanics for metaphysics.

I have not worked out a full interpretation of quantum mechanics in terms of Deleuze's metaphysics. Instead, I have shown an isomorphism between Deleuze's account of modality, along with the processes of individuation and actualization, and quantum decoherence along with measurement. What is compelling about this isomorphism is that it accounts for all the stages we see in applying quantum mechanics: the mathematical description of the system in terms of a wave function (whose extensive interpretation is confounding), the prediction of probabilistic outcomes via the application of the Born rule, and the determinate results of taking measurements. If one takes these three moments all to be equally real, but two of them to be non-actual, then many of the common interpretations of quantum mechanics are inconsistent. Importantly, Deleuze's metaphysics provides ontological reasons why these interpretations are insufficient. However, the question of what a Deleuzian interpretation of quantum mechanics would be is still open. This requires further work, and this work should begin by seeing whether it has to reinvent the wheel. Multiple different interpretations may turn out to be cut to fit for Deleuze's metaphysics. By my estimation, there are several good candidates, including objective collapse interpretations and Carlo Rovelli's recent relational quantum mechanics (see Calamari 2021 for a brief exploration of the latter). However, this problem requires much work yet, as quantum theory stubbornly refuses to provide its own interpretation.

Notes

1. There is very little work on Deleuze and quantum mechanics. The most sustained engagement is the work of Arkady Plotnitsky (see especially Plotnitsky 2006). To my knowledge, there is no published straightforward attempt to apply Deleuze's metaphysics to the measurement problem.
2. For a recent example of a modal interpretation of quantum mechanics, see da Costa et al. (2013).
3. For more, see *Deleuze and Guattari Studies* 11, no. 2, especially Bowden (2017) and Mader (2017).
4. In this article, Maudlin separates the measurement problem into three distinct, but related, problems. This is the first formulation he provides. In only referring to this formulation, I do not reject the other two measurement problems he outlines. Rather, it is this problem in particular that we are concerned with in this chapter.
5. For a technical introduction, see Zurek (2003).
6. For a full account of the history of intensity which Deleuze is drawing from, see Mader (2014).
7. For an exploration of Schrödinger Logics in relation to Deleuze's metaphysics, see Ardoline (2024).
8. It is worth noting that French and Redhead (1988) point out that the violation of PII is only interesting if what violates it can be said to be an individual. In non-individual cases, the PII is simply not applicable rather than violated. If one is imagining individuation as a process proceeding from a situation in which no individuals exist to the existence of the first individuals, then this would seem to be the latter inapplicable case. This is not a correct understanding of individuation, however. Deleuze claims that "all differences are borne by individuals" (1994, 247). Since intensive differences are differences, they are borne by individuals, though they are not characterized as individual differences (i.e., as merely differences between individuals). The intensive, then, is not

non-individual, but the aspect of individuals that functions without identity, a characterization of their self-differing and what that makes possible.

9. We have so far looked at a relatively static picture of a single measurement. If we are to extend this account to the change of quantum systems over time, then we must add a further process. This is wave function evolution, the change in the probabilities of the system over time. If the isomorphism argued for holds, then when looking for a counterpart to wave function evolution, we are looking for a process of change in the virtual. The obvious candidate is differen*t*iation. We will leave this fuller picture aside for a future project as it is not necessary to the further arguments of this chapter.
10. Wallace's account differs from a more classical interpretation of Many Worlds in which every case of a superposition is seen as a branching event rather than every decoherence event. This reduces the expected number of branches quite a bit (see Maudlin 2019, 176–9).
11. Villani remarks that "the 'real' in Deleuze would be the inclusive disjunction of the actual and the virtual" (2007, 51); however, as we supported the view of the intensive as a unique mode above, we are committed to including the intensive in that disjunction. There is argument about whether in being covered over, the intensive should be understood to be included or inhering within the virtual and actual or not, but we leave this issue aside. Again, see Bowden (2017).
12. For instance, Many Worlds raises difficult obstacles for making sense of probabilistic reasoning, rational decision making, and even the confirmation of scientific theories through experiment. See Albert (2015) and Maudlin (2019).
13. For this reason, measurement is perhaps a poor choice of words in that it seems to imply a human practice, when it actually refers to a wider range of interactions of which human measurements made in a lab are a subset (a similar problem applies with the term "observer" in scientific theories).
14. See Williams (2006) and Mader (2017).

References

Albert, David Z. 2015. *After Physics*. Cambridge, MA: Harvard University Press.

Ardoline, Michael J. 2024. *Deleuze, Mathematics, Metaphysics: Difference and Necessity*. Edinburgh: Edinburgh University Press.

Bowden, Sean. 2017. "The Intensive Expression of the Virtual: Revisiting the Relation of Expression in *Difference and Repetition*." *Deleuze and Guattari Studies* 11, no. 2: 216–39.

Bowden, Sean, and Dale Clisby, eds. 2017. "The Virtual, the Actual, and the Intensive: Contentions, Reflections, and Interpretations." *Deleuze and Guattari Studies* 11, no. 2, special issue.

Calamari, Martin. 2021. "The Metaphysical Challenge of Loop Quantum Gravity." *Studies in History and Philosophy of Science Part A* 86: 68–83. https://doi.org/10.1016/j.shpsa.2021.02.003.

da Costa, Newton C. A., and Krause Décio. 1994. "Schrödinger Logics." *Studia Logica* 53, no. 4: 533–50.

da Costa, Newton C. A., Olimpia Lombardi, and Mariano Lastiri. 2013. "A Modal Ontology of Properties for Quantum Mechanics." *Synthese* 190, no. 17: 3671–93.

Dawid, Richard, and Karim P. Y. Thébault. 2015. "Many Worlds: Decoherent or Incoherent?" *Synthese* 192, no. 5: 1559–80.

DeLanda, Manuel. 2002. *Intensive Science and Virtual Philosophy*. London: Bloomsbury Academic.

Deleuze, Gilles. 1994. *Difference and Repetition.* Translated by Paul Patton. New York: Columbia University Press.
Deleuze, Gilles, and Félix Guattari. 2009. *What is Philosophy?.* Translated by Hugh Tomlinson and Graham Burchell. London: Verso.
DeWitt, Bryce S., and Neill Graham. 1973. *The Many-Worlds Interpretation of Quantum Mechanics.* Princeton, NJ: Princeton University Press.
French, Steven. 1989. "Identity and Individuality in Classical and Quantum Physics." *Australasian Journal of Philosophy* 67, no. 4: 432–46. https://doi.org/10.1080/00048408912343951.
French, Steven, and Michael Redhead. 1988. "Quantum Physics and the Identity of Indiscernibles." *The British Journal for the Philosophy of Science* 39, no. 2: 233–46.
Friebe, Cord. 2014. "Individuality, Distinguishability, and (Non-)Entanglement: A Defense of Leibniz's Principle." *Studies in History and Philosophy of Science Part B: Studies in History and Philosophy of Modern Physics* 48, no. 1: 89–98. https://doi.org/10.1016/j.shpsb.2014.08.002.
Hemmo, Meier, and Orly Shenker. 2022. "The Preferred Basis Problem in the Many-Worlds Interpretation of Quantum Mechanics: Why Decoherence Does Not Solve It." *Synthese* 20. https://doi.org/10.1007/s11229-022-03713-y.
Mader, Mary B. 2014. "Whence Intensity? Deleuze and the Revival of a Concept." In *Gilles Deleuze and Metaphysics*, edited by Alain Beaulieu, Edward Kazarian, and Julia Sushytska, 225–48. London: Lexington Books.
Mader, Mary B. 2017. "Philosophical and Scientific Intensity in the Thought of Gilles Deleuze." *Deleuze and Guattari Studies* 11, no. 2: 259–77. https://doi.org/10.3366/dls.2017.0265.
Maudlin, Tim. 1995. "Three Measurement Problems." *Topoi* 14, no. 1: 7–15. https://doi.org/10.1007/BF00763473.
Maudlin, Tim. 2019. *Philosophy of Physics: Quantum Theory.* Princeton, NJ: Princeton University Press.
Plotnitsky, Arkady. 2006. "Chaosmologies: Quantum Field Theory, Chaos and Thought in Deleuze and Guattari's *What is Philosophy?.*" *Paragraph* 29, no. 2: 40–56.
Villani, Arnaud. 2007. "I Feel I Am a Pure Metaphysician (interview with G. Deleuze)." Translated by Robin Mackay. *Collapse* 3: 45–62.
Wallace, David. 2010. "Decoherence and Ontology (or: How I Learned to Stop Worrying and Love FAPP)." In *Many Worlds? Everett, Quantum Theory, and Reality*, edited by Simon Saunders, Jonathan Barrett, Adrian Kent, and David Wallace, 53–72. Oxford: Oxford University Press.
Wallace, David. 2012. "Decoherence and Its Role in the Modern Measurement Problem." *Philosophical Transactions: Mathematical, Physical and Engineering Sciences* 370, no. 1975: 4576–93.
Williams, James. 2006. "Science and Dialectics in the Philosophies of Deleuze, Bachelard and DeLanda." *Paragraph* 29, no. 2: 98–114. https://doi.org/10.3366/prg.2006.0019.
Zurek, Wojciech H. 2003. "Decoherence and the Transition from Quantum to Classical—Revisited." https://arxiv.org/abs/quant-ph/0306072v1. Accessed September 10, 2025.

Chapter 10

The Chaosmos of the Quantum, with Gilles Deleuze

Arkady Plotnitsky

1. Introduction

This chapter primarily concerns quantum physics. Its argument is, however, shaped by an *engagement* with Deleuze's philosophy. The main juncture of this engagement is Deleuze and Guattari's (DG) view of creative thought (*la pensée*) as a confrontation between thinking and chaos, combined with their concept of concept, as the main vehicle of thought in philosophy, as in *What is Philosophy?* (hereafter "thought" will be used in this sense; otherwise "thinking" will be used).

As concerns thought, while also seeing it as a confrontation between thinking and chaos, I adopt a different conception of chaos. For DG:

> Chaos is defined not so much by its disorder but by the infinite speed with which every form taking shape in it vanishes. It is [...] a *virtual*, containing all possible particles and drawing out all possible forms, which spring up only to disappear immediately. (DG 1996, 84)

By contrast, I understand chaos as that which is ultimately beyond all conception, beyond the reach of thinking, including thought, in affinity with the ancient Greek concept of "chaos as the alogon," the term I shall adopt, along with "reality without realism" (RWR) (following Plotnitsky 2021; 2022). The latter term indicates that chaos as the alogon is a form of reality, as is DG's chaos as a virtual. The conception of RWR presupposes more general conceptions of reality and existence, assumed here to be primitive and not given analytical definitions. By "reality" I refer to that which is assumed to exist, without making any claims concerning the *character* of this existence, claims that define realism or ontology.[1] I understand existence as a capacity to have effects on the world, effects that manifest another form of reality, designated, following DG, as "actual." The assumption that anything exists is always inferred from such effects. The concept of RWR, or chaos as the alogon, may apply to both the ultimate reality responsible for quantum phenomena and to the ultimate (mental) reality responsible for thinking (Plotnitsky 2022).

Beyond the reach of thinking, including thought, chaos as the alogon is also beyond the idea of chaos as a virtual, or any of DG's concepts of virtual, from *Difference and Repetition* (Deleuze 1995) onward, which define ontologies of the virtual. I shall, however, retain a version of the virtual, considered as an effect of chaos as the alogon, with qualifications when it applies in quantum theory (QT), because QT contains its own, different concept of virtual (Plotnitsky 2021). Thus, I shall refer to "the DG virtual" and "the quantum virtual," respectively. In DG's argument in *What is Philosophy?*, reality, whether the DG virtual or the actual, is primarily mental, even though it is ultimately created by the material reality of the body and the brain (DG 1996, 201–18). Thought itself might concern material reality, as it does in QT or all modern, mathematical-experimental physics from René Descartes and Galileo onward. By contrast, the reality of matter cannot be considered apart from the reality of thinking. No form of reality, including RWR, is possible otherwise. In QT, at least in RWR interpretations, the actual is defined by quantum phenomena registered in observational instruments. As such, the actual allows for and, in the present view following Niels Bohr, entails a representational ontology, by means of classical physics, in contrast to the ultimate nature of the reality responsible for these phenomena, which is, in RWR interpretations, beyond the reach of thought.[2] Classical physics, however, cannot predict the actual of quantum phenomena; it can only describe this actual. This incapacity led to the introduction of QT by Max Planck in 1900, and eventually, in 1925–27, to quantum mechanics (QM) and quantum field theory (QFT). QFT is necessary in high-energy regimes, when one must consider the effects of special relativity, which deals with electromagnetism, rather than with gravity, as general relativity does. As currently constituted, QFT cannot handle gravity, and there is no quantum theory of gravity at present. Low-energy quantum regimes are handled by QM (or non-relativistic forms of QFT).

Chaos as the alogon is also productive of randomness or chance, and the corresponding form of disorder. DG's statement that "chaos is defined *not so much* by its disorder" leaves space for disorder in chaos as a (DG) virtual, but DG do not say in what sense. The disorder considered here will, however, be defined by randomness or chance and assumed to be an effect of a reality that is an RWR or chaos as the alogon. This does not imply assuming this reality itself to be random or mixed, because randomness and order are *conceptions*, creations of thinking, while, as an RWR, this reality is beyond the reach of thinking. This reality allows for predicting, probabilistically, quantum events, and thus for a degree of order in our interaction with it by means of experimental technology and QT.

My understanding of concepts follows more closely that of DG, while extending the sphere of its application. DG understand a concept not as a generalization from particulars (commonly assumed to define concepts) or a general or abstract idea, but as a multicomponent structure, defined by the organization, composition, of its components. DG limit their concept of concept to philosophical concepts, in juxtaposition to mathematical and scientific ones. I argue that creative mathematics and science are also primarily defined by the invention of new concepts in a sense analogous, although not identical, or in the language of mathematics, homomorphic but not isomorphic, to DG's concept of concept in philosophy.

Werner Heisenberg's invention of QM was enabled by the invention of several new concepts—physical, mathematical, and philosophical. *Mathematically*, Heisenberg used mathematics never previously used in physics. *Physically*, he no longer considered the motion of quantum objects, such as electrons and photons, but only the effects of their interactions with observational instruments. *Philosophically*, this approach implied a new, RWR-type epistemology, because the ultimate nature of reality was treated as an RWR or chaos as the alogon. The approach transformed the very nature of theoretical physics in QT. It made the creation of a new theory an invention of *abstract* mathematical schemes, in the sense both of higher forms of mathematical abstractions and of being divorced from mathematically representing physical reality. QM and QFT are, however, capable of predicting probabilistically (no other predictions are experimentally possible) the outcomes of quantum experiments observed as quantum phenomena. By *quantum* phenomena I refer to physical phenomena, and in considering which, one must consider the Planck constant, *h*, which has a very small magnitude ($6.62607015 \times 10^{-34}$ joule-hertz^{-1}), introduced by Planck in 1900 and essential to the ultimate constitution of nature in QT.

All modern physics only deals with idealized mathematical models, as do QM and QFT. What changed was the nature of the connections between a physical theory and physical reality. The formalism of these theories, in RWR interpretations, no longer represents the ultimate nature of physical reality *responsible* for observed phenomena, but only relates to the data observed by means of probabilistic predictions.

Physical concepts only apply to the description of observed quantum phenomena. As noted, in these interpretations, quantum phenomena and the observable parts of the instruments used are described by classical physics, which can neither describe how quantum phenomena come about nor predict them. While, however, QM or QFT can predict these phenomena, they cannot describe how they come about either, at least in these interpretations. In classical physics or relativity, the description of how the phenomena considered come about is possible because the interference of observational instruments can be neglected, so that the phenomena observed are not affected by observation and thus can, for all practical purposes, be identified with the objects considered, such as planets moving around the sun. By contrast, it is never possible to bypass the role of observational instruments in the constitution of quantum phenomena. This fact precludes identifying observed quantum phenomena, described by classical physics, with quantum objects, which are responsible for quantum phenomena by interacting with observational instruments, but which cannot be described by classical physics. These circumstances still allow for realist interpretations of QM or QFT or alternative realist theories of quantum phenomena. In RWR interpretations, however, the mathematics of QT has no representational connections to quantum phenomena. The only connections between them are probabilistic predictions of data contained in quantum phenomena.

An abstract mathematical scheme of QM or QFT still needed to be found—whether invented or borrowed from already existing mathematics. Matrix algebra, used by Heisenberg in his discovery of QM, was developed in mathematics by then. Heisenberg was famously unaware of its existence and reinvented it. This mathematics had several key features that had never been used in physics previously, especially complex, rather than real, numbers, infinite-dimensional spaces, and noncommutative algebra of new mathematical variables, the so-called operators, initially as matrix variables. These features still define all currently standard quantum theories, including the Yang–Mills theory (a form of QFT) that grounds the standard model of elementary particle physics. The standard model handles all fundamental forces (electromagnetic, weak, and strong) of nature apart from gravity, for which there is no quantum theory thus far. Just like QM, all these theories, regardless of interpretation, only predict the probabilities of the outcomes of quantum experiments. In RWR interpretations, these theories do not represent and even preclude a conception of how what is so observed comes about. This changes the character of theoretical physics by making theories the inventions of new (mathematized) conceptual schemes, enabling one to predict the outcomes of experiments, rather than representing the *ultimate nature* of reality responsible for these outcomes. The character of experimental physics changes as well, by becoming more essentially technological, as the invention of new technological devices able to register the data observed, instead of technology designed to observe what would have happened even if one had not observed it. The above qualification concerning "ultimate" applies because we observe phenomena at classical macrolevels. Experimental technology, such as the Sudbury neutrino detector (Fig. 10.1), enables the *creation* of new phenomena as the effects of the interactions between nature and technology, such as those shown in Fig. 10.2, effects the emergence of which is longer represented by QT.

Nobody has ever seen a moving electron, photon, neutrino, or any other type of particle. It is only possible to observe traces of their interactions with measuring instruments. The emergence and structure of these traces make it difficult and, in RWR interpretations, impossible

Fig. 10.1 Photograph: Sudbury Neutrino Detector

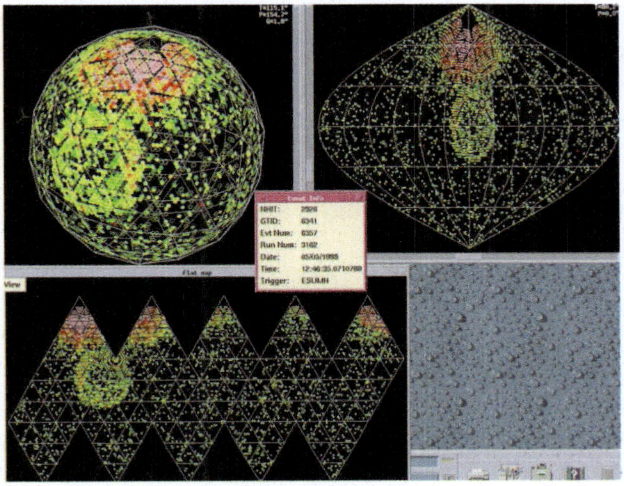

Fig. 10.2 Screen capture images taken from XSNOED (the x-windows SNO Event Display)

to know or even to conceive of how they come about. One can still allow for the statements in Fig. 10.2, such as that "the muon interacting with the detector produced a secondary electron," in RWR interpretations if one assumes that both the muon and the electron in question are "detected" in sequence, with each assumed to be responsible for the trace observed, rather than being observed as such. Consider, however, one of the photographs obtained in the Large Hadron Collider (LHD) at CERN, which served as the basis for establishing the existence of the Higgs boson (Fig. 10.3).

It is not only that the Higgs boson has not been observed as such (no quantum object could be), but that no *traces* of the Higgs boson have

Fig. 10.3 The Higgs boson effect registered in Cern, open-source photograph

been observed. The existence of the Higgs boson has been established on the basis of probabilistic calculations of the distributions of the traces of other particles, with the Higgs boson assumed to be responsible for these traces.

2. Confronting Chaos: Thought and Reality in Quantum Physics

As noted, DG define thought as a confrontation between thinking, or the brain, and chaos, as a DG *virtual*, a confrontation in which thought creates an order manifested in the *actual* (DG 1996, 200). On the surface, this view might appear unsurprising, but actually it is. First, commonly, chaos, in this juxtaposition, is assumed to be some type of disorder, such as that introduced by randomness or chance. DG instead see chaos as something different from disorder, or solely disorder: "Chaos is defined not so much by its disorder but [...] by the infinite speed with which every form taking shape in it vanishes [and as] containing all possible *particles* and drawing out all possible forms, which spring up only to disappear immediately, without consistency or reference, without consequence" (DG 1996, 84). (There is, thus, some disorder, a point discussed below.) The second, greater surprise is that DG define thought as a special kind of this confrontation, manifested especially in philosophy, art, and mathematics and science. While at war with chaos, thought also works together with, instead of protecting itself from, chaos, as do certain forms of thinking—in particular opinion, "which claims to protect us from chaos":

> The struggle with chaos is only the instrument in a more profound struggle against opinion, for the misfortune of people comes from opinion. [...] But art, science, and philosophy require more: they cast planes [of thought] over chaos. These three disciplines are not like religions that invoke dynasties of gods, or the epiphany of a single god, in order to paint the firmament on the umbrella, like the figures of an Urdoxa from which opinion stem. Philosophy, science, and art want us to tear open the firmament and plunge into chaos. (DG 1996, 203, 206, 202)

Philosophy, science, and art are different, if sometimes interrelated, primary modes of this cooperative confrontation—by creating new concepts in philosophy, new formal and propositional structures in mathematics and science, and new compositions in art.

As a DG virtual, chaos is a form of reality that ultimately escapes thought, which is, however, capable of interacting with chaos *actualizing* this interaction (DG 1996, 84). The actual is, thus, a form of reality available to thinking and manifested in mental formations created by *thought* in philosophy, art, and mathematics and science. DG's concept of chaos is unusual and does not appear to have been previously used in philosophy. As suggested by their invocation of "particle" and "virtual," this conception appears to be loosely borrowed from QFT. Its provenance appears to be the concept of virtual particle formation—the birth and disappearance of particles from a vacuum, technically the false vacuum (as there is no empty space). In QFT, however, these transformations involve two kinds of particles, designated in QFT as "real" and "virtual," although the existence of virtual particles is not universally accepted. The reason is that real particles can be observed or, in RWR interpretations, in which they cannot be observed as such, registered in observational instruments, even though one cannot say anything about them apart from their interactions with these instruments. By contrast, virtual particles, even if assumed to exist, are never so registered, and have other properties distinguishing them from real particles.[3] In any event, the concept of the virtual in QFT, the quantum virtual, is not the same as DG's chaos as a virtual, or any concept of DG virtual, even apart from the fact that in physics everything can only happen with a finite speed, limited by c (the speed of light in the vacuum), by relativity. In their argument, however, this infinite speed is in the realm of thought, to which their concept of chaos and this infinite speed refers.

DG's concept of chaos as a DG virtual is more akin to what happens *between* quantum experiments in general (rather than dealing with the quantum virtual), in high-energy (QFT) regimes, as opposed to what is observed, as actual, in experiments. This affinity is limited if one adopts an RWR interpretation of QFT. This limitation arises because DG's concept of chaos as a virtual is ontological, conceiving of this reality as a form of continuous becoming, "containing all possible *particles* and drawing out all possible forms, which spring up only to disappear immediately, without consistency or reference, without consequence" (DG 1996, 84). This conception is not applicable, any more than any other, to the ultimate *physical* reality responsible for quantum phenomena in RWR interpretations. There are ontological interpretations of this reality, some akin to (and possibly among the sources of) DG's chaos as the virtual, such as those based on John A. Wheeler's concept of "quantum foam" (Wheeler and Ford 1998). These interpretations assume the existence of a "sea" of particles continuously transforming into each other, except that, because one deals with physical reality, there is no infinite speed. In RWR interpretations, no concept of any kind, even that of "something happens," can apply to the reality ultimately responsible for quantum phenomena, and thus to anything apart from experiments.[4]

This is how what happens *happens* in QFT in RWR interpretations. Any well-defined physical experiment requires at least two observations and the corresponding measurements: the initial one, M_1, at time, t_1, defining the initial condition of the experiment and enabling predictions concerning future experiments by means of the corresponding theory; and the second one, M_2, at time t_2, that confirms (or possibly refutes) the theory's prediction. In low-energy (QM) regimes, if M_1 registers an electron, M_2 will also register an electron. It is not possible to be certain that it is the same electron, but it will be an electron, a quantum object of the same type as registered in M_1, with the same permanent characteristics, such as its mass, charge, or spin. It is also possible that M_2 will register nothing, even if the experiment is isolated from outside interferences. In classical physics or relativity, one always deals with strictly the same object, always found in any experimental sequence, if isolated from outside interferences (which can change what happens), and for individual or simple (non-chaotic) systems, our predictions of the outcome of M_2 are ideally exact.

The situation is, however, different in high-energy (QFT) regimes. There M_2 can, with a comparable probability, register an object, such as an elementary particle, different in type from the one registered in the previous observation, M_1, within a single quantum experiment, or under the same initial conditions. The identity of the type of the object observed in M_1, say, an electron, can no longer be guaranteed in M_2. This was a revolutionary discovery, a consequence of Dirac's equation, which implied the existence of an antiparticle to the electron, the positron, which could be registered by M_2, following that of an electron registered by M_1. In QED, the simplest form of high-energy QFT dealing only with electrons, positrons, and photons, when the first observation registers an electron, the second may register a positron, or a photon, or an electron–positron pair. Once one moves to higher energies, the panoply of different particles becomes greater, and experiments can register many elementary particles, different in type, in successive observations.

The question becomes how these transformations come about; specifically, whether they conform to a realist or ontological interpretation, akin to DG's ontological view of the ultimate nature of reality, as the DG (chaos as a) virtual, or in accord with RWR interpretations. In either type of interpretation, as actual, each quantum phenomenon, either in low-energy (QM) or high-energy (QFT) regimes, is an unrepeatable, unique effect of an RWR reality, discrete relative to any other such effect. As such, these effects also entail the irreducible element of randomness, or chaos as randomness, but only in the actual, because, as beyond thinking, this reality can no more be thought of as random than as causal. What is more radically transformational in QFT vs. QM is *the actual*—quantum phenomena, as effects of this reality. These effects are associated with different types of particles in successive measurements even within the same experimental arrangement. At the same time, there is no assumption of a uniform nature of the ultimate reality responsible for these effects, an assumption precluded in RWR interpretations. These interpretations, however, equally preclude assuming this reality to be transforming or multiple. As beyond conception, this reality can no more be single than multiple, no more permanent than transforming, no more continuous than discrete. According to Heisenberg:

> There is no description of what *happens* to the system between the initial observation and the next measurement. [...] The demand to "describe what happens" in the quantum-theoretical process between two successive observations is a contradiction in adjecto, since the word "describe" refers to the use of classical concepts, while these concepts cannot be applied in the space between the observations; they can only be applied at the points of observation. (1962, 57, 145)

The same would apply to the word "happen" or "system," or any word we use, including reality, insofar as there is any concept associated with this word, which is not the case of "reality" in "reality without realism." It is, accordingly, impossible to assume a continuous ontology, including that akin to the DG virtual, of the ultimate nature of reality responsible for quantum phenomena in RWR interpretations. These interpretations only assume an ontology of the actual, in which all quantum phenomena are discrete relative to each other. Deleuze's general definition, "What we call virtual is not something that lacks reality but something that is engaged in a process of actualization following the plane that gives it its particular reality" (2005, 31), can still apply to the ultimate reality responsible for quantum phenomena in RWR interpretations, but only under the assumption that this "reality" is beyond the reach of thinking, rather than assuming the immanent ontology of the virtual.[5] In RWR interpretations, QFT-type transformations only pertain to the actual, observed in instruments. Nothing can be said or even thought about what happens between quantum events, including in terms of the mathematical formalism of QM or QFT.

A quantum observation replaces measurement in the classical sense of measuring some preexisting properties of the objects considered with establishing, through the interaction between quantum objects and observational instruments, quantum phenomena. The latter can then be treated classically, by measuring what is observed in instruments, without measuring the properties of quantum objects. A quantum observation is a unique act of creation, an outcome of a *complex* process involving a *complex* technology created in the course of a *complex* history of experimental physics. Thus, the radically transformational features of QFT in RWR interpretations pertain strictly to the actual, as effects of the chaos as the alogon, and as such are fundamentally different from the transformations of DG's chaos as a DG virtual, defined ontologically. There is no chaos as a virtual in DG's sense in QFT in RWR interpretations, but only transformational effects of chaos as the alogon, effects defining the actual. The quantum virtual can be defined accordingly (Plotnitsky 2021, 273–7).

Quantum phenomena also exhibit mathematizable regularities. While strictly probabilistic in character, these regularities allow for mathematical handling by means of QM or QFT. QED is the best experimentally confirmed theory ever. It is difficult to expect anything close to this degree of rigor and precision in dealing with human thinking, although there have been attempts to approach it through models borrowed from QT (Plotnitsky 2023a; 2023b). Nevertheless, thought, or even thinking, is impossible apart from randomness or chance and probability, which, accordingly, need to be addressed in considering thought as related to the idea of chaos, either as a DG virtual or as the alogon.

Randomness and chance can designate different concepts, but they are close and can be used interchangeably for the present purposes. I shall speak of randomness. Randomness is a manifestation of the unpredictable as concerns either future events or events that have already happened, even though they might have hidden causes, in which case they are at bottom not random. Randomness introduces disorder into our interactions with the world. On the other hand, the use of probability or statistics helps to restore a degree of order to these interactions. I define probability, following Bruno De Finetti (2010), as a degree of belief, which implies a degree of subjectivity, and statistics as estimating frequencies of possible events and as such more objective.

By "causality," I refer, ontologically, to the architecture of the physical reality considered, which determines, by law, the behavior of the systems associated with it, thus establishing the corresponding form of order. This architecture is causal if the state of this reality, A, at a given point determines, by a given law (such as Newton's law of gravity), the state of this reality at any future point, X, and A is itself determined by the same law by any previous state. (This definition does not require A to be a cause of X.) By determinism, I refer, epistemologically, to the possibility, at least ideally, of exact predictions concerning the system or the architecture of reality considered, which need not allow for determinism, even when they are causal. In the case of individual or small (non-chaotic) systems in classical physics, causality and determinism are coextensive. On the other hand, while classical statistical physics and chaos theory are, ontologically, causal, they are not, epistemologically, deterministic, because the complexity of the systems considered makes probability unavoidable in predicting their behavior. This is, however, merely due to our insufficient knowledge of the assumed causal reality considered. The same is true in ontological or realist probabilistic theories elsewhere. In fact, before quantum physics, these were the only reasons commonly (exceptions, such as Hume, are few) considered philosophically, expressly by Kant, who even excluded probability from the domain of philosophy vs. practical utility (Kant 1997, 661–2). Quantum physics changed this, in RWR interpretations irrevocably. In these interpretations, the recourse to probability in QT is a consequence of placing the ultimate nature of the reality responsible for quantum phenomena beyond conception. This placement precludes the assumption of causality. Thus, it is not a matter of a practical lack of knowledge, possible or at least conceivable in principle, of how things happen, but of the impossibility in principle of conceiving of how they happen.

As noted, DG's statement that "chaos is defined *not so much* by its disorder" leaves space for disorder in their conception of chaos as a virtual. They do not explain what they mean by "disorder," perhaps referring to any feature preventing the ascription of a complete order to chaos. Randomness is a form of disorder, but not all disorder is randomness. Regardless of how DG saw this disorder, the concept of (chaos as) disorder as defined by randomness or chance is fitting in QT, and I would argue in considering thinking and thought. While, however, randomness is inherent in quantum physics, quantum phenomena are not entirely random. First, in certain situations, they exhibit statistically correlated patterns (not found in classical physics); and second, they allow for predicting, probabilistically, quantum events, including these correlations, by means of QT. As such, they allow for a degree of order in our interaction with nature. This order vis-à-vis that of classical physics reflects an essentially different type of randomness and probability.

Thus, chaos as randomness and probabilistic order are found in both classical and quantum physics. In classical physics, it is underlain by an ontological order, and the recourse to probability is merely practical, due to our inability to access this order. In quantum physics, it is underlain by chaos as the alogon, as an RWR-type of reality, which makes randomness and, if possible as it is in QT, the recourse to probability irreducible, because no knowledge or even conception concerning this reality in possible. Hence, no specific order or even a general concept of order could be assigned to this reality, any more than randomness or disorder, correlatively to the fact that this reality can be no more assumed to be discrete than continuous, and no more single than multiple. Randomness or probability, or other features just mentioned, are only effects of this reality.

An element of disorder remains present when only probabilistic or statistical estimates are possible, because one cannot control what will *actually* happen. Probability, again, manifests the interplay of chaos and order. It is "chaosmic," in Joyce's famous coinage in *Finnegans Wake* (1939), favored by DG, and, in Joyce's case, influenced by QT, to which Joyce expressly refers in the novel: "I am working out a quantum theory about it, because this is the most tantumising [a play on tantalizing] state of affairs," he says (Joyce 2012, 149). The sentence also describes the novel itself, on the model of "chaosmology" of QT, resulting in the linguistic, informational, and other features of its super-complex, chaosmic structure (Plotnitsky 2023a; 2023b). The term "chaosmos" has, however, also been used in describing classical theories, especially chaos theory, to which Joyce could not have referred because it was only introduced some thirty years after his death in 1941. (DG's chaos as a DG virtual is not connected to chaos theory, to which they refer in different contexts, e.g., DG 1987, 486–8.) Hence, there are two fundamental types of chaosmos associated with randomness and probability, correlative to the nature of the reality considered in its ultimate constitution—that amenable to an ontology and that, as RWR or chaos as the alogon, beyond all ontology, including that of chaos as the DG virtual.

Where, then, does the *cosmos* in the chaosmos of quantum events or effects come from, if it does not emerge from some underlying causal order, precluded in RWR interpretations? This question becomes even more pressing given that this quantum chaosmos not only allows for the use of probability, but also exhibits the existence of correlations between certain sets of quantum events, while each individual event involved is random, which makes quantum physics as much about order as randomness. That, in certain circumstances, random individual events can combine to form a statistically correlated order, in the absence of any underlying causal order, appears to be especially enigmatic, as is the fact that QT correctly predicts these correlations. Where does this order come from? The existence of an underlying causal order would provide an answer akin to that of classical physics when the latter uses randomness and probability. While this type of answer cannot be excluded, to assume it poses major difficulties, in view of Bell's and the Kochen–Specker theorems, and related findings (Plotnitsky 2021, 253–7). In RWR interpretations, we not only do not but also cannot know or even conceive of how this order, or this randomness, comes about. They are effects of that which is neither ordered nor random, any more than anything else. It is possible to assume that the ultimate nature of reality is random or mixed; and, while continuously causal conceptions of reality have been dominant, random ones have been around since the pre-Socratics, as in Democritean atomism. Short of the RWR view, however, there remains the question of how discrete random events come about, assuming that one can conceive of any emergence of events that is not continuously causal (Wittgenstein 1985, 179). The problem is circumvented in RWR interpretations by in principle precluding even a conception of how quantum events, random or correlated, come about.

3. Concepts: The Art of Composition

Before discussing how the mathematics and technology of quantum physics work, I shall consider the concept of concept, by extending DG's concept of concept to mathematics and science. DG reserve their concept of concept for philosophical concepts, in juxtaposition to scientific or mathematical concepts. "The concept belongs to philosophy and only to philosophy," they contend, possibly only referring to the concept as they define it (DG 1996, 11–12, 33–4). In mathematics and science, thought is defined by planes of reference (versus planes of immanence, giving rise to concepts in philosophy), logical propositions and formal mathematical or scientific structures, such as functives. These features are necessary for mathematics and science, and they must be considered in defining mathematical and scientific concepts. Conversely, there are aspects of philosophical concepts that are not found in mathematical and scientific concepts. This is why *the concept of a mathematical or scientific concept* that I adopt is analogous, homomorphic, but not isomorphic, to DG's *concept of a philosophical concept.* Einstein's concept of space curved by gravity, Heisenberg's concept of matrix variables, or Schrödinger's concept of the wave function are concepts akin to philosophical concepts. So are earlier concepts of classical physics, or purely mathematical concepts, from ancient Greek mathematics onward. Each such concept is, as is each philosophical concept, defined by the specific structure of its elements and by an association with a problem, two key aspects of DG's concept of concept.

A concept is a unique multicomponent entity, defined by the organization or composition of its components, and some of these components may be concepts in turn. As DG say, "there are no simple concepts. Every concept has components and is defined by them [...] There is no concept with only one component" (1996, 16). A concept may also contain parts that cannot be specified, and hence are not concepts, such as RWR or chaos as the alogon, as part of the reality considered in quantum physics, other parts of which, such as those found in quantum phenomena, are specified. All concepts, however innovative, have a history (DG 1996, 17). Concepts are constructed out of elements of earlier concepts, or earlier concepts used as elements.

What defines a concept is, then, the *specific* individual character of its composition. This composition is unique in a new concept and makes this concept unique. A mathematical variable is a concept to which an equation (also a concept) applies, but both could be very different. Some components of a concept may be unique as well, but they need not be. Only its composition must be new for a concept to be new. In this respect a concept is akin to a work of art, defined by its composition, in accord with the ancient Greek meaning of *poiein* (making), from which the term poetry originates. "Composition is the sole definition of art," DG say (1996, 191). A composition can borrow elements from earlier compositions, and in this sense remain a partial *mimesis* of earlier works. But, as a composition, it must be new. The same type of mimesis defines concepts. A theory or a disciplinary field becomes an organized assemblage of concepts. In the present view, however, or that of DG, all *creative* conceptual mimesis is a *technical* mimesis that borrows parts of previous concepts, rather than a Platonist mimesis of a preexisting independent reality.

It may be argued that, insofar as art represents the compositional capacity of human thinking, and transforms this capacity into thought, art ultimately governs thought in all human endeavors. Art qua art, as a culturally defined practice, especially manifests this capacity, and works of art are different from concepts or theories. Thus, a work of art need not contain concepts, although sometimes it does, for example in conceptual art. Philosophy, as defined by DG, may be associated with the invention of concepts as an individual practice. In this respect, this practice is closer to that of art, which, while it may use a tradition, is compelled to be innovative, but only in this respect, because the nature of the conceptual and artistic composition remains different. It is not too far off the mark, however, to say that Heisenberg's QM was a kind of abstract composition of concepts, divorced from representing physical reality, akin to those of Wassily Kandinsky or Piet Mondrian in painting.

Another key aspect of DG's concept of a philosophical concept is its character as positing and even *being* a problem, and remaining a problem, which persists rather than disappears in its solutions. This view, initially advanced in *Difference and Repetition*, has a mathematical genealogy, specifically Niels Henrik Abel's and Évariste Galois's mathematics (Deleuze 1995, 163–4, 179–82). A problem in this sense is not something that, like a theorem, is derived from assumed axioms by means of logical rules, but is something that is posed, created, along with and as a concept. In DG's understanding, however, insofar as it remains a (philosophical) concept, a problem persists rather than disappears in its solutions. This fact enables such a concept to remain "always new" (DG 1996, 5). This is a complex conception that is not easy to interpret, and I shall not attempt to do so (see Plotnitsky 2022, 59–76, 161–6). I shall merely note that it implies two different approaches to solving a problem. The first is technical: it aims to solve a problem by means of manipulations of previously existing concepts. The second is conceptual: it develops new concepts, although a technical manipulation of earlier concepts remains necessary. Heisenberg introduced an entirely new concept of physical variables in confronting the problems facing QT. His concept, however, did much more than solve these problems. It defined a new field in which new types of problems emerge and are being solved, while this concept itself continues to remain in play.

4. Mathematics as Physics: Abstract Compositions of Quantum Theory

We might be lucky that nature allows our mathematics to handle quantum phenomena and to have this mathematics. It was, nevertheless, still necessary to create this mathematics, and to find a way of using it in physics. Heisenberg was able to do both. I shall describe the mathematics of QM in its currently standard Hilbert space version.

There are alternative formalisms, such as those of C* algebras, all thus far mathematically equivalent to Hilbert space formalism, still most common. Introduced later by John Von Neumann (1932), it was not used by Heisenberg. The key features defining it were, however, contained in Heisenberg's version, as exemplified by his matrices for a toy model he considered in his paper introducing QM (Heisenberg 1925), containing the coordinate, q, and the momentum, p, measurements, respectively (ω is the frequency of the radiation, emitted by electrons) (Figures 10.4.-10.5)

$$pq - qp = i\hbar I \left(\hbar = \frac{h}{2\pi}\right)$$

I is a unit matrix:

$$\begin{array}{cccccc} 1 & 0 & 0 & 0 \ldots. & & \\ 0 & 1 & 0 & 0 \ldots & & \\ 0 & 0 & 1 & 0 \ldots & & \\ . & . & . & . & & \\ . & . & . & . & . & \\ . & . & . & . & . & . \end{array}$$

$$\mathbf{q} = \sqrt{\frac{\hbar}{2\omega}} \begin{bmatrix} 0 & \sqrt{1}\,e^{-i\omega t} & 0 & 0 & \cdots \\ \sqrt{1}\,e^{i\omega t} & 0 & \sqrt{2}\,e^{-i\omega t} & 0 & \cdots \\ 0 & \sqrt{2}\,e^{i\omega t} & 0 & \sqrt{3}\,e^{-i\omega t} & \cdots \\ 0 & 0 & \sqrt{3}\,e^{i\omega t} & 0 & \cdots \\ \vdots & \vdots & \vdots & \vdots & \ddots \end{bmatrix},$$

Fig. 10.4 Heisenberg's matrix for coordinate, q.

$$\mathbf{p} = i\sqrt{\frac{\hbar\omega}{2}} \begin{bmatrix} 0 & -\sqrt{1}\,e^{-i\omega t} & 0 & 0 & \cdots \\ \sqrt{1}\,e^{i\omega t} & 0 & -\sqrt{2}\,e^{-i\omega t} & 0 & \cdots \\ 0 & \sqrt{2}\,e^{i\omega t} & 0 & -\sqrt{3}\,e^{-i\omega t} & \cdots \\ 0 & 0 & \sqrt{3}\,e^{i\omega t} & 0 & \cdots \\ \vdots & \vdots & \vdots & \vdots & \ddots \end{bmatrix}$$

Fig. 10.5 Heisenberg's matrix for momentum, p

These elements, which contain imaginary quantities ($i = \sqrt{-1}$), do not commute, $qp \neq pq$, as such canonical variables for the position and momentum, as functions of real variables, $f(x, t)$, would do in classical physics. Correlatively to this commutativity, in classical physics it is always possible to make the simultaneous exact measurements of both, thus also enabling causality and, for individual or simple systems, ideally deterministic predictions. This is never possible in quantum physics, as is manifested in the Heisenberg uncertainty relations, $\Delta q \Delta p \geq h$ (where q is the coordinate, p is the momentum in the corresponding direction, h is the Planck constant, and Δ is the standard deviation, which measures the amount of variation of a set value). Here p and q are *measurable quantities* observed in instruments, while p and q above are abstract mathematical entities (operators), never measured, but used to predict the values for p and q, respectively. Both p and q can never be *exactly* measured simultaneously, which makes predictions concerning quantum phenomena probabilistic. The measurement of either can be exact.

In the case of Schrödinger's equation, the main equation of QM, the key mathematical innovation (when time was considered) was the complex wave function (ψ), a vector in an infinite-dimensional complex Hilbert space, which enables one to predict the probabilities of quantum events (H is an energy operator also known as a Hamiltonian). In one dimension:

$$i\hbar\frac{\partial\psi(x,t)}{\partial t} = \left[-\frac{\hbar^2}{2m}\frac{\partial}{\partial x^2} + V(x,t)\right]\psi(x,t)$$

Dirac's equation, which describes the relativistic (high-energy) electron by using the same types of variables, is, in its compact form, reproduced on the plate in Westminster Abbey commemorating him:

$$i\gamma \cdot \partial\psi = m\psi.$$

This simplicity is deceptive: the equation encodes a complex Hilbert space machinery. As introduced by Dirac, the equation was:

$$\left(\beta mc^2 + \sum_{k=1}^{3}\alpha_k p_k c\right)\psi(x,t) = i\hbar\frac{\partial\psi(x,t)}{\partial t}$$

$$\alpha_i^2\beta + \beta^2 = I_4$$

(I_4 is the identity matrix)

$$\alpha_i\beta + \beta\alpha_i = 0$$
$$\alpha_i\alpha_j + \alpha_j\alpha_i = 0$$

This single equation unfolds into four coupled linear first-order partial differential equations for the four quantities that make up the wave function. p is the momentum operator, but in a more complicated Hilbert space than in QM. This mathematical architecture allowed one to predict the probabilities of high-energy (QED) events, in which, as explained, the identity of the type of the particles considered is no longer preserved. The same type of mathematical structure in the same probabilistic relation to experiments define all QFT, all the way to the Yang–Mills theory.

Beginning with Heisenberg, then, the mathematics of QT was a radical change from all mathematics previously used in physics in the following aspects:

1. While all previous physics used mathematics over real numbers, $\boldsymbol{R}$, and was finite-dimensional, QM uses Hilbert spaces over complex numbers, $\boldsymbol{C}$, which are abstract vector spaces of finite and infinite dimensions. Classical physics or relativity might use complex numbers, but only as a calculational tool. They do not figure in the final solutions of the equations used nor are they related to what is observed, and everything that is observed is represented by real numbers.
2. The second key feature is the noncommutativity ($xy \neq yx$) of Hilbert space vectors and especially operators, like p and q above, known as "observables," which are complex *mathematical* quantities, as opposed to classical physics or relativity, where all observable quantities are represented by commuting ($xy = yx$) functions of real variables. These quantities are related to *physically* observable real quantities, not in terms of representation, impossible because the observed quantities are real, but in terms of probabilistic predictions by using (3).
3. Born's (or a functionally similar) rule establishes the relation between the so-called "quantum amplitudes," which are complex quantities, and probabilities, which are real numbers.

The probabilities involved are non-additive, or non-Kolmogorovian, as they would be in classical physics: the joint probability of two or more mutually exclusive alternatives in which an event might occur is not equal to the sum of the probabilities for each alternative. Instead, the probability of alternative future events is derived by using the addition of "amplitudes." The term "amplitude" is adopted from classical wave theory, but, at least in RWR interpretations, it no longer has any physical meaning and serves strictly as part of the mathematical procedure for predicting probabilities. Quantum amplitudes are complex quantities, vectors in a Hilbert space, which are connected with possible alternative events, associated with variables as operators, say p, when predicting a future momentum measurement. Operators designate *operations* that transform vectors, such as amplitudes, according to linear rules, $p\ (x + y) = p\ (x) + p\ (y)$ and $p\ (\alpha x) = \alpha p\ (x)$, for any constant, α, and some additional rules, keeping in mind that in general operators do not commute ($pq \neq qp$). For an individual event, say, when ψ is a wave function for a particle associated with a Hilbert space, Born's rule says that the probability density function $p\ (x, y, z)$ for predicting a measurement of the position at time t_1 (after measuring it at time t_0) is equal to $|\psi(x, y, z, t|^2$, a square of the modulus of a complex quantity, which is always a real quantity.[6] When one estimates the probability for two or more alternative events, one adds amplitudes and then applies Born's rule.

Beyond finding the mathematical scheme of QM, Heisenberg's approach revolutionized the very practice of theoretical physics. It also redefined experimental physics. Taking advantage of and bringing together two meanings of the word "experiment" (as a test and as innovative creation), in the wake of Heisenberg's invention of QM the practice of quantum physics became the first practice of physics that is both, and jointly, *generatively* experimental and *generatively* mathematical. Experiments now engender, create new reality, rather than observing the preexisting one. By the same token, the practice of theoretical physics in QT is *generatively* mathematical because it no longer consists, as in classical physics or relativity, in developing an idealized mathematical representation of quantum objects and behavior. Instead, it consists in inventing abstract mathematical schemes enabling us to predict, probabilistically, the outcomes of quantum events. Just like classical physics or relativity, quantum physics is an experimental science because of its mathematical project, as Heidegger said (1967, 93). Quantum physics, however, gave this conjunction a new form by establishing a new physical reality and its future course in each experiment, and by defining a mathematical reality, as a reality of thought, by creating new mathematical compositions divorced from representing how physical phenomena come about. In fact, one also experiments with mathematics. It is experimenting with abstract mathematics, divorced from physical concepts that, with Heisenberg, became the way to new theories in quantum physics.

As it has done throughout the history of physics, from pre-Socratic physics to relativity and QT, philosophy can help our thought in finding ways of discovering new concepts and connections between them *in the art of composition* of fundamental physics. In modern physics, mathematics, added to experimental technology, became a technology of thought driving new physics, including by overcoming established opinion, doxa, as Heisenberg did against a powerful doxa of realist thinking. RWR thinking can, however, become a doxa too, standing in the way of thought and new physics. This can happen to any creation of thought. Thought can only avoid this by keeping alive its friendship with chaos by creating new compositions, which continue to remain alive, to remain new, by engendering other new compositions. Thought is a fight against the doxa of opinion, a fight changed by a chaosmic *poeien*, building a cosmos open to the future of thought in each new composition, a unique act of creation, just like each quantum experiment. As each time unique, a *poeien* of thought, wherever it occurs, is quantum-like. This *poeien* is, however, only possible through an interaction with chaos, by keeping the cosmos it creates part of a chaosmos that never surrenders to opinion or other means of programming us by shielding us from chaos. Chaos is a precious friend of thought, which makes thought possible and keeps it alive.

Notes

1. The terms "realist" and "ontological" may designate more diverging concepts. Still, these terms are close in their meaning and will be used, as adjectives, interchangeably. I adopt "realism" as a general term, and refer by "ontology" to a given representation or conception of the reality considered.
2. By classical physics I refer to classical mechanics (including chaos theory), classical statistical physics, and classical electromagnetism.
3. For a discussion of QFT and virtual particles in RWR interpretations, see Plotnitsky (2021, 273–307).
4. Although DG's chaos as a virtual and their other concepts of the virtual allow for different interpretations, they are forms of ontology; moreover, a continuous ontology, which appears to underly all discreteness, important as the latter is in some of their arguments. While the primacy of continuity in their philosophy has been occasionally challenged, its ontological nature has not, to my knowledge, ever been questioned. As beyond the reach of thinking, "chaos as the alogon" or RWR precludes any ontology, continuous or discrete. Not coincidentally, Deleuze's concept of singularity, one of his key concepts, is that of a singular (unique) continuous line, not that of a singular (unique) discontinuous event, as in quantum physics. Another key concept, that of the fold, inspired by Gottfried Leibniz, is also continuous (Deleuze 1993).
5. Even in considering a mental virtual, one deals with the real, which, while the *condition* of the actual, may never be actualized, rather than something possible or potential, eventually actualized. The virtual is the *condition* of the actual and not the *condition of the possibility* of the actual, in accord with the well-known juxtaposition between Deleuze's and Kant's philosophy.

6. The probability density function specifies the probability of a random variable falling within a particular range of values. This probability is given by the integral of this density function over this range.

References

De Finetti, Bruno. 2010. *Philosophical Lectures on Probability*. Translated by Hykel Hosni. New York: Springer.

Deleuze, Gilles. 1993. *The Fold: Leibniz and the Baroque*. Translated by Tom Conley. Minneapolis: University of Minnesota Press.

Deleuze, Gilles. 1995. *Difference and Repetition*. Translated by Paul Patton. New York: Columbia University Press.

Deleuze, Gilles. 2005. "Immanence: A Life." Translated by Anne Boyman. In *Pure Immanence: Essays on a Life*, 25–34. New York: Zone Books.

Deleuze, Gilles, and Félix Guattari. 1987. *A Thousand Plateaus: Capitalism and Schizophrenia*. Translated by Brian Massumi. Minneapolis: University of Minnesota Press.

Deleuze, Gilles, and Félix Guattari. 1996. *What is Philosophy?*. Translated by Hugh Tomlinson and Graham Burchell. New York: Columbia University Press.

Heidegger, Martin. 1967. *What is a Thing?*. Translated by William B. Barton Jr. and Vera Deutsch. South Bend, IN: Gateway.

Heisenberg, Werner. 1925. "Quantum-Theoretical Re-interpretation of Kinematical and Mechanical Relations." In *Sources of Quantum Mechanics*, edited by Bartel Leendert van der Waerden, 261–77. Mineola, NY: Dover.

Heisenberg, Werner. 1962. *Physics and Philosophy: The Revolution in Modern Science*. New York: Harper and Row.

Joyce, James. 2012 [1939]. *Finnegans Wake*. Oxford: Oxford University Press.

Kant, Immanuel. 1997 [1781]. *Critique of Pure Reason*. Translated by Paul Guyer and Allen W. Wood. Cambridge: Cambridge University Press.

Plotnitsky, Arkady. 2021. *Reality without Realism: Matter, Thought, and Technology in Quantum Theory*. Cham: Springer Nature.

Plotnitsky, Arkady. 2022. *Logos and Alogon: Thinkable and Unthinkable in Mathematics, from the Pythagoreans to the Modern*. Cham: Springer Nature.

Plotnitsky, Arkady. 2023a. "The No-Cloning Life: Uniqueness and Complementarity in Quantum and Quantum-like Theories." *Entropy* 25, no. 706. https://doi.org/10.3390/e25050706.

Plotnitsky, Arkady. 2023b. "'The Agency of Observation not to be Neglected': Complementarity, Causality and the Arrow of Events in Quantum and Quantum-Like Theories." *Philosophical Transactions of the Royal Society A* 381, no. 2256. https://doi.org/10.1098/rsta.2022.0295.

Von Neumann, John. 1932. *Mathematical Foundations of Quantum Mechanics*. Translated by Robert T. Beyer. Princeton, NJ: Princeton University Press.

Wheeler, John A., and Kenneth Ford. 1998. *Geons, Black Holes, and Quantum Foam: A Life in Physics*. New York: W. W. Norton.

Wittgenstein, Ludwig. 1985 [1922]. *Tractatus Logico-Philosophicus*. Translated by Charles K. Ogden. London: Routledge.

Chapter 11

Immanence, Process, and Relationalism in Loop Quantum Gravity

Martin Calamari

Fundamental physics faces a significant challenge: to formulate a theory that combines quantum theory and general relativity. Incorporating quantum field theory (QFT), the foundation of the standard model of particle physics, and general relativity (GR), the theory of gravity and spacetime geometry underlying the standard model of cosmology, into one theoretical framework is the quantum gravity (QG) problem.

The quest for a QG theory is motivated by conceptual and physical reasons. Conceptually, while QFT and GR are extremely successful theories, they rest on incompatible presumptions. For instance, whereas quantum theory, including non-relativistic quantum mechanics (QM) and relativistic QFT, assumes preferred space and time or fixed spacetime structures, GR posits spacetime as a dynamical entity. However, while GR assumes continuous metric and matter fields, quantum theory introduces discreteness and probabilistic behavior. Physically, the need for a QG theory is because GR is no longer valid for extreme cosmological regimes, like black holes or the very early universe. At the fundamental level of the Planck scale (Planck length $l_P \approx 10^{-33}$ cm; Planck time $t_P \approx 10^{-44}$ s; Planck energy $E_P \approx 10^{19}$ GeV), where quantum effects cannot be neglected, GR simply breaks down.

Despite nearly a century of efforts, no fully developed or empirically tested QG theory exists. Instead, several tentative approaches exist, including string theory, loop quantum gravity, causal dynamical triangulations, causal set theory, and group field theory (Smolin 2001; Oriti 2009; Murugan et al. 2012). Although there is no universal consensus, many physicists distinguish the QG problem from the purpose of unifying all fundamental forces—such as "grand unified theories" or a putative "theory of everything"—a task that has proven unsuccessful so far. String theory has been the prominent unification program since the mid-1980s. In contrast, loop quantum gravity and other approaches do not pursue a unified theory, but a theory that explains the *quantum* properties and dynamics of gravity, and hence *spacetime*, and its interactions with other quantum fields.[1]

The absence of a QG theory is primarily due to two reasons. First, the Planck scale lies just beyond the reach of existing high-energy experiments or direct astrophysical and cosmological observations. Consequently, over the past few decades, researchers have proposed various low-energy or tabletop experimental and observational testing strategies (Amelino-Camelia 2013; Wallace 2022; Huggett et al. 2023). Second, all candidate QG theories not only raise severe technical concerns but also challenge fundamental physical and metaphysical assumptions. As a result, a growing number of philosophers are contemplating the profound implications that a QG theory might have for our understanding of the fundamental nature of the physical world (Callender and Huggett 2001; Huggett et al. 2020; Wüthrich et al. 2021; Rickles et al. 2006; Weinstein and Rickles 2019; Knox and Wilson 2022, part VI; Vassallo 2022).

This chapter explores the metaphysical and ontological implications of loop quantum gravity (LQG) and their alignment with certain aspects of Deleuze's philosophy. Rather than delving into LQG's technical features or cosmological applications (which fall within the domain of quantum cosmology),[2] my focus centers on three ontological claims underpinning the philosophical foundations of the full LQG theory (Rovelli 2004; Rovelli and Vidotto 2015; Ashtekar and Pullin 2017; Rovelli 2020; Rovelli and Vidotto 2022):

1. The physical world is fundamentally *non-spatiotemporal.*
2. The physical world comprises non-spatiotemporal quantum *events* or *processes.*
3. The physical world involves only *relative*, ontologically interdependent, quantum events or processes.

I argue that these claims resonate profoundly with Deleuze's ontology of becoming, difference, and chaosmology. In section 1, I show how the first claim challenges absolute or transcendent structures as space and time or spacetime in favor of *immanence*. Section 2 illustrates the second claim's shift from substance ontology to *process* metaphysics and ontology. In section 3, I explain how the third claim rejects ontological independence, the traditional primary substancehood criterion, advocating instead for ontological *relationalism*.

1. Immanence: The Disappearance of Spacetime

Many QG approaches, including LQG, suggest that classical space, time, and spacetime from GR disappear at the Planck scale. This implies that the putative fundamental entities (or degrees of freedom) constituting the physical world might be non-spatiotemporal, rendering ordinary spacetime emergent rather than fundamental. Recovering its topological and geometrical structures at appropriate classical ($\hbar \to 0$) and continuum ($N \to \infty$) limits poses a significant challenge. While "spacetime emergence" is a central issue in QG philosophy (Oriti 2014; Huggett et al. 2013; Huggett and Wüthrich 2018; Oriti 2020; 2021; Huggett 2022), this chapter will not cover it. Instead, we shall examine the nature of the presumed non-spatiotemporal fundamental entities in LQG. To achieve this, we must first understand how LQG conceives the disappearance of spacetime.

1.1. The principle of background independence

According to LQG, the absence of spacetime at the fundamental level stems from the principle of background independence (Belot 2011; Smolin 2006; Read 2023). Although no universally accepted definition of background independence exists,[3] for my purposes it will be defined as the principle that a QG theory must avoid relying on absolute, fixed, non-dynamical, and ontologically independent entities, such as space, time, or spacetime. It should not postulate preexisting spacetime topologies, geometries, or a preferred time variable t to describe the dynamics of fundamental degrees of freedom. Instead, it should only involve interacting dynamical entities, *including spacetime*, independent of background structures. Consequently, in QG, spacetime disappears as it becomes an integral part of the fundamental quantum dynamics. Let me explain this argument in more detail.

The principle of background independence is inherited from GR, the paradigm of a background-independent theory.[4] The core of GR is that "there is no distinction between non-dynamical background and dynamical physical variables" (Rovelli 2007, 1310). GR describes spacetime geometry as a four-dimensional (pseudo)-Riemannian manifold (a differentiable manifold M with a pseudo-Riemannian metric $g_{\mu\nu}$) able to affect *and be affected by* matter. This is because "spacetime and the gravitational field are the same entity" (Rovelli 2004, 9). Consequently, spacetime is not a fixed background but a dynamical (classical), continuous, smooth metric field interacting with and altered by other matter fields. Gravity manifests as spacetime curvature due to the matter–energy distribution. This implies that absolute, preferred, global, fixed space and time or spacetime coordinates, reference frames, event locations or spacetime points are not physically observable quantities. Therefore, the physical world comprises interacting dynamical fields (Rovelli 2007, 1312), rather than matter fields propagating on a spacetime background or evolving according to a preferred global time.

1.2. Background independence and immanence

Deleuze appears to grasp GR's background independence, crucially relating it to the idea of immanence. In *A Thousand Plateaus*, Deleuze and Guattari resume Lautman's account of Riemannian geometry (Deleuze and Guattari 1987, 485), an integral part of GR's mathematics.[5] Lautman (2011, 110) distinguishes between intrinsic and extrinsic properties of mathematical entities. Intrinsic properties are inherent to the entity, while extrinsic properties refer to the surrounding "environment" or the "universe" in which it is "embedded." Riemannian geometry exemplifies the intrinsic viewpoint as it "studies the intrinsic properties of a manifold," disregarding "any space in which this manifold is embedded" or "any reference to a universal container or to a center of privileged coordinates" (Lautman 2011, 112). The main example is the curvature of a surface or an *n*-dimensional space, which Riemann proved to be intrinsically defined, independently of a higher-dimensional space, in his famous 1854 *Habilitation* lecture (Riemann 2007). Lautman (2011, 112) stresses that the Riemannian intrinsic perspective is a major feature of GR. It dissolves the distinction between "the container and the contents" by positing the interdependence of "geometry and physics," making it impossible "to separate space, the Riemannian manifold and matter." This conveys GR's background independence, treating spacetime as a dynamical field rather than an absolute or fixed background.

Deleuze and Guattari adopt this central idea of Riemannian geometry and GR to differentiate between transcendent (arborescent) and immanent (rhizomatic) multiplicities. Arborescent multiplicities define over-coding embedding spaces or global reference frames constituting a transcendent absolute as "the horizon or background," or "the Encompassing Element" (*l'Englobant*), which always presumes an "outside observer" (Deleuze and Guattari 1987, 491–5). Instead, rhizomatic multiplicities involve immanent processes similar to Riemann's intrinsic viewpoint, exploring a manifold by following a path on it, as opposed to considering it as "embedded in a Euclidean space to a sufficient number of dimensions" (Lautman 2011, 113; cf. Deleuze and Guattari 1987, 371, 373, 556 n. 39). Furthermore, the plane of immanence "never has a supplementary dimension" and is not embedded in a higher-dimensional "ambient space" (Deleuze and Guattari 1987, 266, 493). Conversely, the plane of transcendence "exists only in a supplementary dimension" (Deleuze and Guattari 1987, 265). Consequently, Deleuze's philosophy associates transcendence with an extrinsic, background-dependent viewpoint, while immanence reflects an intrinsic viewpoint and background independence.

1.3. *The background dependence of non-general-relativistic physics*

A significant challenge in developing a QG theory is that all physical theories, except for GR, are background-dependent.

Classical mechanics
In Newtonian mechanics, space and time are absolute, real substances, possessing specific topological and geometrical properties like infinity, continuity, and a Euclidean metric. They act as independent universal reference frames for the dynamics of physical bodies. Consequently, globally simultaneous events can be defined at any location and for any observer. Furthermore, while they influence the position and paths of material bodies, space and time remain unchanged. As Newton famously states: "Absolute, true, and mathematical time, in and of itself and of its own nature, without reference to anything external, flows uniformly"; "Absolute space, of its own nature without reference to anything external, always remains homogeneous and immovable" (Newton 1999, 54).

Special relativity
Special relativity (SR) eliminates absolute simultaneity, making spatial and temporal locations relative to the observer's reference frame. This results in well-known relativistic effects like length contraction and time dilation, which are captured by the light cone structure of Minkowski spacetime $(M, \eta_{\mu\nu})$, a flat four-dimensional (pseudo)-Euclidean space. An *event* is any physical occurrence at a spacetime point. A continuous succession of events forms a spacetime curve, known as a *world-line*. However, despite discarding preferred spatial and temporal distances, SR retains an invariant four-dimensional spacetime distance (given by the Minkowski metric $\eta_{\mu\nu}$), according to the Poincaré group, SR's global symmetry group. Thus, Minkowski spacetime still represents a fixed,

non-dynamical background independent of the dynamics of physical systems.

Quantum theory
The fundamental dynamical equation of standard quantum mechanics is the Schrödinger equation, which describes the dynamics of quantum systems as the evolution of the quantum wave function (or quantum state) $|\psi\rangle$ with respect to *t*. However, QM is inadequate for particle physics as it ignores relativistic effects and only treats systems with a finite and fixed number of degrees of freedom. Quantum field theory overcomes these limitations by incorporating QM and SR, providing the basis of the standard model of particle physics. QFT addresses high-energy physics' relativistic phenomena, infinite degrees of freedom, and particle creation and annihilation. Nevertheless, standard QFT remains background-dependent since it disregards gravitational interactions, thus describing quantum dynamics within SR's Minkowski spacetime.

As a result, all non-general-relativistic physics defines the dynamical evolution of classical and quantum systems against the background of some preexisting space, time, or spacetime.

1.4. How LQG meets the challenge of quantum gravity: quantum spacetime

The dichotomy between background-independent and background-dependent theories reveals the incompatibility between GR and quantum theory. However, GR is only *partially* background-independent, as it still allows, albeit arbitrarily, the definition of specific spacetime coordinates or spacetime geometries and assumes particular topological and geometrical properties like manifold dimensionality, continuity, and smoothness. These classical features are expected to vanish or significantly alter in the QG regime, since the physical world described by QFT is quantum-mechanical. This is the QG challenge: "to understand what is a general-relativistic QFT, or a background independent QFT" (Rovelli 2004, 7).

This is precisely what LQG tries to accomplish.[6] According to LQG, spacetime, and hence the gravitational field, is not a classical (continuous, smooth) field at the fundamental level, but exhibits all the quantum properties that characterize any quantum dynamical system: discreteness, superposition, probability, quantum fluctuations, uncertainty, randomness. In short, "spacetime is a quantum object" (Rovelli 2001, 110). LQG describes *quantum spacetime* as a quantum dynamical field consisting of "quanta of space" or "atoms of space," meant as quanta of the gravitational field. The "quanta of space" display quantum discreteness akin to the discrete energy spectrum of quantum particles.[7] Furthermore, they can be in quantum superposition, hence exhibiting quantum interference and entanglement, like generic quantum states. Their quantum dynamics are probabilistic, defined by transition amplitudes for quantum process probability, thus revealing the inherent indeterminism and randomness of any quantum system. The "quanta of space" are considered as quantum excitations of the quantum gravitational field, analogously to how particles in QFT are conceived as elementary excitations of quantum fields. However, whereas standard QFT is background-dependent, LQG is background-independent.

This is the ultimate reason why the physical world is fundamentally non-spatiotemporal. The "quanta of space," Rovelli emphasizes, "are not quantum excitations *in* space: they are quantum excitations *of* space itself" (2004, 264). Quantum spacetime is neither embedded *in* space and evolving *in* time nor an embedding background for the dynamical evolution of matter fields. The "quanta of space" constitute space itself, while spacetime results from their quantum interactions. So, spacetime is expected to emerge from the coarse-grained collective dynamics of non-spatiotemporal entities.

This description corresponds in LQG to the spin-network and spin-foam formalism (Rovelli and Vidotto 2015; Rovelli 2020; Perez 2013). Roughly, spin networks define the three-dimensional spacetime quantum geometry. Their states $|\Gamma, j_l, v_n\rangle$ are represented by abstract graphs Γ,[8] the nodes n of which are intuitively understood as "quanta of space," while each link l defines a contiguity relation corresponding to the space quanta local interactions. Spin foams encode the quantum dynamics of the four-dimensional quantum geometry of spacetime as spin networks evolve from an initial to a final state. The quantum dynamics is defined by transition amplitudes for the probability of interaction processes between "quanta of space," similar to how particle interactions are described in the Feynman path-integral formulation of QFT.

2. Process: The Disappearance of Substance

According to LQG, we have seen, the physical world comprises immanent (background-independent) non-spatiotemporal, interacting quantum dynamical entities. This raises the question: What is the nature of such entities? In other words, what is the ontology of the physical world according to LQG?

2.1. A Parmenidean "Block Universe"?

The spacetime disappearance in QG has led many philosophers to suggest a contemporary "Parmenides's revenge" (Huggett et al. 2013), often referred to as the "problem of time" (Weinstein and Rickles 2019; Thébault 2022; Callender 2017). Roughly, the argument is that since, commonly, change involves time and QG suggests a non-spatiotemporal world, this implies a static "Block Universe," a Parmenidean world of being without becoming.[9] This view is epitomized by Barbour's timeless quantum cosmology (Barbour 1999; 2020). According to Barbour, "the quantum universe is static. Nothing happens; *there is being but not becoming*. The flow of time and motion are illusions" (2009, 2, my emphasis). This recalls Plato's contrast between "that which always is, without ever becoming" and "that which always becomes, without ever being" (*Timaeus* 27d–28a), representing the Parmenideanism/Platonism and Heracliteanism divide.

Rovelli's view, and LQG more generally, have been often interpreted as advocating such a Parmenidean/Platonic stance (Belot and Earman 2001; Healey 2002; Huggett et al. 2013), an understanding undoubtedly reinforced by Rovelli's claim that "time does not exist." This interpretation is entirely mistaken, however. The spacetime disappearance in LQG does not entail a static Parmenidean "Block Universe" without dynamical evolution, change, or becoming.[10] Indeed, the opposite is the case. Rovelli insists that fundamental physics concerns becoming, as happening of events, rather than being: "*LQG describes reality in terms of processes*" (Rovelli 2020, 124). Therefore, rather than a form of Parmenideanism/Platonism, LQG upholds a form of Heracliteanism, or rather, *process metaphysics and ontology*, for which *there is becoming but not being* (see Calamari 2021 for an in-depth analysis). Let me explain.

2.2. The process metaphysics and ontology of LQG

Rovelli (2018a, 86–7) argues that the QG's spacetime disappearance involves rejecting the traditional notion of permanent, self-subsistent objects as concrete particulars bearing properties and, ultimately, any substance metaphysics and ontology. Fundamental physics dismisses substantial beings—"something that *is*"—suggesting an ever-changing, impermanent world of events and processes. Rovelli closely links here to Smolin, despite their otherwise different views.[11] Smolin agrees that if the world consisted of enduring objects or substances like point-mass particles, then change would merely be "alterations in how something is" (2001, 53). He argues that relativity, quantum theory, and especially QG suggest that the world fundamentally consists of change, events, and processes, compelling us to think a world where "process is more important than, and prior to, stasis" (2001, 53). Rovelli and Smolin strongly insist that the world or universe comprises events or processes, not "things" (Rovelli 2018a, ch. 6; Smolin 2001, ch. 4).[12]

As a result, according to LQG, the world's fundamental non-spatiotemporal nature is distinct from its dynamical nature. Despite being non-spatiotemporal, the world is dynamical because change, becoming, events, and processes are ontologically primary and independent of spatiotemporal determinations. Thus, whereas spacetime is emergent, change is fundamental.[13] The "problem of time" in QG posits that the absence of time entails the disappearance of change (and vice versa). However, this argument is flawed because it assumes the background-dependent view from non-general-relativistic physics, which does not hold in the QG context.[14] The fundamental disappearance of spacetime does not negate dynamical evolution, as the world consists of non-spatiotemporal, interacting quantum events or processes, rather than enduring substance-like entities moving in a background spacetime. Events and processes are not attributes of a substance or occurrences unfolding in space and time. They are non-spatiotemporal change or becoming *simpliciter*: "the world is nothing but change"; it is "ceaseless changing" (Rovelli 2018a, 85–6, translation modified).[15]

The rejection of any being or substance ontology has also an *epistemological* implication. Traditionally, defining something involves understanding its substantial being or essence (Aristotle 1984b, *Metaph.* Z 3, 1028b34–36; Z 7, 1032b1–2). If substance disappears, its explanatory power also vanishes. Rovelli makes this point explicitly: "we understand the world by studying change, not by studying things," since physics concerns "how *events happen*, not how things *are*"; in other words, "we understand the world in its becoming, not in its being" (Rovelli 2018a, 89, 91, 92).

In sum, LQG endorses a process metaphysics and ontology that sharply contrasts with the traditional metaphysics of being and substance. Rather than positing "something that *is*," LQG describes the physical world as composed of transient becomings, quantum events, and a "network of quantum processes" (Rovelli 2020, 129).

2.3. Being is becoming

This process metaphysics and ontology aligns with Deleuze's ontology of becoming and difference, as well as the broader tradition of process-oriented philosophers like Nietzsche, Bergson, and Whitehead (see Rescher 1996; Robinson 2009). Consider Bergson's description of "real becoming": "*there are changes, but there are underneath the change no things which change: change has no need of a support*" (Bergson 2002, 259; see Saint-Ours 2008, 2015 for a comparison with Rovelli). This idea challenges the Aristotelian conception that movement and change are accidents of an unchanging, underlying substance taken as *substratum* (Aristotle 1984a, *Cat.* 5, 4a10–21). Bergson posits that reality is "pure mobility" or change, whereas "things," "substances," or "states" represent abstract generalizations derived from concrete becoming or duration. "There do not exist *things* made, but only things in the making, not *states* that remain fixed, but only states in process of change" (Bergson 2002, 274). So, real becoming is "substantial," but only insofar as "substance is movement and change" (Bergson 2002, 265), not an eternal, unchanging being or essence. As Deleuze first demonstrated, Bergson's duration, change, and becoming are alteration—"*what differs from itself*," the "difference of self with itself." For Bergson, and indeed for Deleuze, becoming is *difference* (Deleuze 2004, 25–6, 37; 1991; see Lundy 2018 for a thorough analysis).

The rejection of the ontological priority of being and substance in favor of that of change and becoming is also crucial for Whitehead. Whitehead's "principle of process" holds that "*how* an actual entity *becomes* constitutes *what* that actual entity *is* [...]. Its 'being' is constituted by its 'becoming'" (Whitehead 1978, 23). Whitehead condemns the reification of abstract concepts, such as the belief in an immutable substantial being as the essence of nature or individual substances. Drawing on Locke's and Hume's empiricism, he rejects the Aristotelian and Cartesian substance–predicate conception of a permanent, independent substance with essential attributes distinct from its changing accidents. Such a substance is not only unintelligible and epistemically inaccessible but is what obstructs the understanding of nature's actual becoming. Deleuze (1993, 59–64, ch. 6) notes a shared rejection of the substance–predicate model among the Stoics, Leibniz, and Whitehead, who provide the three main event ontologies in the history of philosophy (see Robinson 2009 for discussion). He insists that the confusion between predicate and attribute is a crucial objection raised by these philosophers to substance metaphysics. Predicates are *events* and relations expressed by verbs (actions and becomings), not abstract universals as attributes of a substance or an infinite being.[16] Deleuze argues that what is ultimately common to these event and process-oriented philosophies is not the pursuit of eternity, but the problem of how the physical world allows *the production of the new* (Deleuze 1993, 89; see Smith 2012).

Deleuze's solution to this problem is his ontology of becoming and difference. From his early works, he argues for a new metaphysics and ontology, which requires a resolute departure from the traditional metaphysics of being and essence. This orthodox metaphysics rests on the presumed dichotomy between a supposedly transcendent "true world" of eternal being and the mere "apparent world" of empirical becoming. However, Deleuze contends that this distinction and opposition is a metaphysical fiction. As Nietzsche first demonstrated, the idea of an "apparent world" depends on the presumption of a "true world," *and vice versa*.[17] Deleuze's way out of this fiction is to argue that being *is* becoming. This implies a double—distinct yet inseparable—affirmation: first, "there is no being," "there is only becoming"; and, second, "being is the being of becoming" (Deleuze 1986, 23). This philosophical move is central to Deleuze's view of the world as immanent process and becoming and his philosophy of difference (Deleuze 1994). The identity of being and becoming ultimately leads to the claim that the only identity is the "identity of that which is different" (Deleuze 1994, 40). In other words, being is becoming, that is, *difference*.

3. Relationalism: The Disappearance of Ontological Independence

Section 1 established that LQG rejects any absolute or preferred spacetime at the fundamental level, leading to the disappearance of classical spacetime and a non-spatiotemporal physical world. However, as section 2 clarified, this does not imply a static world. Instead, the world comprises dynamical events or processes. Any object or substance-like entity vanishes, resulting in a fundamentally immanent, non-spatiotemporal network of interacting quantum dynamical processes. This raises a last question: how can we understand a dynamical world without spacetime? More precisely, how can we conceive a non-spatiotemporal quantum dynamics?

3.1. The relational view and the relationalism of fundamental physics

The answer to this question lies in the *relational view* of dynamics. The relational view is the logical conclusion of the background-independence principle and carries a crucial ontological implication. The key to understanding it is to realize the status of physical interactions between dynamical entities in a background-independent context like QG. In non-general-relativistic physics, the dynamics of a physical system is defined as the time evolution of its degrees of freedom, that is, how the physical variables change over time. However, this approach is ill-suited for QG, where no preferred space and time or spacetime background exists. Therefore, to define dynamical evolution consistently in QG, we must consider the interaction between physical systems, that is, how the physical variables change and evolve *in relation to each other* rather than as a function of *t*. This is the idea of relational dynamics, which we can informally define as follows:

> **Relational dynamics**: dynamics is the relative change of physical variables as a function of each other, corresponding to physical interactions between physical systems.

While this conception is necessary in QG, according to Rovelli, it is the result of the convergence between the *relationalism* present in GR and quantum theory, according to his relational interpretation of quantum mechanics. Therefore, to fully grasp the relational view and its ultimate ontological implications, we must first understand the relationalism underlying fundamental physics.

A) The relationalism of general relativity. The background independence of GR, we have seen, involves the gravitational field, and so spacetime, resulting in its being itself a dynamical field. So, according to GR, the physical world consists of interacting dynamical entities, including spacetime. Consequently, in contrast to all non-general-relativistic physics, the dynamics of GR is not defined as "the evolution of physical variables with respect to an independent variable 'time,'" but as "*relative evolution of the physical variables with respect to one another*" (Rovelli 2021, 76). GR employs a relational dynamics, where variables A, B, C, do not change as a function of t, $A(t)$, $B(t)$, $C(t)$, and so on, since t is an unphysical quantity. Instead, they change as a function of other variables, $A(B)$, $B(C)$, $C(A)$, and so on (Rovelli 2004).[18] In other words, GR does not describe the evolution of dynamical entities as a function of a global time variable or against a fixed background spacetime, but as the mutual change of the physical variables describing the interactions between dynamical entities. Consequently, there are no absolute, non-dynamical physical variables, but only *relative* dynamical ones. This is the relationalism of GR.

B) The relationalism of quantum theory. Rovelli believes that the relationalism of GR is connected with the relationalism of quantum theory. This is the primary motivation in view of a QG theory for his proposal of an alternative interpretation of QM, called relational quantum mechanics (RQM) (Rovelli 1996; 2005; 2022; Laudisa and Rovelli 2019; Dorato 2015; 2016; Van Fraassen 2010; Dorato and Morganti 2022).[19] RQM endorses an event—as opposed to an object/substance—ontology, according to which "the actual elements of reality" are *quantum events* (Rovelli 2004, 214).[20] A quantum event is what happens in the physical interaction between systems and corresponds to the values taken by the physical variables in the interaction process. Note that quantum events or processes and physical systems are not ontologically different entities. Physical systems are either elementary or composite, more or less stable, quantum events (Di Biagio and Rovelli 2021).[21] Therefore, on Rovelli's view, the physical world consists of "relational quantum events happening at interactions between physical systems" (Rovelli 2018b, 7), which ultimately means quantum events occurring in *relation* to other quantum events. This sharply contrasts with classical mechanics. Classical mechanics presumes that physical variables always have definite and context-independent values; that is, physical variables refer to absolute states or intrinsic, preexisting properties of a system at any given *t*, independent of any other system (for example, a measurement device). In contrast, in quantum theory, "rather than independent entities with definite properties, there are entities that have properties and characteristics only with regard to others, and only when they interact" (Rovelli 2022, 75). As a result, in quantum theory "*all* physical variables are relational" (Rovelli 1996, 6) as "the actual value of *all* physical quantities of *any* system is only meaningful in relation to another system" (Rovelli 2018b, 6). Therefore, quantum theory is concerned with how "physical systems *affect one another* when they interact" (Rovelli 2004, 215). Consequently, since properties can only be ascribed to interacting entities and through interaction processes, there are no non-interacting entities in the physical world.[22] Furthermore, since quantum theory violates the value definiteness and non-contextuality of classical mechanics (as proved by the Kochen–Specker theorem), the physical world ineluctably involves *contextuality*, that is, interdependence or relationality. This is the core relationalism of quantum theory.

C) The relationalism of loop quantum gravity. The relationalism of GR and quantum theory converge into the relationalism of LQG. Since there is no space and time or classical spacetime at the fundamental level, the "quanta of space" *are* space itself. More precisely, as Rovelli puts it, "the spatiality of the world consists of the web of their interactions," since the "quanta of space" "exist only as terms of ceaseless interactions" (2018a, 108). Like all quantum events or processes, the space quanta exist only through physical interactions and thus in relation to each other. The "quanta of space" are not spatiotemporal occurrences in or with respect to a spacetime background. They are non-spatiotemporal quantum events or processes whose relational dynamics yields quantum spacetime and, in the classical and continuum coarse-grained limit, classical spacetime. Therefore, the dynamics of quantum spacetime is the relative evolution of the dynamical degrees of freedom describing interactions between "quanta of space." In other words, a spacetime region is equivalent to a process, and vice versa (Rovelli and Vidotto 2015, 52). Ultimately, according to LQG, the physical world consists of the relative quantum dynamics of interacting quantum events or processes.

3.2. Chaosmology

The relational view has a far-reaching ontological implication, which, I suggest, resonates significantly with Deleuze's philosophy. According to the traditional foundationalist view of the hierarchy of being, the primary criterion defining a being or substance is *ontological independence*. A primary or fundamental substance does not depend on anything else for its being and existence (O'Conaill 2022). In resolute contrast to this view, the relationalism of modern and contemporary fundamental physics implies, first, an ontology of events or processes, rather than objects or substances, and, second, a fundamental *ontological interdependence*, where only *relative* events and processes exist, *pace* (neo)-Aristotelian metaphysics (Tahko 2012; Simpson et al. 2018).[23] Therefore, since events and processes, as well as their properties, exist only in relation to one another as a result of interaction processes, there are no ontologically independent, self-subsistent, or free-standing entities with essential, permanent, and preexisting properties. Instead, there are only interrelated, transient events and processes.

This view, which we might call ontological relationalism, is at the core of Deleuze's philosophy. In Deleuze and Guattari's theory of multiplicities, rhizomatic multiplicities are defined by their connectivity and mutual immanence; that is, they exist through their immanent interactions and variable connections rather than any presumed self-identical independence and permanence (Deleuze and Guattari 1987). Deleuze further elaborates this view in his book on Leibniz, which involves what he describes as the objective relativism or perspectivism of modern and contemporary philosophy—"perspectivism as a truth of relativity," he writes, "not a relativity of what is true" (1993, 23)—and that, finally, leads to its own form of neo-Leibnizianism, or rather, *chaosmology*.[24] According to Deleuze, one of the main features of the new relative perspectivism of the contemporary world in both philosophy and science[25] is the denial of the condition of absolute closure that characterizes Leibniz's monadology in favor of an *open* world in becoming, that is, a "chaosmos" of interacting and interdependent events and processes (see Deleuze 1993, 81, 137, 154; Deleuze and Guattari 1987, 616). The world is a chaosmos of purely immanent events and processes in relative becoming. This is the fundamental relationalism of the new chaosmology.

Conclusion

In loop quantum gravity, the principle of background independence involves the absence of any absolute, fixed space and time or spacetime at the fundamental level. This view is consistent with Deleuze's idea of immanence. LQG endorses a process metaphysics and ontology, where the physical world consists of dynamical yet non-spatiotemporal quantum events or processes, rather than static objects or substances. This aligns with Deleuze's ontology of becoming and difference. LQG maintains a relational physics view, which entails a form of ontological relationalism. This view suggests that quantum events or processes are interdependent and result from interactions, rejecting the notion of ontological independence. This view is consistent with Deleuze's theory of multiplicities and relative perspectivism, which characterizes his view of the world as chaosmos. As a result, LQG's philosophical foundations and Deleuze's philosophy support the view that the physical world is an immanent chaosmos of processes existing only in relation to each other.

Notes

1. See Armas (2021) for an overview of various perspectives.
2. See Ashtekar and Singh (2011), Agullo and Corichi (2014), and Rovelli and Vidotto (2015, ch. 11) for an introduction to and review of loop quantum

cosmology. See Vidotto (2017) for a philosophy-oriented presentation of loop quantum cosmology.

3. Read (2023) offers the most systematic and comprehensive analysis.
4. GR technically encodes background independence through "general covariance" or, more precisely, "diffeomorphism invariance" (Rovelli 2021).
5. For the significance of Riemann for Deleuze, see Plotnitsky (2006a; 2009), Calamari (2015; 2017), Jedrzejewski (2017), Widder (2019), and Duffy (2013).
6. In the following, I will not provide a formal exposition of LQG. Specifically, I will not distinguish between the canonical (Hamiltonian) and covariant (path-integral) formulations of LQG, as they are basically complementary. For a technical account and a comprehensive understanding of background independence in LQG, see Rovelli (2004) and Rovelli and Vidotto (2015).
7. In LQG, spacetime discreteness is not an *a priori* assumption, as in other QG approaches (such as causal set theory), but a physical prediction that could be tested empirically.
8. In this context, the term "abstract" denotes its mathematical meaning of "not embedded," indicating independence from any embeddings into a background spacetime manifold.
9. The four-dimensional "Block Universe" view already arises at the classical level due to the relativity of simultaneity, which denies a global temporal order. However, this view lacks empirical support. First, SR and GR reject a universal present ("now"), but they allow for *local* presents, enabling a *partial* order of spacetime events. Second, we inhabit a variably curved, dynamical spacetime described by GR, not the flat, fixed, Minkowski spacetime of SR. Third, the Block Universe theory presumes a background-dependent view *external* to the universe itself, which is evidently fictional. Finally, in GR and, *a fortiori*, in QG, change and dynamics can be defined independently of time, undermining the notion of a static world. Therefore, as Rovelli (2020) insists, the presentism/eternalism (or "Block Universe") dichotomy is a false alternative.
10. The argument suggesting LQG's non-dynamical worldview, often associated with QG's "problem of time," stems from a technical interpretation of LQG's canonical formulation. Roughly, the argument is that the canonical quantization of Hamiltonian GR leads to the Wheeler–DeWitt equation, $\widehat{H}\,|\psi\rangle = 0$, where the vanishing of the Hamiltonian constraint $\widehat{H}$ and the absence of the time variable suggest a "frozen," static universe. See Rovelli (2015) for a stringent refutation of this interpretation.
11. Rovelli and Smolin now hold divergent views on foundational physics. Smolin advocates *temporal naturalism*, asserting the reality and fundamentality of time (Smolin 2013; 2015). This contrasts with Rovelli's view, which aligns with the spacetime disappearance in LQG. Additionally, their interpretations of QM diverge (Rovelli 2022; Smolin 2019).
12. Some philosophers have argued that LQG and Rovelli's rejection of object-based ontology support ontic structural realism (OSR), which views the world as relational structures like symmetries and laws (Ladyman and Ross 2007; Rickles et al. 2006; Candiotto 2017). However, Dorato and Morganti (2022) recently refuted this in the context of Rovelli's relational interpretation of QM. I concur with them. While Rovelli, like structural realism, is eliminativist about "things" or substance-like entities, he does not endorse the ontological priority or ontic status of structures. For instance, LQG's spin network and spin foam mathematical structures are theoretical descriptions of quantum processes, not actual physical entities (Rovelli and Vidotto 2015, 140–1).

13. This view, as recently suggested (Saint-Ours 2008; 2015; Calamari 2021), suggests the possibility of *change without time*, challenging the traditional metaphysics and ontology of time.
14. This background dependence is the reason why Rovelli (2001, 113) insists that only the Heisenberg picture is viable in QG, not the Schrödinger one, which assumes the evolution of the wave function *over* time $|\psi(t)\rangle$, while operators $\widehat{Q}$ (representing observables) remain time-independent. This is also part of the reason why the Wheeler–DeWitt equation in LQG, a generalization of the Schrödinger equation, is often misinterpreted. The Wheeler–DeWitt equation reduces to the standard QM Schrödinger equation only when a *particular t* can be chosen, which is not the case in QG.
15. In the English translation, the phrase "the world is in a ceaseless process of change" might lead to an Aristotelian interpretation, suggesting that the world is a preexisting substance undergoing property changes. However, in the original Italian version, Rovelli employs the gerund "changing" instead of the noun "change," emphasizing that "the world is ceaseless changing." This choice underscores the world's processual nature and becoming. As is well known, Deleuze's ontology of becoming and event highlights the significance of infinitive verbs as "infinitive-becomings" in dismissing the substance–predicate relation (and more precisely the confusion between predicate and attribute, as we shall see shortly) of substance-based metaphysics; see Deleuze (1990), Deleuze and Parnet (1987, 63–4), Deleuze and Guattari (1987), and Deleuze (1993).
16. In consequence of this objection, Deleuze (1993, 48, 61) stresses that the classical concept definition as *ens rationis* is refuted for the event concept, which, in turn, provides the view of philosophy as creation of concepts where the concept expresses "the event, not the essence or the thing" (Deleuze and Guattari 1994, 21).
17. As Nietzsche famously put it: "The real world—we have done away with it: what world was left? the apparent one, perhaps? ... But no! *with the real world we have also done away with the apparent one!*" (1998, 20).
18. This means that even the evolution of physical time *T* in the sense of clock time (more precisely, as the proper time of a clock's world-line) is measured not in relation to *t*, $T(t)$, but in relation to other physical variables, $A(T)$, $B(T)$, or other clock times, $T_1(T_2)$, $T_2(T_3)$, and so on; see Rovelli (2004, 29–31).
19. Rovelli considers quantum mechanics as a framework theory that includes all (non-relativistic and relativistic) quantum theories.
20. The ontology of QM is notoriously controversial and falls outside the scope of the present chapter. Yet note that the claim here is distinct from the realism/anti-realism debate about the ontological status of the quantum wave function (or quantum state) $|\psi\rangle$. In this respect, RQM is metaphysically anti-realist about the wave function and realist about quantum events or processes. According to RQM, the wave function is not an objective and independent actual physical object; it is an abstract mathematical tool used to compute the probabilities of quantum events.
21. Unlike in Copenhagen-like interpretations, there is no quantum/classical world divide in RQM, for quantum events or processes result from interactions between *any* system, either microscopic or macroscopic. Thus, according to RQM, there is only the quantum world, the difference between events or systems being a matter of scale and approximation.
22. Thus, ordinary notions in physics such as "non-interacting," "isolated," or "free" system are only useful approximations or idealized models.
23. This refutation also involves the notion of "metaphysical grounding" and, more generally, grounding-based conceptions, insofar as they assume independence

as a definitional criterion of fundamentality, as Bliss (2024, 13–14) corroborates: "that whatever is fundamental is *in some important sense* independent is a core, if not the core, feature of fundamentality as it has been historically conceived. It is also the core feature of most accounts of fundamentality contemporarily."

24. See Plotnitsky (2006b) and Beaulieu (2022) for philosophical work on Deleuze's, and Deleuze and Guattari's, chaosmology and "chaosmological thinking."
25. Deleuze references Riemann and modern advances in differential geometry and algebraic topology in GR and particle physics, such as fiber bundle topology, following Châtelet's analyses: "Modern mathematics has been able to develop a fibered conception according to which 'monads' test the paths in the universe and enter in syntheses associated with each path. It is a world of captures instead of closures" (Deleuze 1993, 92). In *A Thousand Plateaus*, Deleuze proposed that this shift marks the passage from "monadology" to "nomadology" (Deleuze and Guattari 1987, 573–4, n. 27; Deleuze 1993, 158). See Calamari (2015) for a comprehensive analysis.

References

Agullo, Ivan, and Alejandro Corichi. 2014. "Loop Quantum Cosmology." In *Springer Handbook of Spacetime*, edited by Abhay Ashtekar and Vesselin Petkov, 809–39. Berlin: Springer.

Amelino-Camelia, Giovanni. 2013. "Quantum-spacetime Phenomenology." *Living Reviews in Relativity* 16, no. 1: 1–137.

Aristotle. 1984a. *Categories*. In *The Complete Works of Aristotle*, vol. 1, edited by Jonathan Barnes. Princeton, NJ: Princeton University Press.

Aristotle. 1984b. *Metaphysics*. In *The Complete Works of Aristotle*, vol. 2, edited by Jonathan Barnes. Princeton, NJ: Princeton University Press.

Armas, Jacome. 2021. *Conversations on Quantum Gravity*. Cambridge: Cambridge University Press.

Ashtekar, Abhay, and Jorge Pullin, eds. 2017. *Loop Quantum Gravity: The First 30 Years*. Singapore: World Scientific Publishing.

Ashtekar, Abhay, and Parampreet Singh. 2011. "Loop Quantum Cosmology: A Status Report." *Classical and Quantum Gravity* 28, no. 21: 1–137.

Barbour, Julian. 1999. *The End of Time: The Next Revolution in Physics*. Oxford: Oxford University Press.

Barbour, Julian. 2009. "The Nature of Time." https://arxiv.org/abs/0903.3489. Accessed September 10, 2025.

Barbour, Julian. 2020. *The Janus Point: A New Theory of Time*. New York: Random House.

Beaulieu, Alain. 2022. "The Role of Poincaré's Study of the Celestial Mechanics in the Development of Deleuze and Guattari's Chaosmological Thinking." *Thaumàzein. Rivista di Filosofia* 10, no. 2: 185–206.

Belot, Gordon. 2011. "Background-independence." *General Relativity and Gravitation* 43, no. 10: 2865–84.

Belot, Gordon, and John Earman. 2001. "Pre-Socratic Quantum Gravity." In *Physics Meets Philosophy at the Planck Scale: Contemporary Theories in Quantum Gravity*, edited by Craig Callender and Nick Huggett, 213–55. Cambridge: Cambridge University Press.

Bergson, Henri. 2002. *Henri Bergson: Key Writings*. New York: Continuum.

Bliss, Ricki. 2024. *Grounding, Fundamentality and Ultimate Explanations*. Cambridge: Cambridge University Press.

Calamari, Martin. 2015. "Riemann–Weyl in Deleuze's Bergsonism and the Constitution of the Contemporary Physico-Mathematical Space." *Deleuze Studies* 9, no. 1: 59–87.

Calamari, Martin. 2017. "The Mathematical Model of Smooth and Striated Spaces and the Connectivity Problem." *Deleuze Studies* 11, no. 3: 328–55.

Calamari, Martin. 2021. "The Metaphysical Challenge of Loop Quantum Gravity." *Studies in History and Philosophy of Science Part A* 86: 68–83.

Callender, Craig. 2017. *What Makes Time Special?*. Oxford: Oxford University Press.

Callender, Craig, and Nick Huggett, eds. 2001. *Physics Meets Philosophy at the Planck Scale: Contemporary Theories in Quantum Gravity*. Cambridge: Cambridge University Press.

Candiotto, Laura. 2017. "The Reality of Relations." *Giornale di Metafisica* 39, no. 2: 537–51.

Deleuze, Gilles. 1986. *Nietzsche and Philosophy*. Translated by Hugh Tomlinson. London: Continuum.

Deleuze, Gilles. 1990. *The Logic of Sense*. Translated by Mark Lester and Charles Stivale. New York: Columbia University Press.

Deleuze, Gilles. 1991. *Bergsonism*. Translated by Hugh Tomlinson and Barbara Habberjam. New York: Zone Books.

Deleuze, Gilles. 1993. *The Fold: Leibniz and the Baroque*. Translated by Tom Conley. Minneapolis: University of Minnesota Press.

Deleuze, Gilles. 1994. *Difference and Repetition*. Translated by Paul Patton. New York: Columbia University Press.

Deleuze, Gilles. 2004. *Desert Islands and Other Texts, 1953–1974*. New York: Semiotext(e).

Deleuze, Gilles, and Félix Guattari. 1987. *A Thousand Plateaus: Capitalism and Schizophrenia*. Translated by Brian Massumi. Minneapolis: University of Minnesota Press.

Deleuze, Gilles, and Félix Guattari. 1994. *What is Philosophy?*. Translated by Hugh Tomlinson and Graham Burchell. New York: Columbia University Press.

Deleuze, Gilles, and Claire Parnet. 1987. *Dialogues*. Translated by Hugh Tomlinson and Barbara Habberjam. London: Athlone.

Di Biagio, Andrea, and Carlo Rovelli. 2021. "Stable Facts, Relative Facts." *Foundations of Physics* 51, no. 30. https://doi.org/10.1007/s10701-021-00429-w.

Dorato, Mauro. 2015. "Events and the Ontology of Quantum Mechanics." *Topoi* 34, no. 2: 369–78.

Dorato, Mauro. 2016. "Rovelli's Relational Quantum Mechanics, Anti-Monism, and Quantum Becoming." In *The Metaphysics of Relations*, edited by Anna Marmodoro and David Yates, 235–61. Oxford: Oxford University Press.

Dorato, Mauro, and Matteo Morganti. 2022. "What Ontology for Relational Quantum Mechanics?." *Foundations of Physics* 52, no. 66, https://doi.org/10.1007/s10701-022-00581-x.

Duffy, Simon. 2013. *Deleuze and the History of Mathematics: In Defence of the 'New'*. London: Bloomsbury.

Healey, Richard. 2002. "Can Physics Coherently Deny the Reality of Time?." In *Time, Reality and Experience*, edited by Craig Callender, 293–316. Cambridge: Cambridge University Press.

Huggett, Nick. 2022. "Spacetime 'Emergence.'" In *The Routledge Companion to Philosophy of Physics*, edited by Eleanor Knox and Alastair Wilson, 374–85. New York: Routledge.

Huggett, Nick, and Christian Wüthrich. 2018. "The (A)temporal Emergence of Spacetime." *Philosophy of Science* 85, no. 5: 1190–203.

Huggett, Nick, Tiziana Vistarini, and Christian Wüthrich. 2013. "Time in Quantum Gravity." In *A Companion to the Philosophy of Time*, edited by Heather Dyke and Adrian Bardon, 242–61. London: John Wiley & Sons.
Huggett, Nick, Keizo Matsubara, and Christian Wüthrich, eds. 2020. *Beyond Spacetime: The Foundations of Quantum Gravity*. Cambridge: Cambridge University Press.
Huggett, Nick, Neils Linnemann, and Mike Schneider. 2023. *Quantum Gravity in a Laboratory?*. Cambridge: Cambridge University Press.
Jedrzejewski, Franck. 2017. "Deleuze et la géométrie riemannienne: une topologie des multiplicités." In *From Riemann to Differential Geometry and Relativity*, edited by Lizhen Ji, Athanase Papadopoulos, and Sumio Yamada, 311–28. Cham: Springer International Publishing.
Knox, Eleanor, and Alastair Wilson, eds. 2022. *The Routledge Companion to Philosophy of Physics*. New York: Routledge.
Ladyman, James, and Don Ross, eds. 2007. *Everything Must Go: Metaphysics Naturalized*. Oxford: Oxford University Press.
Laudisa, Frank, and Carlo Rovelli. 2019. "Relational Quantum Mechanics." In *The Stanford Encyclopedia of Philosophy*, edited by Edward N. Zalta. https://plato.stanford.edu/entries/qm-relational/. Accessed September 10, 2025.
Lautman, Albert. 2011. *Mathematics, Ideas and the Physical Real*. Translated by Simon B. Duffy. London: Bloomsbury.
Lundy, Craig. 2018. *Deleuze's Bergsonism*. Edinburgh: Edinburgh University Press.
Murugan, Jeff, Amanda Weltman, and George F. R. Ellis, eds. 2012. *Foundations of Space and Time: Reflections on Quantum Gravity*. Cambridge: Cambridge University Press.
Newton, Isaac. 1999 [1687]. *The Principia: Mathematical Principles of Natural Philosophy*. Berkeley: University of California Press.
Nietzsche, Friedrich. 1998 [1889]. *Twilight of the Idols or How to Philosophize with a Hammer*. Translated by Duncan Large. Oxford: Oxford University Press.
O'Conaill, Donnchadh. 2022. *Substance*. Cambridge: Cambridge University Press.
Oriti, Daniele. 2009. *Approaches to Quantum Gravity: Toward a New Understanding of Space, Time and Matter*. Cambridge: Cambridge University Press.
Oriti, Daniele. 2014. "Disappearance and Emergence of Space and Time in Quantum Gravity." *Studies in History and Philosophy of Science Part B* 46: 186–99.
Oriti, Daniele. 2020. "The Bronstein Hypercube of Quantum Gravity." In *Beyond Spacetime: The Foundations of Quantum Gravity*, edited by Nick Huggett, Keichi Matsubara, and Christian Wüthrich, 25–52. Cambridge: Cambridge University Press.
Oriti, Daniele. 2021. "Levels of Spacetime Emergence." In *Philosophy Beyond Spacetime: Implications from Quantum Gravity*, edited by Christian Wüthrich, Baptiste Le Bihan, and Nick Huggett, 16–40. Oxford: Oxford University Press.
Perez, Alejandro. 2013. "The Spin-Foam Approach to Quantum Gravity." *Living Reviews in Relativity* 16, no. 1: 1–128.
Plotnitsky, Arkady. 2006a. "Manifolds: On the Concept of Space in Riemann and Deleuze." In *Virtual Mathematics: The Logic of Difference*, edited by Simon Duffy, 187–208. Manchester: Clinamen.
Plotnitsky, Arkady. 2006b. "Chaosmologies: Quantum Field Theory, Chaos and Thought in Deleuze and Guattari's *What is Philosophy?*." *Paragraph* 29, no. 2: 40–56.
Plotnitsky, Arkady. 2009. "Bernhard Riemann." In *Deleuze's Philosophical Lineage*, edited by Graham Jones and Jon Roffe, 190–208. Edinburgh: Edinburgh University Press.

Read, James. 2023. *Background Independence in Classical and Quantum Gravity*. Oxford: Oxford University Press.
Rescher, Nicholas. 1996. *Process Metaphysics: An Introduction to Process Philosophy*. New York: SUNY Press.
Rickles, Dean, Steven French, and Juha T. Saatsi. 2006. *The Structural Foundations of Quantum Gravity*. Oxford: Clarendon Press.
Riemann, Bernhard. 2007. "On the Hypotheses Which Lie at the Foundation of Geometry." In *From Kant to Hilbert: A Source Book in the Foundations of Mathematics*, vol. 2, edited by William Ewald, 652–61. Oxford: Oxford University Press.
Robinson, Keith. 2009. *Deleuze, Whitehead, Bergson. Rhizomatic Connections*. New York: Palgrave Macmillan.
Rovelli, Carlo. 1996. "Relational Quantum Mechanics." *International Journal of Theoretical Physics* 35, no. 8: 1637–78.
Rovelli, Carlo. 2001. "Quantum Spacetime: What Do We Know?." In *Physics Meets Philosophy at the Planck Scale: Contemporary Theories in Quantum Gravity*, edited by Craig Callender and Nick Huggett, 101–22. Cambridge: Cambridge University Press.
Rovelli, Carlo. 2004. *Quantum Gravity*. Cambridge: Cambridge University Press.
Rovelli, Carlo. 2005. "Relational Quantum Mechanics." In *Quo Vadis Quantum Mechanics?*, edited by Avshalom Elitzur, Dolev Shahar, and Kolenda Nancy, 113–20. Berlin: Springer.
Rovelli, Carlo. 2007. "Quantum Gravity." In *Philosophy of Physics*, edited by Jeremy Butterfield and John Earman, 287–329. Amsterdam: Elsevier.
Rovelli, Carlo. 2015. "The Strange Equation of Quantum Gravity." *Classical and Quantum Gravity* 32, no. 12. https://iopscience.iop.org/article/10.1088/0264-9381/32/12/124005/pdf. Accessed September 10, 2025.
Rovelli, Carlo. 2016. *Reality is Not What it Seems: The Journey to Quantum Gravity*. London: Penguin.
Rovelli, Carl. 2018a. *The Order of Time*. London: Penguin.
Rovelli, Carlo. 2018b. "Space is Blue and Birds Fly Through It." *Philosophical Transactions of the Royal Society* 376, no. 2123. https://doi.org/10.1098/rsta.2017.0312
Rovelli, Carlo. 2020. "Space and Time in Loop Quantum Gravity." In *Beyond Spacetime: The Foundations of Quantum Gravity*, edited by Nick Huggett, Keizo Matsubara, and Christian Wüthrich, 117–32. Cambridge: Cambridge University Press.
Rovelli, Carlo. 2021. *General Relativity: The Essentials*. Cambridge: Cambridge University Press.
Rovelli, Carlo. 2022. *Helgoland: Making Sense of the Quantum Revolution*. Translated by Erica Segre and Simon Carnell. London: Penguin.
Rovelli, Carlo, and Francesca Vidotto. 2015. *Covariant Loop Quantum Gravity*. Cambridge: Cambridge University Press.
Rovelli, Carlo, and Francesca Vidotto. 2022. "Philosophical Foundations of Loop Quantum Gravity." https://doi.org/10.48550/arXiv.2211.06718.
Saint-Ours, Alexis de. 2008. "Time and Relation in Relativity and Quantum Gravity: From Time to Processes." In *The Ontology of Spacetime II*, edited by Dennis Dieks, 25573. Amsterdam: Elsevier.
Saint-Ours, Alexis de. 2015. "Does Time Differ from Change? Philosophical Appraisal of the Problem of Time in Quantum Gravity and in Physics." *Studies in History and Philosophy of Modern Physics* 52: 48–54.
Simpson, William, Robert C. Koons, and Nicholas J. Teh, eds. 2018. *Neo-Aristotelian Perspectives on Contemporary Science*. New York: Routledge.

Smith, Daniel. 2012. *Essays on Deleuze*. Edinburgh: Edinburgh University Press.
Smolin, Lee. 2001. *Three Roads to Quantum Gravity*. New York: Basic Books.
Smolin, Lee. 2006. "The Case for Background Independence." In *The Structural Foundations of Quantum Gravity*, edited by Dean Rickles, Steven French, and Juha T. Saatsi, 196–239. Oxford: Oxford University Press.
Smolin, Lee. 2013. *Time Reborn: From the Crisis in Physics to the Future of the Universe*. London: Penguin.
Smolin, Lee. 2015. "Temporal Naturalism." *Studies in History and Philosophy of Science Part B: Studies in History and Philosophy of Modern Physics* 52: 86–102.
Smolin, Lee. 2019. *Einstein's Unfinished Revolution: The Search for What Lies Beyond the Quantum*. London: Penguin.
Tahko, Tuomas E. 2012. *Contemporary Aristotelian Metaphysics*. Cambridge: Cambridge University Press.
Thébault, Karim. 2022. "The Problem of Time." In *The Routledge Companion to Philosophy of Physics*, edited by Eleanor Knox and Alistair Wilson, 386–400. New York: Routledge.
Van Fraassen, Bas C. 2010. "Rovelli's World." *Foundations of Physics* 40, no. 4: 390–417.
Vassallo, Antonio. 2022. *The Foundations of Spacetime Physics: Philosophical Perspectives*. New York: Routledge.
Vidotto, Francesca. 2017. "Relational Quantum Cosmology." In *The Philosophy of Cosmology*, edited by Khalil Chamcham, Joseph Silk, John D. Barrow, and Simon Saunders, 297–316. Cambridge: Cambridge University Press.
Wallace, David. 2022. "Quantum Gravity at Low Energies." *Studies in History and Philosophy of Science* 94: 31–46.
Weinstein, Steven, and Dean Rickles. 2019. "Quantum Gravity." In *The Stanford Encyclopedia of Philosophy*, edited by Edward N. Zalta. https://plato.stanford.edu/entries/quantum-gravity/. Accessed September 10, 2025.
Whitehead, Alfred N. 1978 [1929]. *Process and Reality*. New York: The Free Press.
Widder, Nathan. 2019. "The Mathematics of Continuous Multiplicities: The Role of Riemann in Deleuze's Reading of Bergson." *Deleuze and Guattari Studies* 13, no. 3: 331–54.
Wüthrich, Christian, Baptiste Le Bihan, and Nick Huggett. 2021. *Philosophy Beyond Spacetime: Implications from Quantum Gravity*. Oxford: Oxford University Press.

Chapter 12

Zig-Zagging Through the Deleuzian Virtual: The Dark Universe of Diagrams, Cosmology, and Quantum

jan jagodzinski

The first quarter of the twenty-first century, as characterized by the unprecedented developments in the bio-geo-sciences with the convergence of NBIC (nano-bio-info-cogno) technologies foremost along digital-algorithmic lines, impacts heavily on the context of the Anthropocene (euphemistically called the crisis of climate change). These developments provide a historical moment to revisit, update, and possibly revise or expand the cosmological vision of Deleuze and Guattari, as it has become obvious that the separation of nature and culture as characterized by modernist thought has collapsed. The awkward nomenclature "natureculture" is now ubiquitous in the literature, as the sciences and humanities search for a new footing in their cross-relations.

Much has changed since Deleuze wrote his prophetic projection on control societies. Learning machines now use neural networks based on the neurological structure of the human brain. The projection of AI is to further such biological developments that continue to integrate living nerve cells with artificial ones, providing for the cross-synthetization of the senses. It is now possible to "see" with the tongue thanks to such technology. The first generation of quantum computers is upon us; genetics at the germ level and cloning are now subject to easy manipulation via technologies such as Crisper-cas9. Gene-editing kits are available for "home" experimentation. The agrammaticality of language, where Deleuze draws on a host of poets and artists (Mallarmé, Péguy, Artaud, cummings, Burroughs, Roussel, Bisset, Dada) has been supplemented by programming languages: COBOL, Python, Java, Swift, Ruby on Rails, and so on. More significant perhaps has been the development of ChatGPT and the AutoML-Zero programming language by Google, which further enable "genetic algorithms" to correct themselves "mathematically," with the continued claims that *now* human bias has been eliminated. Such transformations inform an attempt to query Deleuze's (at times with Guattari) cosmology in relation to recent developments in physics and information theory.

It is well known that Deleuze was deeply worried by the way the cybernetic instrumentalization of information theory was at the root of "control societies," which he had just begun to develop by way of a postscript, indicating that such an event had already taken place. Equally well explored have been Deleuze's comments, found in the closing pages of his cinema book on the time-image, which reference early electronic images made cinematically possible through digitalization. And, in the last pages of an appendix in his book on Foucault, Deleuze refers to the projected future developments of the Nietzschean *Übermensch* in relation to the *superfold*; the superfold being the "unlimited finity" that is generated from a finite number of components. His example at that time was the double helix of DNA. Deleuze refers to this development as possibly the "eternal return" itself of infinite variation, or creativity itself in its evolutionary sense, meant not solely as a reference to biological capacity, but in the *fundamental sense* of (Bergson's) "creative evolution" or "evolutionary expansiveness" as to the mutability of the laws of nature that are subject to change.

This is a cosmological position currently held by a number of astrophysicists who posit an autodidactic universe, especially Lee Smolin (Smolin 2013; Unger and Smolin 2015), as initially advanced by John Webb et al. (2011). Such a direction questions the current standard model of cosmology known as the concordance or Lambda-CDM model, which is a linear model with a single origin (the Big Bang), and an expanding universe (inflation) as initiated by Edwin Hubble in the 1920s and 1930s, where general relativity provides precise predictions within the context of our galaxy. Dark *matter* was conceptualized as an "auxiliary hypothesis" in the 1960s and 1970s to deal with the anomalies in energy that kept on appearing with this model (the rotational-curve anomaly of distant galaxies). This was followed by the "invention" of dark *energy* to explain the acceleration of space.

The problem is that this model was then universalized for the entire universe, which is currently being questioned through different conceptions of cosmic duration, such as Roger Penrose's conformal cyclic cosmology (CCC), which still embraces the framework of general relativity, while dark matter is now challenged by "modified Newton dynamics" (MOND) as first proposed by astrophysicist Mordehai Milgrom (Merritt 2020). The term "multiverse" has become a catchphrase in the popular media to lump together various approaches that have begun to question both the "cosmological principle" and the "cosmological constant" that are bedrock ideas of the standard model, although it remains evident that an ontogenetic origin did take place due to the evidence of cosmic microwave background radiation discovered in 1976 by Penzias and Wilson. The First Cause remains, in what I refer to throughout this chapter as the *radical* Real of a Dark Universe as it will be developed.[1] To note: this Big Bang is a violent, non-propitious change, an explosion, a *metastasis* as a madness to acquire order from disorder that paradoxically stabilizes a meta-state for a duration. Deleuze's cosmology is characterized by such a fundamental displacement as the disturbance of a constant metastasis. The unthought of the Dark Universe plunges thought into forms of "madness," into paradoxes, strangeness, absurdity, and senseless forms of creativity that dwell at the edges of reason.

For Deleuze the metaphysician, a theory of "everything" is questionable. More so, symmetry, equilibrium, and the Second Law of Thermodynamics that calculates entropy and overlooks the incalculable micro "free" energy of intensity are always questioned. This is a "transcendental illusion" (Deleuze 1994, ch. 5; Clisby 2017). The cold and still universe of absolute zero or heat death that the Third Law of Thermodynamics projects also does not sit well with him (or Guattari). Both dispute such conservation laws. Following Anaximander's *apeiron*, if there is complete symmetry, there is insufficient reason for anything to happen. Deleuze constantly forwards *sufficient-reason-as-dissymmetry* (Ross 2024), contra Leibniz, whose cosmological influences were fundamental and had to be surpassed. Deleuze's "calculating God" on a cosmological scale

> makes the world by calculating, but his calculations never work out exactly [*juste*], and his inexactitude or injustice in the result, this irreducible inequality, forms the condition of the world. [...] The world "happens" while God calculates; if the calculation were exact, there would be no world. (Deleuze 1994, 222)

There is always an *unknown remainder* that belongs to the Dark Universe. The disparity between forces, diversity, the heterogeneity of a series, as well an ineradicable inequality of the remainder itself in mathematical calculation, are fundamental to chaosmos. "Disparity—in other words, difference or intensity (difference of intensity)—*is the sufficient reason of all phenomena*, the condition of all that appears [...] Intensity is difference, but this difference tends to deny or cancel itself out in extensity and underneath quality" (Deleuze 1994, 222, 223, emphasis added).

In *The Logic of Sense*, Deleuze's God plays an "Ideal Game" of chance with the dice-throw—the "unassignable time" of the event, attributed to the Lucretian swerve of the *clinamen*, which, for Deleuze, becomes a figure of absolute contingency (Deleuze 1990, 307), assigning the power of constant creation with each throw. An event has no existence beyond the complex exchange between virtual and actual. Each dice-throw bears paradoxically on its own outcome and the whole of chance at once. This is contra quantum Qbism, which uses Bayesian formalization for chance events of probability. Deleuze's dice-throw makes for an interesting retort to Einstein's "God does not play dice" and Niels Bohr's reply: "Stop telling God what to do." Limitation as "hesitation" and "retardation" (as developed by Bergson), and its corollary "openness," characterize Deleuze's cosmos. "Chaos does not exist; it is an abstraction because it is inseparable from a *screen* that makes something—something rather than *nothing*—emerge from it" (Deleuze 1993, 76–7, emphasis added). The "given *as given*" is already order by the empirical "fact" that it is "given" (that of the screen). "Disorder" for Deleuze does not exist. It is illusionary, as "nonbeing" (or nothing) is illusionary, as these concepts depend on negation. Differentiated order is always affirmative. Limitations are a real factor in the passage of nature, which are subject to continual replacement. Such limitations (infinitely mutable) as laws and rules are never overcome in the sense of negation or opposition, but are subject to contingent events.

As Deleuze muses in the "last" letter Z of the alphabet of his own making: "at the origin of things there is no big bang, there is the Z [...], the big bang should be replaced with the Z, which in fact is Zen, which is the path of the fly [...] to invoke the zig-zag [...] no universals, but sets of singularities" (2001, n.p.). As he further writes in the closing of his meditation on Foucault: "The *forces* within man enter into a relation with *forces* from the *outside*, those of silicon which supersedes carbon, or genetic components which supersede the organism, or agrammaticalities which supersede the signifier" (Deleuze 1988, 131–2, emphasis added). The "superman" (*Übermensch*) is "the formal compound of the forces within man [*sic*] and these new forces" (Deleuze 1988, 132). Z belongs to the radical Real, to the Dark Universe. It is this mention of the "outside" where questions of cosmology and immanent vitalism are confronted. The genesis of thought begins with an intensive encounter with this "outside," which is external and independent of thought itself. Deleuze presents a "*transcendental* sensibility" that consists of discontinuous instants of material impression in what is an intensive zone that appears continuous, as opposed to what is an "*empirical* transcendentalism" in the Kantian system. Deleuze speaks of "infinite speed" as things come and go in the virtual, as assemblages form that undergo organization and disorganization. More concisely:

> Chaos is defined not so much by its disorder as by the infinite speed with which every form taking shape vanishes. It is a void that is not a nothingness but a *virtual*, containing *all possible particles and drawing out all possible forms*. [...] Chaos is an infinite speed of birth and disappearance. (Deleuze and Guattari 1994, 118, emphasis added)

Infinite speed in this context refers to the limitations of the speed of light (*c*), the photon that has no mass. Chaos is the medium where what is virtual awaits its actualization. As Rudolphe Gasché (2014, 57) maintains, with the affirmation in 2012 of the Higgs boson, the Higgs field introduces inertial mass to other fundamental particles such as electrons and quarks. By doing so, particles "slow" down on the plane of immanence as mass determines how much a particle resists changing its speed or position when encountering a force. The three planes that "cut" through chaos—via concepts, functives, and percepts/affects—actualize different shapes and concepts as particles are then imbued with "reality" (mass). Matter offers its own resistance, its own retardation (slowing down), complexifications and foldings with the world; in effect, a dialectical movement between problems and solutions as singularities are constantly redistributed in such events.

It is these forces from the "outside" that link Deleuze and Guattari to their cosmology, to the question of chaosmos, and to issues of information, data, physical laws, and quantum, where "the given *as given*" (that is, the given of difference in-itself) is the contemporary ontological challenge that is raised in this twenty-first century of critical, no less apocalyptic change, where the extinction of *Homo sapiens* is not off the table. As Émile Torres notes, in the 1850s the discovery of the Second Law of Thermodynamics led to a new hermeneutics "according to which entropy death of our solar system and/or the entire universe posed an actual threat to our long-term survival," leading to "cosmic doom and existential vulnerability" (2023, 71). These developments require a return to the virtual-actual metaphysical philosopheme that remains the heart of Deleuzo-Guattarian metaphysical thought.

In what follows I develop two lines of argumentation for such an undertaking. The first is to revisit the current status of information theory, especially concepts that have to do with entropy, data, and the algorithmic state of the art in relation to the virtual, and to target the diagram as to its significance for the imaginary in relation to quantum theory. How do these new developments in digitalization change or modify Deleuze's perhaps too blasé position in his last writings, which are by and large dismissive of cybernetic developments? The second trajectory is to query issues of contemporary cosmology that raise tensions between quantum theory and general relativity, where the writings of Roger Penrose become especially interesting, with his theory of conformal cyclic cosmology. I try to show how Penrose's "minor" position shares a number of touchstones with Deleuzian cosmology.

Information

Information is immanent to the world, as it facilitates actualization. The usual scientific account of information as an encoded transmission between sender and receiver subject to statistical rules of probability is transformed in Deleuze's cosmology. Singularities are distributed with information processing as an unfolding of individuation of inexhaustible potentiality. Information storage is now simply the segmentation or blockage of flows, and its articulation is synonymous with transduction (as developed by Gilbert Simondon). Noise, signal, and meaning are reconfigured. The resulting assembled meta-state already holds the "seeds" of its own destruction, a displacement that leads to a new organization, which is also what the Derridean supplement at the "heart" of things claims. Information as energy or force is the "hidden" causal agent for such iteration and transformation. It is virtual information, allowing for the "difference that makes a difference," intensity that transmits across the virtual to the actual, which is a constant and perpetual reorganization of reality.

This is first-order information (Simondon) emerging from the virtual as both "anterior" and "immanent" to what is individuated. The second-order is situated as an expression of the first-order information in the actual; together, they are a product of Becoming (there is no "master design" implied by the first-order). Becoming is always different to itself, as every manifestation of Becoming never entirely resembles what came before it; nor is it determined by what it is not—as an apophatic approach would have it. The apparent stasis of Being always has a beyond; it is always in flux, vibrating with pushes and pulls, an undulation of repulsions and attractions of differences. The metastasis of affect is forcefully at work, through intensities whose order is not defined by scale (small or large), nor by distance, which already is an actualized entity. Deleuze evokes "depth" but as topological depth of intensity. "Extensity can emerge from the depths only if depth is definable independently of extensity. [...] The original depth, by contrast, is indeed space as a whole, but space as an intensive *quantity*: the pure *spatium*" (Deleuze 1994, 230, emphasis added). Pure *spatium* is what produces space, whereas extension is *location* in metric space, while *non-locality* refers to the intensive virtual realm of potential energy in relation to empirical energy. Information is strongly related to entropy. Hence, Deleuze maintains that information, which is extensively calculable, overlooks the virtual non-locality of energy. The resultant informational pattering is the tension between the internal logic of the "bit," which is to say its spatial "locality," in its co-extensive relations with "other" bits (the "non-locality" of an intensive virtual realm) within the logic of a milieu where the phenomenon occurs. "Bits" ("stuff" or "things" in a non-technical sense) are not fully individuated prior to their relations, upon which they then take on forms of signification. The "how" and "what" of expression cannot separate the syntactical from the semantic as related to the context of a larger problematic.

The processes in the virtual "appear" to move faster than the speed of light, as if "particles" communicated at a distance through worm holes, by instant contagion; this is the non-locality of virtual intensity. Speed is yet another term for Becoming, which, in-itself, is undirected and non-teleological; it is its own "absolute." This absolute, as a radical Real or Dark Universe, is much like an envelope (a cosmic egg) that holds a more "secret" depth from which material emerges and repeats with difference. These typological "envelopes" of intensification emerge from the non-locality of the Dark Universe; they result in retardation or "cancelation," a slowing down, due to the limits imposed by relative speeds. Such envelopes are "islands" of order; negentropy (after Erwin Schrödinger). If information as datum, the given *as given*, is the relative degree of organization (entropy), if it "emerges" or plays a role in the organizing of these virtual assemblages or desiring machines, then Kane Faucher asks whether this rethought definition of information "in effect, assigns to information the role of the differenciator, as the *dark precursor* that is difference-in-itself, which is disguised in every series, and is always a displacement of itself in those series" (2013, 183, emphasis added). Information becomes "the pre-individual actuator or quasi-cause that allows for the individuation of matter" (Faucher 2013, 184). The process of metastasis is one of communicative interaction that happens between the series of metastable and semi- or fully stable system states, through the tensions and permutations of de- and re-territorialization. This is a dynamic dephasing of Being, the processes "from *em-phasing* to *emphasizing* (both etymological roots of 'to show') the object to consciousness; that is to the domain of representation" (Faucher 2013, 139, emphasis added). This is a *contagious* communication from the radical Real of the Dark Universe, a reciprocal determination of individual and milieu in the process of individuation (Simondon) wherein "a metastasis of affect is produced" (Faucher 2013, 139). Contagion takes hold and makes a difference as an assemblage becomes a "bloc of becoming" or a pack. Metastasis as a function of immanent information belongs to the realm of the "dark precursor," which Deleuze endows with the capacity to facilitate force relations. It could just as well be the "Z" Deleuze quips about at the end of his interview with Parnet.

The dark precursor conceptually shares some aspects of the "radical Real" as developed by François Laruelle (2013) in his theory of non-standard philosophy. Vitalist pre-consciousness emerges from this "radical Real" *unilaterally*, which is to say the radical Real is foreclosed to conscious thought entirely, as it is with Deleuzian virtuality. Laruelle calls this "unilateral duality," which is closer to quantum superposition; more like a "collapse" that happens when the virtual is actualized: a non-relation becomes relational. Hence a Dark Universe is suggestive of the chaosmos, a concept that sidesteps the antinomy of order and disorder, cosmos and chaos. Deleuzian cosmology recognizes order but also the contingency of chance and the surpassing of limits that characterizes an open evolution.

Diagrams contra Images

To return to the "outside." Deleuze, at first alone and then with Guattari, embraces Bergson's problematic quest for a "living unity" that links life, consciousness, and the material universe—a cosmological unity which Nietzsche, Whitehead, and Tarde sought in their own ways to formulate. Deleuze and Guattari make this quest explicit when they write in *Anti-Œdipus*: "we make no distinction between man and nature: the human essence of nature and the natural essence of man become one within nature in the form of production or industry" (1983, 4). Matter becomes the stumbling block in this problematic, caught by representation as a form of substance, a passive "material" subject to manipulation. The connection between the cosmos and the living subject is the continuity of duration, the imperceptible flow of life passing between the past and the now into an open future that has yet to exist. The encephalized brain alone, in isolation, produces nothing. No thought "without a body," as Lyotard (1988) made firm.

The "appearance in-itself" of the "outside" to consciousness presents the paradox of a "frame" that takes on many dimensions: as a methodology, as a technology, as a mediation and a delimitation in a pragmatic sense. A perceptual subtraction takes place to delimit the incoming flux of information, of chaos itself, otherwise our particular species becomes overwhelmed, like the blind person who can suddenly "see" but is unable to process what she or he suddenly "sees," remaining paradoxically blind. Overwhelmed by what? By intensities and flows of materiality that our bodies must process. "Screens" of all kinds are required to mediate the outside, screens as both protection and *translucency*. This sits well with Deleuze's notion of "the formation of the third eye, a haptic eye, a haptic vision of the eye [...] as if the duality of the tactile and the optical were surpassed visually in this haptic function born of the *diagram*" (Deleuze 2003, 129, emphasis added). This is also nicely articulated by Deleuze and Guattari in their comment that "We require just a little order to protect us from chaos" (1994, 201). And further is their remark that the protective "umbrella" (screen) can become "opaque" through opinion avoiding the chaos of existence: "poets, artists make a slit in the umbrella, they tear open the firmament [a biblical reference to the "expanse" of the heavens] itself, to let in a bit of free and windy chaos and to *frame* in a sudden *light* a vision that appears through the rent" (1994, 203, emphasis added). To recall: "Chaos does not exist; it is an abstraction because it is inseparable from a *screen* that makes something—something rather than nothing—emerge from it" (Deleuze 1993, 76–7). Deleuze's universe of open evolution is founded on the underlying conditions of the virtual; this is where the *order* of nature lies. To embrace evolution is not to abandon order, rather it requires seizing its processes.

Musica universalis (the Music of the Spheres), as initially proposed by Pythagoras, recovers the model of the spheres (mechanosphere, biosphere, technosphere, "bubbles," ecological spheres, Econosphere, Datasphere, and so on), which are conceptual diagrams rather than framed representations of the world (Birat 2017). The states of a quantum system form a complex projective space when it comes to the irresolvable paradox between wave and particle. This fundamental superposition in quantum theory resolves itself only through decoherence, which brings together diagrams, laboratory apparatuses based on those diagrams, and the limits of measurement, as well as the potentialities that representation presents due to interface limitations with consciousness being as-it-is. It is the creative imagination at work, as intuition, or what Penrose (1997) refers to as "understanding," which enables such diagrams to be thought of in the first place. Diagrams refer to reductive "models" to help explain a mechanism more concisely, and to apparatuses that are pivotal and groundbreaking in their utility for grasping quanta behavior, the best-known and most iconic being various "split-screen" apparatuses.

In quantum physics the diagram can be grasped as that process that happens between the *indetermination* of the virtual, that is, the wave superpositions that are in play, and their *overdetermined* or actualized collapse (decoherence). What is taking place in this zone between the invisible (virtual) and the visible (actual) is Lucretius' *clinamen* or "swerve," that is only subject to diagrammatic speculation in distinction from corpuscular or particle thinking. It is the *underdetermination* of vectors or the immanent "tensor field" with its manifold variations that provide speculation for future actualization. The "swerve" carries with it a contingency of trajection, as discussed earlier. Quantum complementarity, in this particular conception, refers to the irresolvable paradox between a particle's position (determination) and its momentum (indeterminism). The more you know about the former, the less you know about the latter, and vice versa. This applies to other "dualistic" paradoxes such as chaos/order, motion/stasis, infinitely big/infinitely small, and especially discontinuity/continuity, requiring imaginary numbers (n*i*) that do not exist apart from real numbers (+ –), but are "laterally hiding" in a perpendicular dimension, extending natural numbers from one dimension to two.

The diagram is the key to understanding Deleuze's use of *image* throughout his work, which comes from Bergson. To recall a well-cited passage from Bergson: matter, in his view, is an aggregate of "images." And by "image" Bergson means a certain existence which is more than that which the idealist calls a representation, but less than that which the realist calls a thing—an existence placed halfway between "thing" and "representation" (Bergson 1991, 9). Deleuze's discussion of Beckett's plays in his essay "The Exhausted" offers an insight into this assertion. The image is "not a representation of an object but *a movement in the world of the mind*" (Deleuze 1997b, 169, emphasis added). In his play... *but the clouds...*, Beckett arrives (through the process of exhaustion) at an *affect-image*, which would be the equivalent of "pure perception" in Bergsonian terms, that is, an image from the "outside" that is not as yet subject to memory, equivalent to what C. S. Peirce would call "firstness." The pheneron (or firstness) simply refers to qualities of the world, their material intensities, independent of any act of human perception, the autonomy of percepts and affects, an a-representational and a-human realm, what Mark Hansen (2015) calls worldly sensibility, which is imbued with aesthetic effects, as Alfred North Whitehead maintained. The difficulty seems to be whether this intuitive grasp of the undifferentiated flow of life—that is, "real" experience, as opposed to "possible" experience—is *mediated*, especially through technologies, including language. The brain is certainly a screen that receives images as information, which are selected (slowed down, "retarded") and acted upon through and by the body below the level of consciousness through what has been variously theorized as "flesh" (Merleau-Ponty), cognitive nonconscious (Katherine Hayles), as well as Hameroff and Penrose's (1996) Orch OR theory, especially at the level of microtubules where superposition still remains in a potential state before collapse. This "brain screen," as mediated by enabling technological apparatuses, presents the challenges that come with the strides in information technologies and artificial intelligence.

Deleuze charts three distinct stages of metalanguage in Samuel Beckett's oeuvre to reach such an "image" of "pure perception"—"as the movement in the world of the mind" (Deleuze 1997b, 169). The stages that exhaust the normative everyday sense of possibility proceed seemingly in a "reverse" direction to reach this pure perceptual point. The process of exhaustion presents the paradox of a breakthrough/breakdown. It follows, in the quantum sense, the superposition (the both/and) of particles and waves that decohere and collapse as an image that emerges from the dark, like Deleuze's example of lightning in relation to the dark precursor that precedes it. Metalanguage I is described as discontinuous, an "atomic, disjunctive, cut and chopped language" (Deleuze 1997b, 156), where quality turns into quantity through number. It rests on a fundamental continuity (meaningful syntax), which can be divided infinitesimally. The "stuttering of *langue*" or vibrating *langue*-machine is the diagram here (Deleuze 1997a). Word-particles or word-atoms are far-from-equilibrium noises that convey non-sense that disrupts and wedges into meaning as linguistic (phonetic) fragments. At this metalanguage level the diagram as abstract machine deterritorializes classical semiotic models and dissipates them into noise, as "atomic decay." Creative stuttering is discontinuous, intervening into "the block of a single expiring breath." It "makes the sentence grow out from the middle, adding particle upon particle" (Deleuze 1997a, 111). To exhaust the possible with words "one cuts and chops the atoms" (Deleuze 1997b, 156). Combinable words (as atoms) proliferate to their infinite possibilities through inclusive disjunctions that end in their inexistence. A new order might be created with just enough noise to break excessive rationality and logic. At the same time, the new emergence could just break down into the formlessness of noise itself. Noise as nonsense intensifies entropy in this case, unlike Shannon–Weiner instrumental information theory, where it is suppressed for clarity of communication. If meaning stops, one might ask, does it confront "a true silence" that provides for a hesitation for thought to emerge? Either way, Deleuze calls on a pure life-energy or life-force for such stuttering.

Metalanguage II shifts from word to voice. Voice ends up grasping language attributed to its flows as stories and "possible worlds" (Deleuze 1997b, 157), and, above all, attributed to a sense of self who utters these words. The curious and paradoxical aspect of metalanguage II can be thought of as Schrödinger's wave equation itself, which happens to be the key to the entire periodic table of elements, each element of which might be considered a voice in-itself; they form the inclusive disjunction of the table. Yet it is the "erasure" as well as the "recognition" of Schrödinger's name of that very equation which presents the "silence" that establishes the standard quantum model. Schrödinger's name and all who use the equation are "emptied" into the equation itself. In other words, there is an "exhaustion" that leads to silence—to the edge of what is known. The table of elements appears complete, but it is not, as artificially created elements are added to it. As a "table," it seems to close off other possible worlds (elements) that challenge it. As much can be said of the Feynman descriptions of how to calculate the scattering processes in quantum field theory. The "sum" of the various diagrams provides a satisfying way that this happens, where locality and unitarity (the idea that information is not "lost") are made manifest in every step of the calculations. Other flows now "dry up" (Deleuze 1997b, 156).

Metalanguage III presents another version of "the dark precursor," that "spark" (or singularity) which opens up the "pure image" as a vital force of genesis and creation; an image in its indefinite state that goes "beyond" word and voice. The (pure) image, like "white noise," has no memory and is subject to random processes, with infinitesimal events occurring in a continuous time series with no causative predictable linearity, providing only a glimpse, a translucent grasp of the "given" or "outside." It can be understood as an image formed when many screens that polarize light are stacked on top of one another. Each screen, like Deleuze's dice-throw, changes what light (as information) can come through, as each polarized light screen introduces new probabilities from the screen before, and hence adds more "noise" to the passage of light. Stacked the right way, no light comes through: total darkness, frozen silence perhaps, with no information available. Yet such a state is not possible.

This flow or spark of energy lasts for a "short duration" (Deleuze 1997b, 161). It might be thought of as decoherence, and follows Stuart Hameroff and Roger Penrose's (1996) Orch OR theory of consciousness as quantum decoherence of quantum mechanics and spacetime geometry, providing a speculation on Deleuze's moment of the image's self-dissipation as "a singular darkness," as "a movement in the world of the brain" (Deleuze 1997b, 161, 169). At the end of "The Exhausted," Deleuze says, "we are faced with 'eternal darkness,' the end of the black night" (1997b, 170). Such a "collapse" (decoherence) occurring in the brain might be likened to what is properly "informative" for Deleuze—as a unique effect of such a "disruption." It is cause for producing a partial and temporary stability—a meta-state.

Penrose with Deleuze

Deleuze and Guattari's cosmology forwards an open evolution of becoming. In brief, their cosmology posits an incomplete reality that is always in the processes of emerging. It is mathematically incalculable. Deleuze draws on the mathematical philosophy of Albert Lautman wherein mathematical thought is confronted by paradoxes, ambiguities, uncertainties, and contradictions, which address the problematic at hand that connects the Idea and the event. Mathematical "solutions" never fully resolve the problem; there is always a "remainder," which is the disruptive power of a paradoxical element within an explanatory framework. "[T]rue problems are Ideas and [...] these Ideas do not disappear with 'their' solutions, since they are the indispensable conditions without which no solution would ever exist" (Deleuze 1994, 168). There is a constantly shifting distribution of singularities from one event into the next into which it morphs. "The moment that the two series resonate and communicate, we pass from one distribution into another. The moment that the two series are traversed by the paradoxical agent, singularities are displaced, redistributed, transformed into one another, and change sets" (Deleuze 1990, 53).

In *Foucault* (1988) Deleuze presents the process of subjectivation in possible holographic terms, in a way that both challenges and supports recent cosmological theorizing regarding black holes and string theory, as first formulated by Leonard Susskind's (2009) concept of a holographic universe. In an early paper, Tim Murphy (1998) points out the close affinity between David Bohm's (1980) holographic universe, which theorizes an "implicate order" (as the "outside"), and an implicate order that is consistent with Deleuzian topological ontology as developed in *Difference and Repetition* (1994). The "undivided universe" is where everything is enfolded in everything else, which might support string theory with its 10+ dimensions of vibrating "strings." But the analogy stops there, as the hologram theory is quite different than recent applications to the physics of black holes by astrophysicists such as Jacob Berkenstein, Gerard 'tHoof, Leonard Susskind, Juan Maldacena, and Nassim Haramein, who propose a generalized "holographic principle" in terms of both the surface and volume entropy of a spherical system. These are recent attempts to "square the circle" between quantum gravity and general relativity. Space does not allow for further discussion of black hole controversies here, except to quickly gloss a number of significant concerns. Unlike quantum theory, where processes are always (in principle) time-reversible, which means the initial state can be traced back from the final state if there is no measurement, information cannot get lost. In contrast, black hole evaporation is irreversible. Its initial state remains unknown. This amounts to the "black hole informational loss" problem. There are many mathematically *consistent* solutions to fudge this initial problem of "loss" that remains irreversible, each speculating on a different set of initial conditions, each making their own claims about conjoining quantum gravity and general relativity theory (Hossenfelder 2020). Secondly, black hole physics as informed by string theory posits anti-de Sitter space with a *negative* cosmic constant of –1; this is a projected imaginary universe. Given the exponential expansion of the universe, we "live" in a quasi-de Sitter space where the *positive* cosmic constant is close to 1. Major disagreements emerge between loop quantum gravity theorists (the foremost being Carlo Rovelli), for whom black hole entropy undergoes different calculations, and string theorists mentioned above. Anti-de Sitter space (AdS_n) is again a construction utilizing imaginary numbers.

Deleuze's own questioning of the Second Law of Thermodynamics' entropic calculability, that is generalized at the expense of negentropic "islands" of "smooth spaces," and Bohm's (1951, ch. 22, sec. 12, 608–9) equal concerns regarding entropic measurement at the micro and macro levels, where measurement issues of thermodynamic irreversibility enter into quantum theory, find an interesting solution through Penrose's evolutionary (cyclical) universe: conformal cyclic cosmology (CCC). Although the Second Law of Thermodynamics is not rejected, it does undergo further modifications. Already in 1969, Penrose developed what became known as the "Penrose process," which compared black hole mechanics with thermodynamics. For Penrose, black hole information *is* lost; the entropy of black holes is much lower as the degrees of freedom of information are swallowed up (lost): useful energy is used up, arguing that as the universe expands exponentially, black holes begin to give off Hawking radiation, which leads to Penrose's CCC scenario as described below. This is a minority position, as the Second Law of Thermodynamics is being modified to fit his hypothesis as supported by his particular equations. While the total entropy of the universe cannot *decrease*, Penrose (like Deleuze) posits black holes as "islands" of negentropy that will begin to release radiation in the remote future. Penrose's position is that there is "a universal tendency [for entropy] to *decrease* as we examine things further and further into the *past*. Accordingly, the situation ought to have been one of exceptionally low entropy" (2018, 1179, emphasis original). Cosmic microwave background radiation presents "a *maximum* entropy state!" (2018, 1179, emphasis original), the uniformity of which is exceedingly low entropy. This scenario leads to a meeting point where the dissipation of black holes into massless protons leads to the Big Bang of quantum fluctuations.

CCC competes with the more accepted theory of the inflationary universe, a position that Penrose (1989) had already questioned regarding its failures to address the Second Law of Thermodynamics some three decades ago. Besides the standard inflation cosmological model there are a number of so-called "bounce cosmologies" of expansion and contraction, where the universe has no beginning nor ending; rather it is cyclical, as the Big Bang is actually a Big Bounce. An earlier phase of contraction connects with the present phase of expansion (see the critical review by Craig and Sinclair 2012). There is also a revised string theory's claim of an "emergent cosmology" (Brandenberger 2023) where there is no inflation. CCC offers an alternative to these as well.

Put in a nutshell, Penrose "stands cosmological inflation theory" on its head by introducing an evolution cyclical theory of the cosmos, which supports Deleuze's own embrace of the Nietzschean "eternal return" hypothesis. Put more bluntly, the Big Bang doesn't exist as we know it. CCC narrates a process of the cosmos changing from one aeon to the next. Evolution in this sense does not mean progress. Whether new laws emerge or new constants are established, that is, whether a new cosmos with "a difference" unfolds, is only a speculation. From a Deleuzian standpoint a "difference" would emerge, as the repetition of sameness, of identity, would not be the case. A "different" order would emerge. The merit of CCC in relation to Deleuze and Guattari's standpoint is that there is an interrelation between the actual and virtual in each event, which results in a new determinability of the multiplicity that CCC also theorizes via the "morphological genetic" change from one aeon to the next that generates new dark matter (erebons). Penrose suggests that each aeon might present a different shift in laws, but that's about it. There is no explanation of why the constants are what they are. They are the given and not random, supporting the idea (like Deleuze) that order/disorder remains a fundamental paradox. Conformal geometry seems to play with both discontinuous and continuous multiplicities during the transition from one aeon to the next. They become the inverse of each other. In the discontinuous multiplicity that characterizes the remote future, all the relations are an effect of division and *degree*, as elements maintain their identity. Super-black holes begin to leak radiation and disperse as massless protons (high entropy), which then undergo change to a continuous multiplicity (low entropy) at the beginning of the next aeon, when gravity takes away their degrees of freedom. Their kinetic energy manifests into the potential gravitational energy that characterizes erebons. The relation between the elements is such that any change to these relations leads to a change in the multiplicity, a change in *kind*.

Hawking's "leaky" black holes (he later called them gray holes, which is closer to my use of Dark Universe) perhaps offer a clue as to why first-generation galaxies were brighter and more massive in earlier times during the reionization transition of gas dynamics, a finding that does not sit well with the standard model of the universe. In the "remote future," Penrose claims that all we have is massless photons, which is the equivalent of the stretched-out Big Bang. The evidence for this are the "Hawking points" that can be found on the cosmic background radiation (CBR) "sky map," five of which are found in the same place by both satellites: Planck (ESA) and WMAP (NASA's Wilkinson Microwave Anisotropy Probe). They are said to be *derivates* of a previous aeon. The statistical evidence from these points is 99.7 percent confidence (Gurzadyan and Penrose 2013, 2). This "observation" (if indeed accurate) is precisely illustrative of the translucency (anexactness), shadow, twilight, membrane, and similar terms that exemplify a sense of intuition as to what might be "beyond" the screen. CBR is the twilight zone that gives clues to the Dark Universe. It illustrates the more accurate or strong sense of "becoming" for Deleuze and Guattari: that of the *creativity* itself of an event in its evolutionary sense; what they mean by allowing a "little bit" of "chaosmosis" to come through.

Perhaps a simplified way of grasping this transitional event happens in the exchange of the future past—that is, the Big Bang that has happened and is about to happen again (with a difference?)—and the future perfect—that is, the "remote future" that has yet to come. This time structure repeats the event as already happened (the Big Bang has already happened but we are not aware of it, we are in its expansion) and as about to happen (the remote future where the boundary is reached). In the remote future we can use Deleuze and Guattari's neonism: n – 1, which they apply to the rhizome, to mark the boundary condition, and n + 1 as the crossing over that boundary into the Big Bang. The superposition of n – 1 and n + 1 leads to the many paradoxes that emerge in *n* itself between quantum, which leans to immanence and the plane of consistency, and the general theory of relativity which leans toward transcendence. The *n* refers to *n*-dimensional manifolds or *multiplicities* of Riemannian surface as space in itself (as first developed by Carl Friedrich Gauss). The proliferation of the "remainder" or imperfections and inconsistencies in the mathematics to reconcile the two singularities, that is, the emergent complexities, are what "justifies" the universe.

This can be read in Deleuzian fashion, where the virtuality of the remote past is the becoming of the transcendental remote future that will be "actualized" paradoxically by the immanence of the Big Bang, now envisioned as a smooth space, the surface of the plane of immanence. The cosmology is that, as beings who must cope with the *necessity of the insufficiency of reason*, we remain placed and suspended in aeon time that splits the arrow of time simultaneously as the past (the Big Bang in this case, the current quantum physics where gravity is not a factor, micro-physics, the immanence of the plane of consistency) and the remote future (the transcendence of what can be actualized, which is the cosmology of the universe). We are planetary creatures caught, as it were, by the "light cone" of Minkowski space where intuition and creativity emerge as the in-between of past-future, which *must* be constantly negotiated (ethico-politically). Becoming imperceptible in this reading requires that we become cosmic, as it requires speculation on precisely the Dark Universe that eludes us, that remains invisible. The idea of translucency, shadow, anexact, becomes the possible guiding way in which "vision" is extended beyond its usual framings, a task that both Deleuze and Guattari have presented for us. It is a question of the problematic of our "species-becoming," given the phase change of the Earth and our responsibility in it.

Note

1. I purposely use the term Dark Universe to avoid the racially charged color implications of calling it a Black Universe, which spirals into issues surrounding Afropressimism, which has its own cosmological position poised against a heliotropic cosmology of Bataille's "accursed share" of any natural or cultural system, for instance where the White/Black binary is in play (Mubirumusoke 2022). Unfortunately, the nomenclature of black holes raises the same issue. However, even Stephen Hawking used the idea of a "gray hole" to maintain that black holes do leak information. Can thermodynamics void such raciality? Bryant's (2013) meditation on "black ecology" would be one such excursion. Dark Universe is more in the direction of shadow, twilight, grayness (e.g., Klee's "gray point"—the fateful point between coming into being and passing away), and translucency, as will be made evident in this chapter. Light as electromagnetic radiation and as materialized uncertainty is the account offered here. For a "darker" existential account of gray, see Cohen (2013).

References

Bergson, Henri. 1991. *Matter and Memory*. Translated by Nancy M. Paul and Scott W. Palmer. New York: Zone Books.

Birat, Jean-Pierre. 2017. "Musica Universalis or the Music of the Spheres." *Matériaux & Techniques* 105, no. 5/6: 1–8. https://doi.org/10.1051/mattech/2018015.

Bohm, David. 1951. *Quantum Theory*. New York: Prentice Hall.

Bohm, David. 1980. *Wholeness and the Implicate Order*. London: Routledge and Kegan Paul.

Brandenberger, Robert. 2023. "Superstring Cosmology—A Complementary." *Review, High Energy Physics—Theory*. https://doi.org/10.48550/arXiv.2306.12458.

Bryant, Levi R. 2013. "Black." In *Prismatic Ecology: Ecotheory Beyond Green*, edited by Jeffrey J. Cohen, 290–310. Minneapolis: University of Minnesota Press.

Clisby, Nathan. 2017. "Scale-free Monte Carlo Method for Calculating the Critical Exponent g of Self-avoiding Walks." *Journal of Physics A: Mathematical and Theoretical* 50: 264003. https://doi.org/10.1088/1751-8121/aa7231.

Cohen, Jeffrey J. 2013. "Grey." In *Prismatic Ecology: Ecotheory Beyond Green*, edited by Jeffrey J. Cohen, 270–89. Minneapolis: University of Minnesota Press.

Craig, William L., and James D. Sinclair. 2012. "On Non-Singular Space-times and the Beginning of the Universe." In *Scientific Approaches to the Philosophy of Religion*, edited by Yujin Nagasawa, 95–142. London: Palgrave Macmillan.

Deleuze, Gilles. 1988. *Foucault*. Translated by Sean Hand. Minneapolis: University of Minnesota Press.

Deleuze, Gilles. 1990. *The Logic of Sense*. Translated by Mark Lester and Charles Stivale. New York: Columbia University Press.

Deleuze, Gilles. 1993. *The Fold: Leibniz and the Baroque*. Translated by Tom Conley. Minneapolis: University of Minnesota Press.

Deleuze, Gilles. 1994. *Difference and Repetition*. Translated by Paul Patton. New York: Columbia University Press.

Deleuze, Gilles. 1997a. "He Stuttered." In *Essays Critical and Clinical*. Translated by Daniel W. Smith and Michael A. Greco, 107–14. Minneapolis: University of Minnesota Press.

Deleuze, Gilles. 1997b. "The Exhausted." In *Essays Critical and Clinical*. Translated by Daniel W. Smith and Michael A. Greco, 152–74. Minneapolis: University of Minnesota Press.

Deleuze, Gilles. 2001. *Gilles Deleuze from A to Z* (video), with Claire Parnet, dir. Pierre-André Boutang. Translated by Charles J. Stivale. New York: Semiotext(e).

Deleuze, Gilles. 2003. *Francis Bacon: The Logic of Sensation*. Translated by Daniel W. Smith. Minneapolis: University of Minnesota Press.

Deleuze, Gilles, and Félix Guattari. 1983 [1977]. *Anti-Œdipus: Capitalism and Schizophrenia*. Translated by Robert Hurley, Mark Seem, and Helen R. Lane. Minneapolis: University of Minnesota Press.

Deleuze, Gilles, and Félix Guattari. 1994. *What is Philosophy?*. Translated by Hugh Tomlinson and Graham Burchell. New York: Columbia University Press.

Faucher, Kane X. 2013. *Metastasis and Metastability: A Deleuzian Approach to Information*. Rotterdam: Sense Publishers.

Gasché, Rodolphe. 2014. *Geophilosophy: On Gilles Deleuze and Felix Guattari's What is Philosophy?*. Evanston, IL: Northwestern University Press.

Gurzadyan, Vahe G., and Robert Penrose. 2013. "On CCC-Predicted Concentric Low-Variance Circles in the CMB Sky." *The European Physical Journal Plus* 22. https://doi.org/10.48550/arXiv.1302.5162.

Hameroff, Stuart R., and Robert Penrose. 1996. "Orchestrated Reduction of Quantum Coherence in Brain Microtubules: A Model for Consciousness." In *Toward a Science of Consciousness. I. The First Tucson Discussions and Debates*, edited by Stuart R. Hameroff, Alfred W. Kaszniak, and Alwyn C. Scott, 507–40. Cambridge, MA: MIT Press.

Hansen, Mark B. N. 2015. *Feed-Forward: On the Future of the Twenty-First Century*. Chicago: University of Chicago Press.

Hossenfelder, Sabine. 2020. "The Black Hole Information Loss Problem is Unsolved. Because It's Unsolvable." http://backreaction.blogspot.com/2020/11/the-black-hole-information-loss-problem.html. Accessed September 10, 2025.

Laruelle, François. 2013. *Principles of Non-Philosophy*. Translated by Nicola Rubczak and Anthony Smith. New York: Bloomsbury Academic.

Lyotard, Jean-François. 1988. "Can Thought Go on Without a Body?." *Discourse* 11, no. 1: 74–87.

Merritt, David. 2020. *A Philosophical Approach to MOND: Assessing the Milgromiam Research Program in Cosmology*. Cambridge: Cambridge University Press.

Mubirumusoke, Mukasa. 2022. "Prolegomena to any Future Cosmology." *Philosophy Today* 66, no. 1: 39–55.

Murphy, Timothy S. 1998. "Quantum Ontology: A Virtual Mechanics of Becoming." In *Deleuze and Guattari: New Mappings in Politics, Philosophy and Culture*, edited by Eleanor Kaufman and Kevin J. Heller, 211–29. Minneapolis: University of Minnesota Press.

Penrose, Roger. 1989. "Difficulties with Inflationary Cosmology." *Annals of the New York Academy of Sciences* (Texas Symposium on Relativistic Astrophysics) 571, no. 1: 249–64.

Penrose, Roger. 1997. "On Understanding Understanding." *International Studies in the Philosophy of Science* 11, no. 1: 7–20.

Penrose, Roger. 2018. "The Big Bang and its Dark Matter Content: Whence, Whither, and Wherefore." *Foundations of Physics* 48: 1177–90.

Ross, Bill. 2024. *Order and the Virtual: The Philosophy and Science of Deleuzian Cosmology*. Edinburgh: Edinburgh University Press.

Rugh, Svend, and Henrik Zinkernagel. 2017. "Limits of Time in Cosmology." In *The Philosophy of Cosmology*, edited by Khalil Chamcham, Joseph Silk, John Barrow, and Simon Saunders, 377–95. Cambridge: Cambridge University Press.

Susskind, Leonard. 2009. *The Black Hole War: My Battle with Stephen Hawking to Make the World Safe for Quantum Mechanics*. Grand Ridge, FL: Black Bay Books.

Smolin, Lee. 2013. *Time Reborn: From Crisis in Physics to the Future of the Universe*. Toronto: Alfred A. Knopf.

Torres, Émile P. 2023. *Human Extinction: A History of the Science and Ethics of Annihilation*. Abingdon: Routledge.

Unger, Roberto M., and Lee Smolin. 2015. *The Singular Universe and the Reality of Time: A Proposal in Natural Philosophy*. Cambridge: Cambridge University Press.

Webb, John K., J. A. King, M. T. Murphy, V. V. Flambaum, R. F. Carswell, and M. B. Bainbridge. 2011. "Indications of a Spatial Variation of the Fine Structure Constant." *Physical Review Letters* 107: 191101. https://doi.org/10.1103/PhysRevLett.107.191101.

PART IV

COMPARATIVE COSMOLOGIES

In classical photography, analog double or multiple exposures are considered an accident, a mistake, or unaesthetic. But by regarding such pictures, you don't regard a photo as a clone of reality, rather an inversion of this order: not that the photo simulates reality, but the accidental blending of different places and times holds the possibility to detect the unknown, to speculate about coincidence or to experience cosmic immanence. Through the interlocking of objects in multiple exposures, the object forms in the photography, creating an epistemological, a sensual in-between, spheres of imagination and experience, lines of flight.

—Stefan Paulus

Stefan Paulus, *Line of Flight*, analog double exposure on 35mm slide film, captured between 1998 and 2003 © Stefan Paulus

Chapter 13

Caught Between a Sufficient Cosmos and a Chaosmos: Laruelle, Deleuze, and Guattari on Philosophy, Cosmology, and the Real

Chantelle Gray

Cosmology Between Science and Philosophy

The pre-Socratic philosopher Anaximander is often said to have been the first cosmologist because of his naturalistic and materialist investigations. His major work, a philosophical poem titled "On Nature," contains "material on cosmogony, cosmology, astronomy, meteorology and zoogony" (Gregory 2016, 2), and he famously postulated that the basic building blocks of life were neither *aer* (air, mist, or vapour), as Anaximenes would have it, nor water, as Thales postulated, nor a mixture of these, as Anaxagoras held, but the *apeiron*, a largely enigmatic substance that has been interpreted as "the boundless, the indefinite, and the eternal" (Shields 2003, 4). The focal point of academic interest in Anaximander of late has not been the *apeiron* as such, but his cosmology, which presents "a description of the stages by which the cosmic architecture was created" (Hahn 2003, 79–80). In contrast to Thales' Earth, "which floats on water, Anaximander's Earth floats on nothing at all," but is, instead, suspended at the midpoint of the cosmos, "equidistant from everything" (Hall 2023, n.p.). More important is that Anaximander reached this model as the result of a firm commitment to rational, naturalistic thought, rather than supernatural musings. Carlo Rovelli argues that it was in fact Anaximander who founded what we understand today as the scientific method: "Not to fear rethinking the world is the power of science," which has shown itself "ever since Anaximander removed the foundations on which the Earth rested, Copernicus launched it to rotate in the sky, Einstein dissolved the rigidity of space and time, and Darwin demolished the separateness of humanity" (Rovelli 2021, 66).

Anaximander's practice was, moreover, firmly rooted in both science and philosophy, though physics and metaphysics would become increasingly differentiated in the work of Galileo, Kepler, and Newton (Rosenberg 2000, 2). Despite this, the early relations between philosophy and science have been taken up again in the contemporary field known as the "philosophy of science," inspired by the work of philosophers like Rudolf Carnap, Gottlob Frege, Bertrand Russell, and David Hilbert. As the names of these philosophers suggest, the philosophy of science is largely the purview of analytic philosophy, though there is a particularly interesting lineage in continental philosophy which Michel Foucault traces via Gaston Bachelard, Georges Canguilhem, Michel Serres, and Martial Guéroult (Foucault 1989, 4–5), though Jean Cavaillès, Albert Lautman, Gilbert Simondon, Foucault himself, Gilles Deleuze, and also François Laruelle conceivably form part of this line of thinkers, whether as adopters or critics.

Bachelard and Canguilhem initiated what has become known as the "philosophies of the concept," which they opposed to the so-called "philosophies of consciousness." This was, as Knox Peden argues, a "decisive moment in French intellectual history in that the fundamentals of Spinozist rationalism were for the first time posited against those of phenomenology" (2014, 61). Bachelard held, for example, that "science *realises* its objects without ever just finding them ready-made," thereby extending phenomenology into a phenomeno-technics (*phénoménotechnique*) (2002, 70). Canguilhem takes this idea a step further by calling attention to the idea that the scientific object of history does not correspond to the object of science (Canguilhem 1994, 17, my translation), meaning that *scientific ideology*, or the "epistemological problem concerning the way in which scientific knowledge is historically constituted," should not be confused with any attempt on the part of scientists "to systematize their research methods and procedures" within culture, because it is philosophers, with "scientific-sounding doctrines," who perform this ideological work (Canguilhem 1988, 28, 38). The point that Canguilhem is making is that there exists a kind of "double" or parallel history of the scientific object: on the one hand, the "concepts that interpret data" and, on the other, the "theories that explain them" (Gutting 1989, 34). This is also a theme that is important for Laruelle, as will become clearer.

Deleuze, as is well known, is a philosopher of the concept *par excellence*, and although I would not call him a philosopher of science *per se*, he was influenced by the sciences, drawing on the works of Leibniz (differential calculus), Riemann (non-Euclidean geometry), Lautman (the concept of genesis in mathematics), and Simondon (ontogenesis and individuation), to name a few. In his formulation of the transcendent dogmatic image of thought, he turns, for example, to non-Euclidean geometry, especially the work of Riemann, a mathematician who furthered Carl Friedrich Gauss's notion of surfaces as spaces, though Riemann addressed "a much more general problem: that of *n*-dimensional surfaces or spaces" (DeLanda 2002, 4). What Gauss argued is that a "surface is a space in itself" (DeLanda 2002, 4), and that this topological space *locally* resembles a Euclidean space near each point. Riemann takes this further by showing that the instantaneous rate of change—the infinitesimal or velocity at *this* instant—can be calculated as *dy/dx* without the need for reference to the container space of a Euclidean object, such as a tetrahedron or cube, which would be expressed in the more traditional Euclidean values of x, y, and z (length, width, and height). Put differently, the rate of change of x with respect to y can be determined without invoking a global embedding or transcendent space. This, evidently, is why Deleuze (and Guattari) sometimes speak of $n - 1$: to show that a transcendent unity is no longer required. This move allows Deleuze to argue that an Idea, or the virtual, "is an *n*-dimensional, continuous, defined multiplicity" (1994, 182)—like a Riemann surface, which is a one-dimensional complex manifold. The implication is that the virtual "structure," or "structuration," does not *precede* actualization but is instantiated as a problem *at the same time*, thus contingently conditioning the eventual actualization or *relative and incomplete resolution*. The virtual, in other words, *is* the actual. The virtual is not some separate "realm" as is sometimes assumed—the distinction is merely a formal one to help us understand something about the granularity of the individuation process (see Deleuze 1994, 246).

These scientific formulations form the basis of Deleuze's conceptualization of difference and repetition, which, in turn, informs many of the concepts he develops with Guattari, for example the body without organs (BwO), the plane of immanence, the plane of consistency—or inconsistency as they sometimes call it—and the Planomenon, though each of these takes on its own specific nuance according to the constellation of concepts it forms a part of. In the chapter on the ritornello in *A Thousand Plateaus*, as in *What is Philosophy?*, these two originary Deleuzian concepts take a more cosmological form. In "Of the Refrain," Deleuze and Guattari make explicit links between chaos, Earth and cosmos to signal not only personal and societal individuation processes, but also cosmological movements of territorialization, deterritorialization, and reterritorialization, or difference and repetition (1987, 310–50). In *What is Philosophy?*, they still think the cosmological in terms of cuts, but instead of territorialization processes, cosmological movement is now also shown to depend on specific modes. They argue, accordingly, that in its philosophical mode, the cosmological is guided by *conceptual personae*, such that philosophy confronts chaos via the creation of philosophical concepts and "the laying out of a plane," which, in this case, is the plane of immanence (1994, 36). When in its scientific mode, the cosmological is steered by *partial observers*, the plane of consistency this time being the plane of reference which constitutes the functions through which science encounters chaos (1994, 119–20). Finally, Deleuze and Guattari argue that the cosmological is directed by *aesthetic figures* in its artistic mode, but instead of creating concepts, artists contend with chaos by constellating percepts and affects as a plane of composition (1994, 65).

Another variation can be found in how Deleuze and Guattari think of the chaos–cosmos relation which, in *A Thousand Plateaus*, is postulated as a continuum—what Guattari would call a *chaosmos*—while in *What is Philosophy?*, they "seem to mark a clearer distinction between chaos and cosmos," such that the "order on the plane of immanence" can be said to push "back the edge of chaos" even though the cosmological (plane) can never quite conquer the chaotic, as Alain Beaulieu has argued (2016, 201). This attenuation of the chaos–cosmos relation means that while there is a continual recursive movement in the earlier text, there is a more unidirectional relation from chaos to cosmos in *What is Philosophy?*, hence Beaulieu's observation that the cosmological has the capacity to confront and contain chaos to a certain degree, but not conquer it. Chaos could thus be said to determine the plane of organization in a specific mode (philosophical, scientific, artistic), though the cosmological does not in turn condition chaos, as it does in *A Thousand Plateaus*. This is an important nuance in Deleuze and Guattari's work, especially when challenged by the non-standard philosophy of Laruelle, who not only critiques "cosmological" thought, but also the "philosophies of difference" of Jacques Derrida and Deleuze in particular. What I will argue in this chapter is that despite the strong critiques made by Laruelle, there are in fact some resonances between what he calls "The Principle of Sufficient Reason"—or what I called a "sufficient cosmos" in the title of this chapter—and what Deleuze calls the "dogmatic image of thought," just as Deleuze and Guattari's theorization of "chaos" in *What is Philosophy?* could be seen as a doppelgänger of what Laruelle calls "the Real." This should not, however, lull the reader into thinking that there are easy links to be made between Deleuze, Guattari, and Laruelle.

Against the Philosophy of Science, Against Cosmological Thought

Laruelle's project on non-standard philosophy, or simply non-philosophy, offers a provocative and theoretically dense injunction against what he calls "standard philosophy"—specifically the image of thought it gives to itself by which it posits its own givenness, or auto-position. From this position, standard philosophy not only assumes itself to be the highest *form* of thought, but also acts as a "device for controlling what counts as thought" in the first place (Ó Maoilearca 2015, 1). Once this position is assumed, anything and everything becomes *philosophizable* (see Laruelle 2013a, 22; 2015a, 8). Laruelle calls this double movement the Principle of Sufficient Reason and the Philosophical Decision. The first part of this movement sees standard philosophy assuming itself *sufficient* for thinking everything, including the Real which, here, does not refer to the Lacanian notion, but to "ordinary" or radical immanent reality *as lived*. The point for Laruelle is that the Real is not philosophizable because it is *a priori* foreclosed to thought—"an absolute untranslatable" (2013d, 14). Because the Real precedes philosophy, philosophy has no traction on it, which is to say, philosophy is not sufficient for thinking the Real. Once the Real is philosophized, standard philosophy completes the movement by way of the Philosophical Decision: a splitting action that results in either a 1/2 or 2/3 mix, both of which add "a second principle" to the Real (Laruelle 2013d, 25). The Principle of Sufficiency and the Philosophical Decision together form what Laruelle calls the *invariant structure* of standard philosophy. His argument, in brief, is that once philosophy assumes itself sufficient for thinking the Real, it must justify this position, which it does by dividing the Real into coordinates, such as immanence and transcendence or thought and being, which are then synthesized into a whole. Non-standard philosophy is thus Laruelle's attempt to think the Real without "various forms of transcendences" (Smith 2016a, 107).

Another way of saying this is that standard philosophy superimposes a photographic framing on the Real, thus announcing the cosmos as "a shot" and itself as the "creative shot of the World" (Laruelle 2011, 1). Standard photography, like standard philosophy, both have decisional powers that provoke "a certain forgetting of the frame they impose on the Real" (Fardy 2018, 3). This is what Laruelle calls standard philosophy's epistemological function: its sufficient epistemology or epistemic violence. Given this, non-philosophy is an attempt to rehabilitate standard philosophy both at the level of practice and at the level of methodology, which it does by challenging its "photographic" model of the world. This does not mean that non-philosophy is an attempt to replace philosophy; its aim, rather, is to render philosophy *ordinary* or *generic*—a weak force. Just as non-Euclidean geometry does away with the fifth postulate of parallelism in Euclidean geometry, *without relinquishing Euclidean geometry*, so the *non-* in non-philosophy indicates a depotentialization of the "semi-circularity" fundamental to the structure of the Philosophical Decision (Laruelle 2013b, 55, 155). Laruelle charges philosophies of difference—including Deleuze's—with this same semi-circularity, arguing that there remains, in these philosophies, "a specular doublet, a double layer, double stratum or double face, either parallel or arranged in a Moebius strip," which makes philosophers believe that they are "critiquing the whole of representation" when in fact they are critiquing only "one stratum" (Laruelle 2012a, 310). Laruelle's critique of Deleuze is thus that instead of conceding that the Real is foreclosed to thought, difference, as substantive, does away with transcendence only at the level of actuality—or virtual-actual individuation—and so ultimately remains ideal because difference still transcends the Real when it *stands in for it*, so *representing* it, despite Deleuze's attempts to avoid the traps of representation and transcendence. I respond to this charge in the conclusion of this chapter. This point, for Laruelle, is that Deleuze does not overcome this limitation because it is part and parcel of the nature of philosophical thought as "auto-positional" or self-positing, "*be it Deleuzian or not, be it philosophical or not*, is as good or as bad an explanation as any other" (Ó Maoilearca 2015, 14).

As may have been surmised, when Laruelle refers to "standard philosophy," he means philosophy *tout court*. Ray Brassier, in *Nihil Unbound* (2007), argues that this claim unnecessarily burdens the non-philosophical project, in turn "forcing" Laruelle to make excessive claims about non-philosophy's own delimitations. Brassier suggests that philosophy *tout court* be reduced to a certain kind of philosophical reasoning, namely *correlationism*, a concept popularized by Quentin Meillassoux in his 2008 book *After Finitude*, where he describes the unavoidable correlation "between thinking and being" initiated by Kant's philosophy (2008, 5). Meillassoux's point is that we cannot *know* reality independently of "our *relation* to" it, which is another way of saying that we cannot know "phenomena without some transcendental operator" grounding the conditions of possibility for those phenomena to "manifest themselves" (Brassier 2007, 51). This circular reasoning inevitably results in either idealism or naïve realism, and it seems to Brassier that it is this bind precisely to which Laruelle refers.

Abandoning notions of philosophy *tout court* is, moreover, a humbler posture, one more in line with Laruelle's project.

While we might therefore accept that Deleuze and Guattari's chaosmic ritornello succumbs, in the end, to a Philosophical Decision, it is equally reasonable to argue that cosmological thought (philosophical, scientific, or artistic), in *What is Philosophy?*, is foreclosed to chaos, which here seems very close to Laruelle's Real. There is, moreover, a sincere attempt by Deleuze to address sufficiency in his work, as he does via the dogmatic image of thought, where he deals not only with the problem of images *in* thought, or representation—for example representations *of* the Real—but also the image of thought that philosophy gives to itself, namely, sufficiency, even if this is not what Deleuze terms it (see Deleuze 1994, 129–67). Similarly, he and Guattari address sufficiency in their theorization of the process of Œdipalization, whereby a decision of sufficiency determines Œdipus as the Real, just as it determines the human as always-already Œdipalized. Besides, Laruelle himself does not step outside of concepts like immanence, transcendence, and the transcendental, and choosing a term like the "Real" for radically immanent lived reality does not come without its own semantic encumbrances, laden with Lacanian overtones as it is, though Laruelle does make a substantial effort to bypass this problem methodologically.

Non-philosophy thus proceeds by subordinating itself to its immanent cause, the One-in-One, because it understands the Real to be *in-One*, rather than *in*-thought or *in*-being. As such, the Real is given-without-givenness. Said otherwise, non-philosophy aims to think *from* immanence or the Real rather than *for* or *to* it, because it recognizes a non-commutativity of being or thought and the Real. This is not simply another philosophical or representational position; rather, it is a performative posture (Laruelle 2016, 12; Ó Maoilearca 2015, 20). Non-philosophy, likewise, proceeds from axioms rather than postulates, not only to avoid philosophical semi-circularity, but also because axioms perform non-philosophy's insufficiency to determine or lay claim to the Real. The two main methods of non-philosophy—by which it produces philosophical fictionings or *philo-fiction*—are *dualysis* and *cloning*. Dualysis "distinguishes itself from analysis, from synthesis and from their combinations," as well as from dialectics, in that it is a performative practice "which treats the world-terms according to a unilateral duality, introducing the determination-in-the-last-identity to it" (Laruelle 2010, xxx). Many important non-philosophical "clones" (rather than concepts, as found in standard philosophy) have been mentioned. One of these is *unilateral duality*, which specifies that instead of dividing the Real into two coordinates, such as thought and being, and then synthesizing both under the aegis of one of the terms or a third one, non-philosophy refuses a synthesis and does not therefore "envelop or englobe its terms into a world or thought-world it takes as the Real" (Smith 2016b, 70). Another is *unilateral thought*, which indicates that thought moves *from* the Real but never *to* it, the implication being that the Real is decisive for thought, but only "in-the-last-instance."

Readers will recognize the concept of determination-in-the-last-instance from Louis Althusser's work, which he himself takes from Friedrich Engels, specifically Engels's observation that "he and Marx only ever held that the economy was decisive for historical and political change in the last instance" (Fardy 2021, 3). Althusser thus "extends the critique of 'economism' inaugurated by Engels and Marx but adds the caveat that even the 'last instance' of the economy 'never comes'" (Fardy 2021, 4). To strip this concept of any philosophical sufficiency, Laruelle clones and radicalizes it. "Just as the clones of biotechnology parallel their progenitors, so the clones of non-philosophy parallel the Real," which is to say, the clones of non-philosophy "take the form of mutations" rather than syntheses (Fardy 2018, 9). Cloning is what non-philosophy holds that thinking *is*: precisely because thinking has no traction on the Real, thought can only ever extract a clone from the vision-in-One; that is, "the One clones thought in the form of the transcendental One" just as philosophy is cloned "by it in the form of unilateral dualities, the essence of which is [the] identity" of the Real or radical immanence (Laruelle 2012b, 241). As Laruelle puts it, cloning "is the surest destruction of the image of thought, of the procedures of projection and reflection at the heart of representation," processes from which philosophy has never been able to free itself (2012b, 242). When Laruelle clones determination-in-the-last-instance, he diminishes the overreach of its standard form because he shows that even though the Real is an immanent causality, it is only ever so in-the-last-instance—an instance that "never comes" because it never enacts a Philosophical Decision. This "non-acting" is what Laruelle calls an impotent operation because it "produces the same result no matter how many times the original application is reapplied" (2014, 28)—there is no doubling of the world via a transcendental synthesis.

Laruelle understands his method in opposition to the philosophy of science, instead referring to it as a "science of philosophy," precisely because science "changes the order of its thoughts without also claiming to change the order of the real" (Laruelle 2016, 70). It is thus non-propositional, non-thetic, and non-decisional (of) the Real. Such a "scientific" reform of philosophy is aimed at removing its epistemic violence by rendering it non-epistemological so that it 1) becomes artificial, or fictional, in the sense that clones fiction or fashion the posture of scientific thought; 2) is determined-in-the-last-instance, meaning it is "necessary-without-sufficiency" (Laruelle 2015b, 47); and 3) is fractal, where a generic or general fractality refuses the philosophization of science by allowing it its own legitimacy (Laruelle 2021a, 70).

Here Laruelle diverges from both Bachelard and Canguilhem, though I continue to think that Laruelle forms part of this lineage, even if in a "non" manner. In terms of Bachelard, the following passage from *The Philosophy of No* (*La philosophie du non*) certainly gives some indication of this in terms of Euclidean and non-Euclidean geometry:

> Generalization by negation must include what it denies. Indeed the whole impetus of scientific thought for a century now stems from dialectical generalizations of this sort, which embrace what has been denied. Thus non-Euclidean geometry embraces Euclidean geometry; non-Newtonian mechanics embraces Newtonian mechanics; wave mechanics embraces relativist mechanics. (Bachelard 1968, 117)

Notwithstanding these resonances and indications of a non-lineage, Laruelle himself declines any equivalence: "Non-philosophy is *de jure* badly formed, interpreted sometimes as a positivist and scientistic negation of philosophy, and sometimes as a new 'philosophy of no' [*philosophie du non*]" (2013d, 41). Canguilhem, as we saw, distinguishes between "genuine science" and "scientific ideology," and is of the opinion that "the two are intertwined in the history of science" so that "an adequate understanding of the progress of science requires an understanding of both" (Gutting 1989, 45). Laruelle, on the other hand, wants philosophy to refuse to decide on either the Real or the object of science. The problem for Laruelle is thus about "*how to render epistemology and the philosophy of sciences themselves inventive, how to have done with auto-epistemology*," or: how to render philosophy *generic* (2013c, 206).

For Laruelle, the generic, like the fractal or any other of his concepts, is decidedly not cosmological because cosmology is too tied to the ordered world of philosophy, whereas he aims to free thought from its philosophical "Cosmos-function" by submitting it to the "Universe-function" of science (Laruelle 2012c, 181). This may seem confusing, as the cosmos is simply another word for the universe. The difference is subtle and indicates a difference in vision—much as Venus is both the morning star and the evening star. "Universe" is how Laruelle distinguishes a more scientific approach to the cosmos, whereas the "Cosmos" is for him something like a "cosmology of meaning"—the "sufficient cosmos" of philosophy. This is also why Laruelle will insist on philo-fiction or a generic science of philosophy, because it "brings back or *brings about* the universe towards generic humanity, looking at the cosmos under the lens of the universe" (2021b, n.p.). Here we find a clear divergence between Laruelle's project and that of Deleuze and Guattari, for whom the cosmological—and particularly the cosmological in its fabulatory mode—is of great importance. How, then, are we to think of Deleuze and Guattari's cosmological thought given these injunctions against it?

Cloned Personae, Victimology, and Generic Counter-Actualization: A Non-Standard Reading of *What is Philosophy?*

In *What is Philosophy?*, Deleuze and Guattari think of cosmological movement in terms of three modes—the philosophical, the scientific, and the artistic—each of which "confronts" chaos by laying out a plane of consistency. Early in the text, they make a claim that would seem to be wholly in line with Laruelle's project: that philosophy is "not contemplation, reflection, or communication" (1994, 6). Likewise, Laruelle states that non-philosophy is impotent or sterile contemplation: the "non-positional contemplation-(of)-immediate givens or (of)-irreflective transcendental experiences" (Laruelle 2012d, 57). Deleuze and Guattari also state that science does not need philosophy, which, as I have shown throughout, is a very non-philosophical posture (see Deleuze and Guattari 1994, 117). They then go on to argue that immanence is only immanent to itself—as non-philosophy itself axiomatically stipulates: "The One-real and it alone will be called Given [...] The Given implies in its essence that it be radically immanent (to) self and immanent to nothing else (being, the World, Being, Substance, Givenness, Calling, etc.)" (Laruelle 2013b, 72). In *A Thousand Plateaus*, the plane of immanence *is* the cosmological, whereas Deleuze and Guattari make a distinction between the plane of immanence or the cosmological and chaos in *What is Philosophy?*. In their words, the plane of immanence is a "cosmic milieu that sections chaos in the form of a plane" (1994, 43). From this point of view, chaos—and not the plane of immanence—seems the closer concept to the Real. I would even venture to argue that there is at least some implication here that the cosmological (plane) bears the immanent "imprint" or "cut" of the chaos-Real. This would seem to be corroborated by Deleuze and Guattari's idea that the philosopher, like the scientist and the artist, brings back from chaos the coordinates they need to lay out the plane of immanence, reference, or composition from which they respectively produce concepts (with the help of conceptual personae), functions (with the help of partial observers), and percepts and affects (with the help of aesthetic figures) (Deleuze and Guattari 1994, 202–3). This being said, they appear to contradict this position when they argue that thought, via philosophy, science, and art, confronts chaos, "laying out a plane, throwing a plane *over* chaos," though whereas science "relinquishes the infinite in order to gain reference" and art attempts to "create the finite that restores the infinite," philosophy "wants to *save* the infinite by giving it consistency" (Deleuze and Guattari 1994, 187). Clearly, "saving" the infinite or the Real implies both the Principle of Sufficient Reason and a Philosophical Decision at play.

The point is that Deleuze and Guattari vacillate between what seems to be a non-philosophical position and a philosophical one. Laruelle is not so ambivalent in his response to Deleuze and Guattari. Recall that they mention Laruelle in the footnotes of *What is Philosophy?*—rather favorably—though Laruelle reads what Deleuze and Guattari write as them being bent on recuperating non-philosophy for philosophical ends by presenting it as a "form of alterity"—the Other—that "is still masterable" by the whole that is philosophy (Laruelle 2012e, 53). I am somewhat more equivocal about this because, as I hope to have shown convincingly, there is, at least in *What is Philosophy?*, a movement toward submitting the cosmos-function of philosophy (the plane) to the universe-function of science (chaos), even if they do not quite succeed in attaining radical immanence consistently.

This leaves us with a difficult question—*What is Philosophy … now?*—because having read this, it might be tempting either to cast aside non-standard philosophy as an obtuse and pointless project, or to abandon what we know as standard philosophy, thereby ridding ourselves of all remnants of cosmological thought. After all, how can philosophy become "capable of thinking the ordinariness of its violence" (Gangle 2012, 60) if its invariant structure is that of the Principle of Sufficient Reason and the Philosophical Decision? Having grappled with this near-impossible position, allow me to offer two perspectives.

According to the first, we would render the philosophy of Deleuze and Guattari non-standard. This is the method followed by Laruelle himself in, for example, his work on Marx. In an attempt to perform radical non-violence, I could proceed, accordingly, by rendering two of Deleuze and Guattari's concepts generic. To transform the first concept, *counter-actualization*, into a non-standard philosophical clone would require that it *not* remain at the level of philosophical individuation. A generic counter-actualization would thus have to counter philosophy itself by way of cloned personae rather than conceptual personae, the second concept I use here. Recall that for Deleuze and Guattari the conceptual personae are the becomings of a philosophy, "on par with" the becoming of a philosopher (Alliez 2004, 10); or, as they put it, conceptual personae act as philosophers' "intercessors" because they aid philosophers in the creation of their philosophical concepts (Deleuze and Guattari 1994, 63, 64). For Laruelle, as we have seen, philosophical concepts are themselves part of the problem. By cloning the personae, however, counter-actualization moves beyond philosophical individuation as it becomes the under-determining-last-instance of intellectual work, thereby "saving" philosophical concepts from the sufficiency of the Philosophical Decision. The cloned personae thus address the violence of philosophy itself by bringing to light what Jacques Rancière has described as the problem of philosophers and their "poor," which situates philosophy not only within the realm of standard photography, but also within the domain of capitalism, because it shows how philosophical endeavors capitalize on, or extract surplus value from, their "poor" by sacrificing their physical needs "to the holocaust of speculation (economic but also philosophical)" (Kolozova 2017, 95). That is to say, the factory conditions by which the proletarian is transformed into the subject (or object) of "intellectual creations" (Rancière 2003, 75) dematerializes "life at the hands of economic and thought systems that deny the radicality of the Real" (Fardy 2021, 77).

For Laruelle, the victim-in-person, or human-in-person, is decidedly not a philosophical subject or object. Rather, as he explains in *General Theory of Victims*, Philosophy-with-a-capital-P is itself the criminal that produces the victim by enclosing the human-in-human within the sufficiency of the Philosophical Decision. Paradoxically, it is the victim, *as cloned persona*, who can "save" or "redeem" intellectuals—and thus philosophy—by helping them become generic; that is, by freeing them from the cosmos-function of philosophy and submitting them, instead, to the universe-function of science. As Jonathan Fardy says: "victimology is a science that diagnoses the violence done by philosophies that claim to know and speak for the reality of victimization" (2021, 86). This implies that there is a structural intimacy between the cloned personae of the Real and the cloned persona of the victim-in-person, as both are "formal names for what abstract formalization forgets, namely, that the Real and victims can only be decided by an act of thought-violence that subjugates them to the dictatorship of philosophical concepts" (Fardy 2021, 85). As ordinary messiahs, the cloned personae of the victim can counter-actualize philosophical auto-sufficiency because, as Deleuze himself says in *The Logic of Sense*, counter-actualization is to mime (or clone) "*what effectively occurs*" (1990, 161), though whereas Deleuze holds that counter-actualization doubles actualization, it would, as a non-philosophical performance, identify the inhuman or non-conceptualizable (non-philosophizable) as the idempotent lived—an undulatory form oriented toward the subject determined-in-the-last-instance, rather than toward the Real as a totality derived from philosophical interpretation that forms a cogent cosmology of meaning, thereby denying the ordinary, generic victim-in-person who, like the Real, is irreducible to the framings, concepts, and alienations of philosophy.

The second perspective I offer here is a subtle one, undertaken with reference to the work of the Chilean second-order cyberneticians Humberto Maturana and Francisco Varela, specifically their theory of autopoiesis. Although the first exposition thereof was published in an article written with Ricardo Uribe (1974), the more developed work was published as *The Tree of Knowledge* (1988). Here Maturana and Varela explain autopoiesis in terms of the self-organization or self-maintenance of a system or organism (1988, 89), structural coupling between a system or organism and its environment (1988, 99–103), and emergence (1988, 39, 51). Of special importance is the idea that although autopoietic systems or organisms are thermodynamically open to the *exchange of energy* from their environments, they are self-referential or recursive and therefore operationally closed to the *exchange of information*, the reason being that this is the only way for its organization to remain invariant. Simply put, a system must be able to distinguish between itself and what is not itself, or system and not-system, otherwise it will have no way of making reliable predictions about its state and generating effective actions to counteract entropy. The concept of *structural coupling* explains how operationally closed systems or organisms can conserve autopoiesis while obtaining (not exchanging) information about the world beyond their own boundaries, both of which "are necessary conditions for the existence of living beings" (Maturana and Varela 1988, 102, 103). Finally, as Maturana and Varela argue, the maintenance of dynamic systems requires "a compatibility of the organisms with their environment," that is, *adaptation* (1988, 102). When an organism, as a dynamic system, can no longer adapt to its environment, or has lost its capacity for emergent properties, it "disintegrates as an autopoietic system" (1988, 102).

To sum up, an autopoietic entity can be seen as having three necessary conditions, the first being operational closure which allows it to distinguish between system and not-system, and so maintain itself. The second is structural coupling and metabolic flux which, together, provide an autopoietic system with ways to thermodynamically exchange energy with and gather information from its associated milieus and environments. Finally, for organizational recursion and structural adaptation to take place, an autopoietic entity must be capable of "self-transcending" by "creating emergent properties that change its internal structure" through metabolic flux—but not its internal invariant organization (Žukauskaitė 2023, 120; see also Maturana and Varela 1988, 75).

To be clear about what this means, it helps to turn to a definition of autopoietic systems given by Maturana and Varela in *Autopoiesis and Cognition*—and it is worth quoting at length:

> The autopoietic system is defined as a unity by its autopoietic organization. The realization of this organization in a physical system requires components which are defined by their role in the autopoiesis and which can only be described in relation to this. Furthermore these components can only be realized by material elements which can exhibit the necessary properties under the conditions specified by the autopoietic organization, and must be produced in the proper topological relation within this organization, by the particular instance (structural realization) of the autopoietic system that they constitute. Accordingly, an autopoietic organization constitutes a *closed domain of relations specified only with respect to the autopoietic organization that these relations constitute, and, thus, it defines a "space" in which it can be realized as a concrete system*; a space whose dimensions are the relations of the components that realize it. (Maturana and Varela 1980, 88, emphasis added)

One of the implications of this is that notions such as "specification" and "order" are "referential notions," which is to say they "do not have meaning outside the context in which they are defined" (Maturana and Varela 1980, 89). *Relations of specification* thus refer solely to "the specification of components in the context of which defines the system as autopoietic," just as *order*, or *relations of order*, refers only to "the establishment of processes that secure the presence of the components" that effectuate autopoiesis (Maturana and Varela 1980, 89). In terms of autopoiesis and cognition specifically, they reiterate the importance of this—and I quote at length again:

> Notions such as coding and transmission of information do not enter in the realization of a concrete autopoietic system because they do not refer to actual processes in it. Thus the notion of specificity does not imply coding,

> information or instructions [...] To say that the system, or part of it, codes for specificity, is not only a misnomer but also misleading; this is so because such an expression represents a mapping of a process that occurs in the space of autopoiesis onto a process that occurs in the space of human design [...] The notion of coding is a cognitive notion which represents the interactions of the observer, not a phenomenon operative in the observed domain. (Maturana and Varela 1980, 90)

These two descriptions provided by Maturana and Varela, especially the final sentence, are to me very close to how Laruelle describes the Real. What, then, might the implications be if we think of the Real and being in terms of, not autopoietic systems *per se*, but at least in terms of two independent systems or orders? This would imply that there is not necessarily a contradiction between non-philosophy and standard philosophy as they are two distinct orders of systems being described. My argument, in other words, is that what Laruelle provides us with via non-standard philosophy is a way to account for the Real *from the position of the Real*—as Maturana and Varela describe autopoietic systems *from the perspective outside of that of the observer*. From this perspective, there is indeed a non-commutativity of being and the Real, just as there is an opacity and non-translatability of the Real, such that the Real constitutes the non-conditions for thought. Moreover, because the Real is idempotent, it is entirely plausible to argue that no event and no synthesis take place—only sterile contemplation, the "non-positional contemplation-(of)-immediate givens or (of)-irreflective transcendental experiences" (Laruelle 2012d, 57).

What Deleuze and Guattari offer us is something entirely different: an account of being and becoming *from the position of beings-in-becomings*. That is to say, we accept the cut of the Real as having already occurred. From this perspective, it is conceivable to argue that Deleuze's theory of difference and repetition—in all its cosmological iterations, including those made with Guattari—pertains *only* to the order of *beings-in-their-becomings* or *individuating emergences and metabolic fluxes* and *not* to the order of the Real. The specifications made by Deleuze and Guattari are therefore to be understood as referential notions that have no meaning "outside the context in which they are defined" (Maturana and Varela 1980, 89). So, when Deleuze and Guattari speak of immanence, individuation, and so on, what they are referring to is something like the structural couplings and thermodynamic or metabolic exchanges between multiplicities within the unity of beings-in-their-becomings and the larger unity or univocity of the cosmos, itself an individuating system in the midst of individuating. Whereas Laruelle provides us with an understanding of the conditions of the Real, Deleuze and Guattari provide us with an understanding of the conditions of *life*. Given this, it seems to me that they only begin to theorize the Real in *What is Philosophy?* when they make the clearer distinction between cosmos and chaos.

This is, of course, not how Deleuze and Guattari frame it themselves—this is how I have personally come to understand the contributions and limitations of "standard" philosophies and non-philosophy. It should also be said that although the distinction I have made allows for a different kind of conception of Laruelle's project and other philosophical endeavors, this is in no way an attempt to simply let "standard" philosophy off the hook. If anything, it makes clearer when and where philosophy *does* overextend its reach. For example, when Deleuze describes his transcendental empiricism, he does actually attempt to lay claim to the Real (the transcendental part). It strikes me, then, that he not so much corrects Kant's error—tracing the transcendental from the empirical—as refines the problem which, to be fair, is what standard philosophy is supposed to do! As Deleuze and Guattari themselves state: "We cannot say in advance whether a problem is well posed, whether a solution fits" (1994, 82). The point is not to imagine that we have the keys to the universe, or that any thought we have is sufficient for thinking the Real, or even that non-philosophy is somehow "better" than standard philosophy or vice versa: they are all fictionings. The point is, like Bartleby, to refuse a (philosophical) decision.

References

Alliez, Eric. 2004. *The Signature of the World: What is Deleuze and Guattari's Philosophy?*. Translated by Elliot R. Albert and Alberto Toscano. London: Continuum.

Bachelard, Gaston. 1968. *The Philosophy of No: A Philosophy of the New Scientific Mind.* Translated by Georges C. Waterston. New York: Orion Press.

Bachelard, Gaston. 2002. *The Formation of the Scientific Mind: A Contribution to a Psychoanalysis of Objective Knowledge.* Translated by Mary McAllester Jones. Manchester: Clinamen.

Beaulieu, Alain. 2016. "Introduction to Gilles Deleuze's Cosmological Sensibility." *Philosophy and Cosmology* 16: 199–211.

Brassier, Ray. 2007. *Nihil Unbound.* New York: Palgrave Macmillan.

Canguilhem, Georges. 1988. *Ideology and Rationality in the History of the Life Sciences.* Translated by Arthur Goldhammer. Cambridge, MA: MIT Press.

Canguilhem, Georges. 1994. *Études d'histoire et de philosophie des sciences*, 7th ed. Paris: Vrin.

DeLanda, Manuel. 2002. *Intensive Science and Virtual Philosophy.* London: Bloomsbury.

Deleuze, Gilles. 1990. *The Logic of Sense.* Translated by Mark Lester and Charles Stivale. New York: Columbia University Press.

Deleuze, Gilles. 1994. *Difference and Repetition*. Translated by Paul Patton. London: Athlone.

Deleuze, Gilles, and Félix Guattari. 1987. *A Thousand Plateaus: Capitalism and Schizophrenia*. Translated by Brian Massumi. Minneapolis: University of Minnesota Press.

Deleuze, Gilles, and Félix Guattari. 1994. *What is Philosophy?*. Translated by Hugh Tomlinson and Graham Burchell. London: Verso.

Fardy, Jonathan. 2018. *Laruelle and Non-Photography*. London: Palgrave Macmillan.

Fardy, Jonathan. 2021. *The Real is Radical: Marx after Laruelle*. London: Bloomsbury.

Foucault, Michel. 1989. *The Archaeology of Knowledge*. Translated by Alan M. Sheridan Smith. New York: Routledge.

Gangle, Rocco. 2012. "Laruelle and Ordinary Life." In *Laruelle and Non-Philosophy*, edited by John Mullarkey and Anthony P. Smith, 60–79. Edinburgh: Edinburgh University Press.

Gregory, Andrew. 2016. *Anaximander: A Re-assessment*, London: Bloomsbury.

Gutting, Gary. 1989. *Michel Foucault's Archaeology of Scientific Reason*. Cambridge: Cambridge University Press.

Hahn, Robert. 2003. "Proportionality in Anaximander's Cosmic Architecture." In *Anaximander in Context: New Studies in the Origins of Greek Philosophy*, edited by Dirk L. Couprie, Robert Hahn, and Gerard Naddaf, 78–89. New York: SUNY Press.

Hall, Claire. 2023. "Pond of Gloop." *London Review of Books* 45: 10. https://www.lrb.co.uk/the-paper/v45/n10/claire-hall/pond-of-gloop. Accessed September 10, 2025.

Kolozova, Katerina. 2017. "The Inhuman and the Automaton: Exploitation and the Exploited in the Era of Late Capitalism." In *Superpositions: Laruelle and the Humanities*, edited by Rocco Gangle and Julius Greve, 91–102. New York: Rowman and Littlefield.

Laruelle, François. 2010. *Future Christ: A Lesson in Heresy*. Translated by Anthony P. Smith. London: Continuum.

Laruelle, François. 2011. *The Concept of Non-Photography*. Translated by Robin Mackay. New York: Urbanomic.

Laruelle, François. 2012a. "From the First to the Second Non-Philosophy." Translated by Anthony P. Smith and Nicola Rubczak. In *From Decision to Heresy: Experiments in Non-Standard Thought*, 305–25. New York: Urbanomic.

Laruelle, François. 2012b. "What is Non-Philosophy?." Translated by Taylor Adkins. In *From Decision to Heresy: Experiments in Non-Standard Thought*, 185–244. New York: Urbanomic.

Laruelle, François. 2012c. "The 'Non-Philosophical' Paradigm." Translated by Nicola Rubczak and Anthony P. Smith. In *From Decision to Heresy: Experiments in Non-Standard Thought*, 173–84. New York: Urbanomic.

Laruelle, François. 2012d. "A Rigorous Science of Man." Translated by Robin Mackay. In *From Decision to Heresy: Experiments in Non-Standard Thought*, 33–73. New York: Urbanomic.

Laruelle, François. 2012e. "'I, the Philosopher, Am Lying': A Reply to Deleuze." In *The Non-Philosophy Project: Essays by François Laruelle*, 41–75. New York: Telos Press.

Laruelle, François. 2013a. *Philosophy and Non-Philosophy*. Translated by Taylor Adkins. Minneapolis: Univocal.

Laruelle, François. 2013b. *Dictionary of Non-Philosophy*. Translated by Taylor Adkins. Minneapolis: Univocal.

Laruelle, François. 2013c. *Anti-Badiou: On the Introduction of Maoism into Philosophy*. Translated by Robin Mackay. London: Bloomsbury.
Laruelle, François. 2013d. *Principles of Non-Philosophy*. Translated by Nicola Rubczak and Anthony P. Smith London: Bloomsbury.
Laruelle, François. 2014. "A Science of [*en*] Christ." *Angelaki* 19, no. 2: 25–33. https://doi.org/10.1080/0969725X.2014.950851.
Laruelle, François. 2015a. *Introduction to Non-Marxism*. Translated by Anthony P. Smith. Minneapolis: Univocal.
Laruelle, François. 2015b. *General Theory of Victims*. Translated by Jessie Hock and Alex Dubilet. Cambridge: Polity.
Laruelle, François. 2016. *Theory of Identities*. Translated by Alyosha Edlebi. New York: Columbia University Press.
Laruelle, François. 2021a. *The Last Humanity: The New Ecological Science*. Translated by Anthony P. Smith. London: Bloomsbury.
Laruelle, François. 2021b. "Selections from *Philosophie non-standard: générique, quantique, philo-fiction* (Paris: Kimé, 2010)." Translated by Jeremy R. Smith. https://endemictheory.wordpress.com/2021/08/29/translation-of-francois-laruelle-from-fiction-to-philo-fiction-and-science-phiction-selections-from-philosophie-non-standard-2010/. Accessed September 10, 2025.
Maturana, Humberto R., and Francisco J. Varela. 1980. *Autopoiesis and Cognition: The Realization of the Living*. London: Reidel.
Maturana, Humberto R., and Francisco J. Varela. 1988. *The Tree of Knowledge: The Biological Roots of Human Understanding*. New York: Shambhala/Random House.
Meillassoux, Quentin. 2008. *After Finitude: An Essay on the Necessity of Contingency*. New York: Continuum.
Ó Maoilearca, John. 2015. *All Thoughts Are Equal: Laruelle and Nonhuman Philosophy*. Minneapolis: University of Minnesota Press.
Peden, Knox. 2014. *Spinoza Contra Phenomenology: French Rationalism from Cavaillès to Deleuze*. Stanford, CA: Stanford University Press.
Rancière, Jacques. 2003. *The Philosopher and His Poor*. Translated by John Drury, Corinne Oster, and Andrew Parker. Durham, NC: Duke University Press.
Rosenberg, Alex. 2000. *Philosophy of Science: A Contemporary Introduction*, 2nd ed. London: Routledge.
Rovelli, Carlo. 2021. *Helgoland: Making Sense of the Quantum Revolution*. Translated by Erica Segre and Simon Carnell. London: Penguin.
Shields, Christopher. 2003. *Classical Philosophy: A Contemporary Introduction*. London: Routledge.
Smith, Anthony P. 2016a. *Laruelle: A Stranger Thought*. Malden, MA: Polity.
Smith, Anthony P. 2016b. *François Laruelle's Principles of Non-Philosophy: A Critical Introduction and Guide*. Edinburgh: Edinburgh University Press.
Varela, Francisco J., Humberto Maturana, and Roberto Uribe. 1974. "Autopoiesis: The Organization of Living Systems, Its Characterization and a Model." *BioSystems* 5: 187–96.
Žukauskaitė, Audrone. 2023. *Organism-Oriented Ontology*. Edinburgh: Edinburgh University Press.

Chapter 14

Biological Philosophy and Geophilosophy in the Shadow of Cosmology: Reading Canguilhem with Deleuze and Guattari

Samuel Talcott

Gilles Deleuze and Félix Guattari came to philosophy at a moment marked by profound alterations in the experience and conception of space and the universe. Through Einstein's relativist theory of gravitation and its confirmation, cosmology had become a field of scientific research after having long been abandoned. This new work in physics implied that human observations of gravity and its uniformity depended upon a particular, simultaneous structuring of space and time, a spacetime shaped by massive bodies like planets, stars, and black holes. The absolute, infinite universe and its scientific laws were thus called into question. Yet the medieval cosmos, whose dissolution into the infinite universe was wrought by early modern natural philosophy, was not thereby given new life. Scientists found themselves, rather, before a novel problem and the possibility of investigating cosmological structure. Though the term universe remains current, it has since been studied as a cosmos, in something like the sense of the ancient Greek, an order, a world. While Einstein had asserted its static nature, other models accepted the implication that the cosmic structure is dynamic, emerging in a vast historical process. These events in scientific research and speculation led to the Big Bang theory and current scientific accounts of the evolving universe. Though the idea that the cosmos had emerged from the explosion of a primeval atom was proposed in the 1920s, the phrase Big Bang would be used first in the late 1940s (see Lecourt 2006).

The Space Race between the USA and the USSR also began in that moment. Sputnik, the first satellite, was orbiting the Earth by 1957. A human being was in orbit by 1961. By 1969 people were on the moon. As the universe ceased to be an objective impossibility for scientific study, outer space ceased to be an absolute obstacle to human endeavor through its technological transformation into a new medium of exploration. Between the new possibilities of space travel and colonizing other worlds on the one hand, and a new scientific basis for investigating the cosmic universe on the other, it also became possible to write the history of the infinite universe. At the beginning of a new adventure in doing and knowing, the finite history of the old idea of an infinite universe could be examined in a study of the conceptual transformations by which it had become possible. Alexander Koyré's work, notably the book *From the Closed World to the Infinite Universe* (1957), can be interpreted as contributing to such a project.

When knowledge and capabilities alter, the very meaning of the world, history, and thought must be reworked. It is as if new configurations are drawn into the light and the old are cast into the obscurity of error. Yet these new configurations are also accompanied by another sort of obscurity: the shadow and darkness that they cast, so many novel unknowns to investigate. These unknowns concern not only scientists, but everyone. For the world is now different than had been supposed and philosophers must therefore question traditional presuppositions and explanations. Gaston Bachelard pursued this obligation, which he set forth as a principle for future thinking, in his attempts to rework philosophical thought through the new physics and its rupture with ordinary experience. An entire host of figures, ranging from Koyré, Canguilhem, and others to younger thinkers like Foucault or Deleuze and Guattari, can be read as pursuing this task of reworking thought, not only in light of new configurations of knowing and doing, but also in view of the shadows they cast. Yet each pursued the task in their own way. This chapter suggests that reading Canguilhem with Deleuze and Guattari with this philosophical obligation in mind highlights their shared methodological and conceptual links. Going further, the chapter takes up this obligation in its own way by reading these three *together* in the shadow of scientific cosmology and its current problems. We begin, therefore, with a central question raised by the new cosmos.

For physicists and astronomers, cosmology now implies an anthropic principle (Lecourt 2006). The existence of the universe as we know it follows from a series of events that could have happened otherwise. Other conditions could have taken hold, conditions within which human beings would not have existed. This leads to the simplest form of the anthropic principle, a question: What *are* we to make of our existence as observers of the cosmos? The question expresses wonder that we are able to observe and study the cosmos, that the cosmos is such that we exist as its observers. What does it mean that the conditions of the universe are just right for our own existence? The principle's strongest versions abandon questioning and wondering to insist that humankind's existence as cosmological observer is the necessary end and goal of the cosmos itself. Weaker versions find, more modestly, that humans are immanent to the universe and should expect to observe, therefore, only that which exists thanks to the same conditions through which we exist. As the physicist Brandon Carter put it, 'What we can expect to observe must be constricted by the conditions necessary for our presence as observers' (1974, 291). The anthropic principle supports, thus, both assertions of human transcendence and cosmological immanence.

To observe the cosmos is not to know it. The great diversity of cultural traditions shows that the observed cosmos has been understood in radically different ways. The past century of physics and astronomy also shows diverse conceptions issuing from a scientific tradition of confronting the unknown in a search for knowledge that presses beyond what is currently observable. We see this in efforts to devise new techniques to collect data on black holes, objects so massive that light cannot escape their gravitational pull. We see this, too, in work on dark matter and energy, which must exist for current scientific observations of the universe to hold. The "darkness" or "blackness" in these phrases speaks both to our lack of knowledge and our current inability to know, that is, the absence of observable data or information. Ingenious efforts to develop new sources of information or give new interpretations of existing information lead to new understanding, but the knowledge produced is, in important ways, beyond the realm of the sensible. These are not phenomena that we can grasp through ordinary sensory experience. For strong versions of the anthropic principle, this would speak to the transcendent power of human techno-science, while simultaneously raising questions about the extent to which this would be a human transcendence. If our knowledge of the universe is accomplished by a break with the sensible, this universe does not appear to be *our* universe, one that we could explore beginning from our embodied life.

In weaker versions of the anthropic principle, the cosmological encounter with the unknown has a different character. Such knowledge still depends on ruptures with the sensible, but such ruptures are now part of an existential adventure, an adventure rooted in the conditions of existence from which it begins. Scientific instruments are invented, thus, for the sake of a novel encounter with the unknown and some new information about it. But they are derived from the technologies that were first invented for quotidian use. Bachelard noted, for example, that commercial products invented for practical purposes like the pure stearic acid candle, later replaced by the electric lamp, were so standardized that they could be deployed as units of scientific measure (1969, 159). Yet such "phenomenotechniques" do not only standardize measurements, they invent and construct novel phenomena for investigation (Bachelard 1970, 19). So too, today. Charged coupling devices, a technology that has revolutionized photography by introducing digital images, have been reworked to count single electrons and thereby enabled new means in the search for dark matter and energy (Royal Swedish Academy of Sciences 2009). The very possibility of that search also depends on the reworking of massive practical innovations. Alterations in the Earth's crust produced by centuries of industrial mining in places like Sudbury, Ontario, have established conditions in which the planet could be instrumentalized in the search for dark matter. The resultant spaces carved out deep underground are shielded from surface radiation by rocky structures two kilometers thick, becoming places in which scientists can search for evidence of theoretical particles that could never be observed in the noise—both technical and cosmological radiation—on the surface (SNOLAB 2024). These exemplifications suggest the rudiments of a philosophical theory of knowledge: thinking emerges immanently from the cosmos and the means we have invented to inhabit our world, means which also enable new ways to explore the unknown. But this is no Cosmism in which humans master life, death, and the universe. Rather, this immanent cosmological philosophy finds reason to attend to the Earth in our ongoing encounter with the unknown.

This chapter endeavors to contribute to such a cosmological philosophy in two ways. First, I consider how the pursuit of knowledge in biology both opened onto and depended on ideas and research involving the cosmos. I do this by considering Canguilhem and his approach to the history of the biological sciences, especially his study of the milieu concept. This, however, is intimately tied to this chapter's second endeavor. Readers of Deleuze and Guattari will quickly see that Canguilhem deploys ideas that they also take up in their own thought, albeit in altered ways. Following some of these connections, I claim that reading Canguilhem with Deleuze and Guattari demonstrates the power of an immanent pursuit of philosophy that embraces and responds to the unknown without seeking some definitive transcendence. Showing connections between thinkers whose affinities are often noted but little studied, this reading offers a foray into a cosmological history of thought with philosophical stakes.

Thinking in the Shadow of the Cosmos with Canguilhem, Deleuze, and Guattari

Georges Canguilhem (1904–95) was a philosopher, medical doctor, and historian of the life sciences. He knew the young Deleuze well and advised his thesis on Hume (Dosse 2010, 111). Deleuze published an edited volume, *Instinct and Institutions* (1953), in Canguilhem's series *Philosophical Texts and Documents*. And they stayed in touch throughout their lives. The following pages initiate a comparative study of their work, showing not only that each philosophized in the shadow of cosmology, but that Deleuze and later Guattari extended and altered Canguilhem's conceptions.

Canguilhem's *Knowledge of Life*, first published in 1952, argues that biological science became autonomous by using concepts to guide careful experimentation and then revising them based on this experimentation. He finds that careful laboratory experimentation on animals produced new knowledge of the living being as a unity of valorized functions; a central element of this knowledge is the concept of the milieu that emerged from the difficulties of laboratory experimentation. Clever biologists had realized that the lab is not an animal's or plant's habitat, but a strange and foreign space that does not facilitate direct knowledge of organisms. As Canguilhem put it elsewhere, "the discovery that observation interferes with the observed phenomenon" was not made by the "contemporary theorists of wave mechanics," but by biological experimentation in the nineteenth century (1989, 147). This entailed the realization that life tends to take the form of an individual endeavor within a milieu (Canguilhem 2008). The living are never indifferent to the place in which they find themselves. Each living being exists in relation to a meaningful milieu, one that is structured and exists in relation to the organism. This milieu is not the mathematical-physical universe; it can only be grasped when the biologist studies the living thing in terms of the diverse relations it maintains with its own milieu. Not simply the physical world, milieus are the different meaningful realities that the living establish in relation to their own organismic needs and possibilities. Experimental knowledge of life followed when biologists created ruses by which to mimic biological milieus or find other means to expose biological functions in terms of their role in the life of the individual.

Canguilhem thought, for a while at least, that biology was moving in the direction of ethology and that the study of comportment was most apt for understanding the lives and minds of nonhuman animals

(Canguilhem 2002a). This is because ethology aimed to study the living being in its own milieu, according to its own norms of life. For example, though they both live and nest in trees, the squirrel and the robin inhabit very different milieus. A tree branch offers possibilities for perching, flying, and landing for the bird, whereas for the squirrel it is a narrow route that it is best not to fall off. But Canguilhem also recognized that biology can investigate the living body of this tree; here too, though, the cells that constitute it are to be understood in relation to their specific milieu within the plant.

Whatever the level of biological investigation, the relation between individual and milieu guides the search for new knowledge. Such knowledge does not reduce to physics, because life is underdetermined with respect to universal laws. The living must be grasped immanently, in their own specific activities and milieu. Yet biology, in its two senses as life and knowledge of life, can only be understood against the background of outside forces that produce effects and events in the milieu and life of the living. A thought of the outside is, thus, at work in Canguilhem. This outside is thoroughly physical and cosmological, as he implicitly acknowledges in an early discussion of circadian rhythms (Canguilhem 1989, 175–7).

Before considering the way in which human activity and intention can alter these rhythms, Canguilhem reviews current knowledge of their function, stating:

> The homeothermic animal's circadian temperature rhythm is the expression of a variation in attitude of the entire organism with respect to the environment [*milieu*]. Even when at rest the animal's energy, if it is stimulated by the environment [*milieu*], is not entirely at its disposal, one part being mobilized in tonic attitudes of vigilance and readiness. The state of wakefulness is a behavior which, even without alarms, does not work without costs. (1989, 176)

Here we see that the explanation of this biological function is immanent to the organism itself. And still, though the milieu is in part the work of the organism, it is also an encounter with what is foreign to the organism. Even if human technology can turn night into day and day into night, Canguilhem implies, it is still the case that organisms regulate themselves according to the play of light and darkness that is first patterned by the Earth's rotation on its axis as it orbits the sun. From the perspective of life, though, these physical and cosmological events appear only as elements with an informative role in an immanent biological regulation.

We see an appeal to the outside of life, too, in a later comment made in the context of the new knowledge of life propounded by molecular biology. Canguilhem here registers a shift in his own thinking and now recognizes that organisms, as physical and chemical systems, are subject to the Second Law of Thermodynamics (increasing entropy and an irreversible orientation toward their own dissolution). If this means that life can no longer be thought as "the set of functions that resists death," he insists that "it is at least permissible to say that living beings are systems whose improbable organization slows a universal process of evolution towards thermal equilibrium, that is to say, towards death" (2002b, 261). Here we see Canguilhem conceiving the organism in terms of its outside, but even that physical entropy can only be figured as death, an event for the living. His recognition of the organism as a physicochemical mechanism of self-reproduction subtly insists thereby on defining its inevitable dissolution in entropy from the position of life. Entropy *is* death. Again, here we have an immanent biology accompanied by a thought of a physical, cosmic outside.

Yet this outside is not only ontological. Canguilhem's most concentrated examination of milieu studies it through the different meanings it has supported over the course of its epistemological history. He can be read as showing that this biological concept is open to elaboration in terms of the outside and, further, that its conceptualization helped in the very opening of a thought of the outside. Work on the milieu concept, he suggests, is at the root of the modern mechanistic conception of the infinite universe. Concluding a discussion of milieus as the atmospheres surrounding specific forms of life as their vital sphere, an understanding with roots in Hippocrates, Canguilhem finds that Comte's novel use of the term in the singular, the milieu, made it equivalent with Jean-Baptiste Lamarck's term *circumstances* and Étienne Geoffroy Saint-Hilaire's *ambience*. All three terms point, Canguilhem insists, to an "intuition of a formation around a center" (2008, 103). Yet the widespread adoption and use of the term by the physical sciences led in an opposing direction, in which the milieu is conceptualized as "an indefinitely extendible line or plane, at once continuous and homogeneous, and with neither definite shape nor privileged position [...] milieu [here] does not evoke any relation except that of a position endlessly negated by exteriority" (Canguilhem 2008, 103).

In this theoretical use, the concept of milieu undermines the very possibility of an absolute center. "The now refers to the before; the here refers to its beyond, and thus always and ceaselessly. The milieu is truly a pure system of relations without supports" (Canguilhem 2008, 103).

Canguilhem's discussion thus alludes to Koyré's work on the historical conceptualization of the limitless universe and shows the importance of the physical sciences' milieu therein. Yet Canguilhem's study also shows that the modern scientific physics inaugurated by Newton remained indebted to an experience of the milieu as cosmic order, that is, to an ancient vision of the world as a single, living individual with a center. Through this history, it shows two senses of externality being contested, an outside that undoes any center by endlessly negating it and opening onto a plane, and an outside that is always relative to its center.

As much as his history might seem to favor the latter, the organismic outside, Canguilhem also relied on the concept of the plane to distinguish different sorts of activities undertaken by the human being. For example, he used the plane to distinguish epistemology, as the study of knowledge, from philosophizing about knowledge. Commenting on biology's historical oscillation between theoretical interest in a single, continuous vital substance and interest in the discrete seeds of life, he writes:

> If we transpose this epistemological observation onto the plane of the philosophy of knowledge, we must say, against the empiricist commonplace often uncritically adopted by scientists when they arrive at a philosophy of their experimental knowledge, that theories never proceed from facts. (2008, 31–2)

The discovery of new facts allows for the identification of false theorizations, theoretical errors. But the discovery of novel and inexplicable facts only provides a reason to elaborate a new theory that would improve upon the current one. Facts do not suggest new theories themselves. Every fact, furthermore, comes to light and exists only through a theoretical orientation, that is, from a particular account of the world. In this sense, today's facts depend upon theory that might be false tomorrow. The theory that comes to replace it might, more or less radically, recast what counts as a fact, though Canguilhem insists that history shows that these theories involve a sort of enduring oscillation around two interests, interests that do not concern knowledge alone. In making this argument, Canguilhem asserts that epistemology and philosophy are two different activities pursued on different planes, planes that do not intersect because they are guided by different aims. While epistemology studies how the sciences work, philosophy seeks to understand the meaning of scientific activity and knowledge for life. That science, epistemology, and philosophy are activities pursued on separate planes implies that each can be pursued immanently, even though each also opens onto and calls for the others.

Canguilhem explicitly incorporated the encounter with the unknown, with an outside, into his account of philosophy. He stated this most clearly in his 1943 medical thesis, writing: "Philosophy is a reflection for which all unknown material is good, and we would gladly say, for which all good material must be unknown" (1989, 33). Deleuze and Guattari, of course, would also understand philosophy in relation to what is foreign to it, what they called the Outside, or the plane of immanence. As Deleuze insisted, echoing Canguilhem, thinking only happens in this encounter: "Thinking doesn't come from within, but nor is it something that happens in the external world. It comes from this Outside, and returns to it, it amounts to confronting it" (1995, 110).

This interest in what is foreign to philosophy is why Canguilhem decided to pursue medical studies: to renovate his philosophical training in abstractions by an introduction to concrete, human problems (Canguilhem 1989, 33–4). He sought, therein, a philosophical response to the way experience ceaselessly presents us with the unknown and with unexpected difficulties. So doing, he adopted the definition of philosophy as the science of solved problems (Canguilhem 1989, 35). While we may invent solutions to the lived, concrete problems of ourselves and others, none of them is definitive: no medical treatment and no biological knowledge, no means of doing or knowing, is infallibly efficacious or unquestionably secure. Experience occurs as adventure and requires philosophy as a means of reflecting on our condition and the practical responses with which we attempt to meet it. Only a philosophy that attends to the adventurous character of our experience by reflecting on our continual encounters with uncertainty and the unknown, only such a philosophy will be philosophical. Only such a philosophy can open a sense of new possibilities and the value of experimentation. In an allusion to Canguilhem among others, Deleuze called this focus on the problem and the "problematic," "one of the most original characteristics of modern epistemology" (1994, 323, n. 22).

Canguilhem's *The Normal and the Pathological* (1943) offers perhaps his most famous foray into solved problems and is worth considering briefly. This inquiry examines the history of these two concepts and their medical and scientific definition in order to argue that they are neither merely technical nor scientific. It begins by problematizing a "scientific" account of pathology as the merely quantitative variation on normal biological function or structure. By pursuing the history of this account's different uses, he shows the unavoidable reference to a qualitative dimension of the distinction between normality and pathology (1989, 39–112). The difference between the normal and the pathological is not an indifferent truth, he argues, but arises from and translates lived experiences of sickness and recovery. His problematization thereby reminds the scientifically informed medical practitioner not only that the experience of sickness is at the origin of scientific conceptualizations of health and illness, but also that medical care is always an attempt to help a particular person recover from a concrete problem in their life, that is, in the specific milieus they inhabit. Biological knowledge, for Canguilhem, is sought in response to concrete problems of human life, above all, in medicine's attempts to help the sick recover. Biological concepts are authentically scientific, therefore, insofar as they remain true to concrete experiences of sickness and recovery.

This point extends to other sorts of knowledge too, for scientific activity is always pursued within specific vital and social milieus. In his problematization of any scientific claim to establish the difference between the normal and the pathological, we see Canguilhem pursuing philosophy as he understood it, returning to problems that have found an apparent solution and opening them anew for critical examination by investigating the history of their formation and alteration. Given the scientific insistence on definitively overcoming past errors for the sake of knowing the truth, history is precisely what is foreign to the sciences. Only, therefore, by studying their history—that is, the problems the sciences have encountered and claim to have solved in the present—can their philosophical significance be realized. In this, Canguilhem suggests that the sciences, in their pursuit of knowledge, are like other ways of living. They are, more or less, errant endeavors.

Canguilhem's understanding of experience as adventure connects to his interest in error. For him, "Experience is the attempt whose outcome is presumed without being foreseen, it is an attempt that risks failure. In this way, error, which is going astray as much as mistake, accompanies experience as its shadow" (Canguilhem 2015, 674). Before it is a mistaken explanation, a theoretical error is an uncertain attempt, like other uncertain activities that the living undertake in the face of the unknown. For Canguilhem, the voyage of exploration, far out on the ocean and the sight of land, is the very image of error. If we feel definitively lost when we realize that we are wrong, this image shows that every error in understanding is actually the outcome of a failed intention—some deviation from the course we had set ourselves. We are not erratic beings doomed to wander in a meaningless chaos or void. Rather, on the chaotic sea of life, we encounter specific unknowns with which our chosen course confronts us. If our organism orients us through its creative realization of a valorized milieu, life in society and technical activities also provide us with values and milieus, and images of our world. If we may at times feel ourselves lost in a void, the reality is that we suffer from the proliferation of discordant ways of valorizing and evaluating life.

Canguilhem made a similar point through his 1957 commemoration of Fontenelle, whom he remembers for opening up an infinite horizon for human endeavor in a limitless universe in his *Conversations on the Plurality of Worlds* (1686). What is more, he concludes, "Fontenelle maintains the glory of having made reasonable and stimulating for Modern thought an idea that was absurd and depressing to the Ancients, that of a Humanity without destiny in a Universe without limits" (2002c, 58). We can read Canguilhem's biological philosophy as similarly insisting on the plurality of proliferating vital milieus against a life lost in the void. It is modern biological knowledge that has introduced a need for sensitivity to the different milieus that distinct life forms occupy, even within the same physical space. And the physical sciences, which assert the universality of their knowledge, have to be interpreted in their relation to the world of ordinary human life even as they attempt to rupture with it (Canguilhem 2008, 119). At the very moment that people began to undertake adventures into outer space and assert a cosmic existence, if not eventual mastery of life, death, and the stars, Canguilhem urges his readers to recognize instead the limits and immanence of our endeavors. Whatever our exploits in outer space, humans have to recognize that the universe is *our* cosmos, that is, linked to our particular manner of living as a species. The philosopher, meditating on biology and its cosmological implications, finds reason for modesty about our own experience, endeavors, and accomplishments. There is not a simple physical universe to be known at the base of reality, but rather proliferating material milieus as these emerge in encounters with the unknown.

Canguilhem affirms thus what Deleuze later called "Chaodyssey or Chao-errancy [*Chao-errance*]" (Deleuze 1994, 57; 1990, 264). When Deleuze introduces the term *Chao-errance* it is as an alternative to the coherence (*cohérence*) that we seek when we believe that thought should represent its objects. He coined this neologism and near homophone while attempting to elaborate a new image of thought that escapes this classical image. Rather than a thought that aims to establish the identity of objects in universal representations, Deleuze proposes a thought of difference. This thought is adventurous precisely because it finds chaos in the positive repetition of difference; all that happens, happens immanently, in its own terms, but within the chaos that complicates everything. He and Guattari later found that

> From chaos, Milieus and Rhythms are born. This is the concern of very ancient cosmogonies. Chaos is not without its own directional components, which are its own ecstasies [...] The milieus are open to chaos, which threatens them with exhaustion or intrusion. Rhythm is the milieus' answer to chaos. What chaos and rhythm have in common is the in-between—between two milieus, rhythm-chaos or the chaosmos [...] Chaos is not the opposite of rhythm but the milieu of all milieus. (Deleuze and Guattari 1987, 313)

In their last book together, they find that chaos is the infinite speed of all determinations, it is the immanent proliferation of unconnected differentiation (Deleuze and Guattari 1994, 42). Not absolute entropy, but infinitely differing immanence. Such formulations, I suggest, must be heard in resonance with the terms and tendencies of Canguilhem's thought.

Canguilhem had identified chaos as an ancient philosophical myth, which is not to say that he dismissed it. Rather, he problematized by arguing that we are not erratic beings lost in the void, but errant ones. The animal errs within its specific milieu. If we humans can feel that we are erratic, lost without settled purpose or pattern in an infinite chaos, it is because we are capable of alienation from our vital world. Our search for universal knowledge plays a role in this, but so too would our technical ability to vary the milieus within which we operate. But philosophy has as its task for Canguilhem to reconnect human beings to our milieus, milieus that exist alongside and intertwined with the milieus of other living beings, other forms of life. The vital milieu consists of enduring transformations and alterations—birth, maturation, sex and reproduction, aging, death—within a specific world of social milieus, but always also in relation to other forms of life and their milieus.

Deleuze, then later Deleuze and Guattari, insisted on attention to chaos. Yet they did not conceive it as a philosophical myth, but rather the milieu that emerges between milieus. Thus, they named it *Chao-errance*. And yet it seems that this conception of chaos expresses the implication of Canguilhem's idea of life as a proliferating confusion of different milieus. Deleuze also employed the image of the ocean, as we saw Canguilhem do. Yet whereas Canguilhem insisted on the sea as an image of the inescapability of experience and adventure in life, Deleuze presented an image of chaos, a metaphysical principle of adventuresome experience for all beings; he famously found that

> "Everything is equal!" and "Everything returns!" can be said only at the point at which the extremity of difference is reached. A single and same voice the whole thousand-voiced multiple, a single and same Ocean for all

> the drops, a single clamour of Being for all beings: on condition that each being, each drop and each voice has reached the state of excess—in other words, the difference which displaces and disguises them and, in turning upon its mobile cusp, causes them to return. (Deleuze 1994, 304)

All three thinkers ask us to problematize ready-made solutions, opinions, for the sake of confronting the adventure of experience. Canguilhem thought that art, science, and philosophy were different ways of confronting the adventure of life (Talcott 2019). Deleuze and Guattari argued that the confrontation with chaos is what provokes thought and gives birth to art, science, and philosophy, what they call the three Chaoid sisters (Deleuze and Guattari 1994, 208). Each sister confronts chaos and creates worlds in so doing, thus the chaosmos. Deleuze and Guattari, I think we can say, produced a new sense of chaos and vitality by drawing on Canguilhem's insights into the proliferation of vital milieus in the world. Where Canguilhem sought to dispel chaos as the nightmare of an errant, living being, Deleuze and Guattari argued for a vital chaos in which there is always an order emergent from this very proliferation, a chaosmos. They did not thereby simply give a new name to a preexistent concept. They varied and altered it immanently by seeking to apply it to new problems.

We see this in their idea of geophilosophy, which is about the Earth in relation to itself and, implicitly, about the Earth in relation to the cosmos. In *What is Philosophy?*, this approach begins from the empirical reality of the territories, introducing them via social and animal territories studied by ethologists. In this, they deploy ideas that were close to Canguilhem's interest in ethology, which he studied so as to distinguish different forms of life by considering their relation to the milieu (Canguilhem 1989, 178; 2002a). Following their earlier distinction of milieu and territory (1987, 314), Deleuze and Guattari now take a special interest in the territorial relation to milieu. They write:

> We already know about the importance in animals of those activities that consist in forming territories, in abandoning or leaving them, and even in re-creating territory on something of a different nature (ethologists say that an animal's partner or friend is "the equivalent of a home" or that the family is a "mobile territory"). (1994, 67)

For them, variations on the three movements of claiming a territory, moving out of that territory, and finding a new territory also define the social fields and give rise to "*real types or personae.*" In addition to physical and mental territories, there are also spiritual or philosophical territories. The customs of our society help to shape the conceptual persona that thinks in us, but when, and if, we think we also contribute to inventing a new persona. Whereas Canguilhem treated concepts as specifically scientific ways of confronting the unknown, Deleuze and Guattari assign science the task of slowing chaos down by studying it according to a particular variable. And they embrace the chaos that Canguilhem treated as a philosophical myth as the defining feature of philosophical work, which "struggles with chaos [... taken] as undifferentiated abyss or ocean of dissemblance" (1994, 207). The concept is, they find, philosophy's "own creation [that consists in] setting up an event that surveys the whole of the lived no less than every state of affairs" (1994, 34). Further, for them, "A concept is therefore a chaoid state par excellence; it refers back to a chaos rendered consistent, become Thought, mental chaosmos" (1994, 208). Philosophy is, thus, the investigation of conceptual personae and the creation of concepts. The conceptual persona, which philosophers seek to diagnose, shows "thought's territories, its absolute deterritorializations and reterritorializations" and enables an ethology of thought (1994, 69). Deleuze and Guattari diagnose in abundance, though they also suggest that we have entered a new territory. And here, everything is in question, the new problem for thought is about the person "who believes in the world, and not even in the existence of the world but in its possibilities of movements and intensities, so as to once again give birth to new modes of existence, closer to animals and rocks" (1994, 75). Thus, they conclude, "believing in this world, in this life, becomes our most difficult task, or the task of a mode of existence still to be discovered on our plane of immanence today" (1994, 75). Whereas Canguilhem had pursued a biological philosophy, they contribute to a geophilosophy.

This philosophy resonates with Canguilhem's interest in the possibilities of place for the living. But rather than develop this into a way of distinguishing plant, animal, and human forms of life, Deleuze and Guattari consider the Earth itself in relation to the three territorial movements. For them, the Earth always surpasses any territory. Animal migrations merge with this movement of the Earth. Thus, they write:

> The earth is not one element among others but rather brings together all the elements within a single embrace while using one or another of them to deterritorialize territory. Movements of deterritorialization are inseparable from territories that open onto an elsewhere; and the process of reterritorialization is inseparable from the earth, which restores territories. (1994, 85–6)

In this, they generalize—that is, extend, vary, and alter—the claims that Canguilhem made about the milieu beyond the world of errant living individuals to argue for a dynamic and vital Earth, one that we have the difficult task of believing in today. Given the planetary reach of human activity, they argue, we must elaborate a new thought of the Earth. But to think in terms of the planet, which derives from the Greek *planētēs*, "wanderer," is to place the Earth in a dynamic relation to the cosmos and our uncertain place within it.

Toward a Cosmological Philosophy Today

Cosmology and the sciences confirm this dynamic vision of the Earth in ways that are difficult to comprehend. The Earth is one of a small number of planets in our solar system. But it is only one of at least one hundred billion planets in our galaxy alone (Waller 2013, 82). If the Earth can be explained with reference to the solar system, cosmology investigates the dynamic existence of our planet within vastly larger and unknown realms. The scientific encounter with unknown and potentially unknowable entities, like black holes or dark matter, rebounds upon our thought of the Earth and, thereby, upon our sense of who we are. If these are scientific objects, they are also philosophical problems that give us reason to question ourselves.

All this makes it difficult, I think, to maintain an anthropic principle in which humankind is the pinnacle of the universe's development. I expect Deleuze and Guattari would concur, diagnosing that belief as a transcendental illusion. While such an illusion might hinder new knowledge of the cosmos, it also keeps us from problematizing our own understanding of ourselves and what we earthlings do. A modest version of the principle can promote, by contrast, an immanent approach in which we test what we think we know against what resists our understanding. Pushing the limits of what we can observe and the formulae we can elaborate to describe these observations enables speculation about and guides the investigation of that which remains hidden. That the history of cosmology is laden with great upsets and reversals reminds us that what we think we know of the Earth and ourselves is likely to alter, not because we discover a transcendent realm, but because the world we find ourselves living in *is* different than we thought. And evolving knowledge of the world itself contributes to this alteration insofar as reality is relational. This does not mean that the sciences necessarily alienate us from the world, but rather that in this very possibility they remain true to the dynamic reality of the cosmos, that is, chaosmos. While we can hope to produce knowledge that is true to the cosmos, such understanding leads toward its own eventual alteration.

A modest version of the anthropic principle also directs us back to the conditions in which we exist such that we are able to continue pursuing new knowledge. And this pursuit requires the cultivation of a care for the Earth, the planet on which our existence as observers continues to depend, and its proliferating confusion of vital, technical, and social milieus. Cutting-edge cosmology, like so many human activities today, depends for its existence on massive industrial operations that mine minerals, petroleum, and other materials from within the Earth. Contemporary machinery and electronics depend on such materials and the energy derived from them. Mining operations have caused severe local environmental degradation and pollution (McNeill and Vrtis 2017); and they have contributed to exploitation, violence, and war (Kara 2023). But the contemporary societies that depend on these operations generate pollution, environmental degradation, human violence, and species extinctions on a *planetary* scale. Such damage threatens both the continued existence of different human societies and, more broadly, the Earth's systems and the vital milieus of other living things. The dependence of cosmological science on industry raises, therefore, difficult questions about how we can pursue such science while also cultivating a care for the Earth. Under the penumbra of that problem, perhaps a modest anthropic principle can guide astronomers and cosmologists to have a care for the identification of the conditions within which Earth-like planets have formed to learn about the possible limits of planetary life, including on our planet. Or perhaps contemporary cosmology and concern for knowledge of black holes and dark matter has simply to be interpreted as a philosophical fascination with the very limits of modern, industrial life, however "advanced," and an acknowledgment of its finitude. Cosmology today takes us to the edge of *our* conditions of existence.

Reading Canguilhem with Deleuze and Guattari, this chapter has also suggested another way to cultivate a care for the Earth's existence and multiplication of vital and social milieus. For their shared commitment to experience and immanence, to a thought of the Outside, enabled the latter to take up ideas that Canguilhem had elaborated along these lines, both continuing *and* altering them in the face of other problems. Our comparative reading demonstrates this: by seeking to think consistently—by seeking the consistency of an idea—these philosophers both continued and altered it. It was not by holding onto a belief in the finality and certainty of a concept, but by testing it, that Deleuze and Guattari made a certain progress. Their very divergences from Canguilhem, however, suggest the creative possibilities that arise from the pursuit of immanence. If this reading of Canguilhem with Deleuze and Guattari offers a sketch of a cosmological history of philosophy, it is not because of human grandeur, nor the grandeur of thought. Instead, this history shows thought as an uncertain effort that has always already been undertaken differently by others before us, that can only ever be undertaken by an ongoing multiplicity of efforts to think the unknown. Thought is and becomes cosmological in the enduring and diverging multiplicity of its encounters with the Outside.

References

Bachelard, Gaston. 1969 [1928]. *Essai sur la connaissance approchée*, 3rd ed.. Paris: Vrin.

Bachelard, Gaston. 1970 [1931–32]. "Noumène et microphysique." In *Études d'histoire et de philosophie des sciences*, 11–24. Paris: Vrin.

Bachelard, Gaston. 1984 [1934]. *The New Scientific Spirit*. Translated by Arthur Goldhammer. Boston: Beacon Press.

Canguilhem, Georges. 1989 [1943]. *The Normal and the Pathological*. Translated by Carolyn R. Fawcett. Princeton: Zone Books.

Canguilhem, Georges. 2002a [1960]. "L'homme et l'animal du point de vue psychologique selon Charles Darwin." In *Études d'histoire et de philosophie des sciences*, 115–25. Paris: Vrin.

Canguilhem, Georges. 2002b [1963]. "La constitution de la physiologie comme science." In *Études d'histoire et de philosophie des sciences*, 226–73. Paris: Vrin.

Canguilhem, Georges. 2002c [1957]. "Fontenelle, philosophe et historien des sciences." In *Études d'histoire et de philosophie des sciences*, 51–8. Paris: Vrin.

Canguilhem, Georges. 2008 [1952]. *Knowledge of Life*. Translated by Stefanos Geroulanos and Daniela Ginsburg. New York: Fordham University Press.

Canguilhem, Georges. 2015 [1956]. "Expérience et aventure." In *Œuvres complètes*, vol. 4, 673–6. Paris: Vrin.

Carter, Brandon. 1974. "Large Number Coincidences and the Anthropic Principle in Cosmology." In *International Astronomical Union, Symposium 63: Confrontation of Cosmological Theories with Observational Data*, edited by Malcolm S. Lonhair, 291–8. Dordrecht: Reidel.

Deleuze, Gilles. 1953. *Instinct et institutions*. Paris: Hachette.

Deleuze, Gilles. 1990 [1969]. "The Simulacrum and Ancient Philosophy." Translated by Mark Lester and Charles Stivale. In *The Logic of Sense*, 253–79. New York: Columbia University Press.

Deleuze, Gilles. 1994. *Difference and Repetition*. Translated by Paul Patton. New York: Columbia University Press.

Deleuze, Gilles. 1995 [1990]. *Negotiations: 1972–1990*. Translated by Martin Joughin. New York: Columbia University Press.

Deleuze, Gilles, and Félix Guattari. 1987. *A Thousand Plateaus: Capitalism and Schizophrenia*. Translated by Brian Massumi. Minneapolis: University of Minnesota Press.

Deleuze, Gilles, and Félix Guattari. 1994. *What is Philosophy?*. Translated by Hugh Tomlinson and Graham Burchell. New York: Columbia University Press.

Dosse, François. 2010. *Gilles Deleuze and Félix Guattari: Intersecting Lives*. Translated by Deborah Glassman. New York: Columbia University Press.

Fontenelle, Bernard de. 1852 [1686]. *Entretiens sur la pluralité des mondes*. In *Œuvres de Fontenelle*. Paris: Eugène Didier.
Kara, Siddharth. 2023. *Cobalt Red: How the Blood of the Congo Powers Our Lives*. New York: St. Martin's Press.
Koyré, Alexandre. 1957. *From the Closed World to the Infinite Universe*. Baltimore, MD: Johns Hopkins University Press.
Lecourt, Dominique. 2006. "Principe anthropique." In *Dictionnaire d'histoire et philosophie des sciences*, edited by Dominique Lecourt, 887–90. Paris: Presses Universitaires de France.
McNeill, John R., and George Vrtis. 2017. *Mining North America: An Environmental History since 1522*, Oakland: University of California Press.
Royal Swedish Academy of Sciences. 2009. Press release. https://www.nobelprize.org/prizes/physics/2009/press-release/. Accessed September 10, 2025.
SNOLAB. 2024. https://www.snolab.ca/about/about-snolab/. Accessed September 10, 2025.
Talcott, Samuel. 2019. *Georges Canguilhem and the Problem of Error*. Cham: Palgrave Macmillan.
Waller, William H. 2013. *The Milky Way: An Insider's Guide*. Princeton: Princeton University Press.

Chapter 15

Modality as Virtuality versus Modality as Possibility: Gilles Deleuze and David Lewis

Ilona Schweitzer

Modality with Gilles Deleuze

Although Deleuze describes himself as an empiricist, he does not take the empirical diversity of spatiotemporal phenomena as primary. Rather, he seeks the conditions of the given, which reside in the virtual: dynamic multiplicities composed of differential relations and intensive potentials. These virtual multiplicities are stable yet generative, and while they themselves remain consistent, their internal tensions and overlaps give rise to new configurations of potentials through individuation, which in turn produce novel actualizations that continually redefine the empirical world.

> I have always felt that I am an empiricist, that is a pluralist. But what does this equivalence between empiricism and pluralism mean? It derives from the two characteristics by which Whitehead defined empiricism: the abstract does not explain, but must itself be explained; and the aim is not to discover the eternal or the universal, but to find the conditions under which something new is produced (*creativeness*). (Deleuze and Parnet 2007, vii, emphasis in the original)

Deleuze analyzes the concept of structure in detail in his 1972 text "How Do We Recognize Structuralism."[1] A structure is a topology of coexistences that are in symbolic relations to one another, whereby the relations are primary, but their empirical occupation is variable. In this sense, structures are symbolic references whose actualization expresses an empirical reality. The term "virtuality" ultimately refers to given, non-contingent dependencies within the basic material equipment, which configure multiple structures and are thus "real without being actual and ideal without being abstract." These structural multiplicities represent the specific relationships between the heterogeneous, which differenciate themselves into the diversity of the empirical when they become actual and concrete. While the structures are timeless because they express reciprocal determinations, they generate time—namely the "multiple internal time" of their actualizations, which are preceded by "static ordinal genesis" within the structures. By differen*t*iating themselves in the virtual (static) and differen*c*iating themselves in the empirical (actual), the ideal/real structures are the only subjects of ontological genesis, quasi a conditioning of differential relations that guides the specificity of their actualizations.[2] Changes in the virtual imply a changed empiricism: what is realized here and now are certain relations and concrete distributions of singularities—others are realized elsewhere or at other times.

Since the structure precedes the empirical and conditions it, the modal as a metaphysical category and the structural as a transcendental category coincide for Deleuze in the concept of the Idea. It stands for virtual multiplicities, that is, continuous multiplicities that persist as structures but change in their spatiotemporal actualizations.

> In all cases the multiplicity is intrinsically defined, without external reference or recourse to a uniform space in which it would be submerged. Spatio-temporal relations no doubt retain multiplicity, but lose interiority; concepts of understanding retain interiority, but lose multiplicity, which they replace by the identity of an "I think" or something thought. Internal multiplicity, by contrast, is characteristic of the Idea alone. A multiple ideal connection, a differential *relation* must be actualized in diverse spatio-temporal *relationships*, at the same time as its *elements* are actually incarnated in a variety of *terms* and forms. The Idea is just defined as a structure. A structure or an Idea is a "complex theme," an internal multiplicity—in other words, a system of multiple, non-localizable connections between differential elements which is incarnated in real relations and actual terms. In this sense, we see no difficulty in reconciling genesis and structure. (Deleuze 1994, 183, emphasis in the original)

When Deleuze addresses modality in *Difference and Repetition*, he locates the term un-thematized in the realm of virtual ideas.[3] This assertion proves to be justified if one considers the isomorphic movement of Spinoza's modes (complication and explication: Deleuze 1992, 175, 184–6) and the Deleuzian ideas (differen*t*iation and differen*c*iation: Deleuze 1994, 209–10). In Spinoza's construction of modality, empirical reality has its origin in a divine reality that Spinoza calls substance (Spinoza 2003, PI, DI–VI, AI; Propositions XV–XVI). All modes of extension and of thought are created there as capacities of God and are only actualized on given occasions, such as in the effect of physical laws, in cognitive processes, or in affective connections between individuals. "When a mode comes into existence, it is determined to do so by a mechanical law that composes the relation in which it expresses itself, which constrains, that is to say, an infinity of extensive parts to enter into that relation" (Deleuze 1992, 212).

The intrinsic differences of the virtual modes in God (their complication) and the gradual differences of the empirical modes (their explication) are not directly related to each other in Spinoza. However, the intensive quantity of a virtual mode and the extensive quantity of its empirical expression are two different modes of existence of one and the same mode (Deleuze 1992, 215). Thus, Spinoza recognizes two types of modal distinction: intensive internal quantities and extensive external quantities of the attributes of God, the latter understood as qualities (Deleuze 1992, 213–14). Deleuze's ideas are structural equivalents of the intensive internal quantities and, as with the extensive expression of the intensive modes, a physics that realizes them belongs to the potentiality of the ideas. Although Deleuze eliminates Spinoza's divine substance as a notion in his concept, he leaves it intact in terms of content as the totality of modes, which he, like Spinoza, conceptualizes twice: virtual as ideas, actual as material manifestations. In this way, Deleuze, like Spinoza, invokes a material continuum with modal conditions.

Deleuze is mainly interested in processes in the virtual; in this conceptually favored place, the problematic complexity of the structural is philosophically addressable for him; here he can show how an ontological becoming could be conceived both locally and empirically (differen*c*iation) as well as structurally and generally (differen*t*iation). The difference of "quantitative intensities"—comparable to the intrinsic modal differences in Spinoza's substance—becomes the condition of structure-internal differen*t*iation for Deleuze. Every actualization (differen*c*iation) of a structure alters its empirical reality, since spatiotemporal dynamics (the dramas of ideas) unfold their power to determine space and time, whereby the "pre-quantitative and pre-qualitative dramatization" realizes the potentiality of the idea (Deleuze 1994, 218, 221).

Leaning into an analytic use of language, it is fair to say that Deleuze necessarily presupposes modality as a structure, but one cannot say that he presupposes it primitively, since a structure consists only of symbolic referential contexts. If one wants to substantiate the necessity of the structure hypothesis, one would need an explanation that does not lie in the realm of the structural. Deleuze offers a concept of difference that refers to potentially differentiated material levels. This is to say that it concerns constitutive relations in a topology of coexistences/co-localizations (Petitot 2004, 59–60). However, the difference hypothesis is not a physical statement, but a metaphysical assertion: Deleuze follows Duns Scotus in the assumption that being must be stated from its difference (Deleuze 1994, 35). In this sense, the difference hypothesis and the structure hypothesis are interdependent: symbolic relations refer both to positional differences and to classes of reciprocal determinations.

Application of the Deleuzian Modality Hypothesis to a Metaphysics of Quantum Physics

Regarding Deleuze's coupled difference/structure hypothesis, it might be interesting to look for structuralist approaches in the contemporary metaphysics of physics and question them about their physical presuppositions. This can be done quickly. Insofar as a particle metaphysics is the basis, it is the physical laws of particle physics that condition the structural relations; insofar as a quantum metaphysics is the basis, it is the laws of quantum mechanics, which are entirely different. In both cases, physicality and structure coincide; that is, mathematical physics conceptualizes the probabilistic operationality of matter at the respective level of abstraction.[4]

Deleuze's structures, on the other hand, express positional relations that differen*t*iate themselves intra-structurally in a first step and only differen*c*iate themselves empirically in a second step. Deleuze's concept of structure is itself a hybrid, namely the fusion of a formalized structuralism that rises above the linguistic, ethnological, and psychoanalytical structuralisms of the late nineteenth and early twentieth centuries, and a biological structuralism using a mathematical calculus.

> Philosophy cannot be undertaken independently of science or art. It is in this sense to constitute a philosophical concept from the mathematical function of differen*t*iation and the biological function of differen*c*iation, in asking whether there was not a statable relation between these two concepts which could not appear at the level of their respective objects. Arts, science and philosophy seemed to us to be caught up in mobile relations in which each is obliged to respond to the other, but by its own means. (Deleuze 1994, xvi)

Regarding the possible applicability to other fields of science, however, the commitment to empiricism remains central for Deleuze—the specific empiricism is primary, its analysis only leads to appropriate conceptualizations:

> One starts with abstractions such as the One, the Whole, the Subject, and one looks for the process by which they are embodied in a world which they make conform to their requirements (this process can be knowledge,

> virtue, history...). [...] Empiricism starts with a completely different evaluation: analysing the states of things, in such a way that non-pre-existent concepts can be extracted from them. (Deleuze and Parnet 2007, vii)

Consequently, it is not possible to apply Deleuze's metaphysics of difference to processes in quantum physics without reflection, as this metaphysics is not based on a solid practice of analyzing the mathematics of quantum physical processes. Deleuze and Guattari have repeatedly pointed out that, although philosophy and empirical science are capable of fertilizing each other heuristically or inspiring each other concretely, philosophy problematizes becoming in virtual space, while the natural sciences prefer metric space. Nevertheless, the two are closely linked in *A Thousand Plateaus*, since the concrete movements of matter constantly undermine the formal matter–form relations, that is, slip away from the algebraic specifications (Deleuze and Guattari 2010, 407–8). Deleuze and Guattari clearly sympathize here with the singular, unpredictable processes of becoming. In this respect, it is plausible to assume that Deleuze's metaphysics of differen*t*iating/differen*c*iating ideas conceptualizes a local, concrete becoming and is not suitable as an umbrella concept for a cosmological becoming.

As a counterpart to the analytical standpoint, however, scientific structuralism, especially ontic structural realism (OSR), is very close to Deleuze's position when it defines modality via structurality:

> Ontic Structural Realism (OSR) is the view that the world has an objective modal structure that is ontologically fundamental, in the sense of not supervening on the intrinsic properties of a set of individuals. According to OSR, even the identity and of individuality of objects depends on the relational structure of the world. (Ladyman and Ross 2007, 130)

In the context of the still young analytical metaphysics, this corresponds to a small revolution, because in contrast to the widespread deterministic modeling of matter as arrangements of physical properties, the indeterminism of matter suggested by quantum physics is now given a structuralist interpretation. Accordingly, the mainstream of analytical metaphysics is accused of cultivating a neo-scholastic worldview, paying homage to common sense and—if at all—only maintaining contact with an already domesticated science. It could therefore not be expected to provide knowledge about the world (see Ladyman 2017, 144).

Nevertheless, in the context of my investigation, the question arises again as to whether modality can be explained in structural terms alone. OSR claims that this is the case and also puts all dispositional strategies for expressing modality in their place—this is still a veiled form of attributive particle metaphysics. However, particles and thus also their dispositions are no longer given in a quantized field, since a quantum field is structured by symmetry laws (cf. French 2014, 264–5, 302). If such a structure is thus determined by intrinsic symmetry laws, then only a symmetry break can change its suchness, its modality. Although potentiality, modality, and structurality are regarded as redundant concepts, a quantum field structure completely determined by symmetry laws has the energy status *zero*, it has no potentiality. Only symmetry breaks caused by interacting structures lead to particle emissions in quantum field theory (cf. Ladyman 2017, 158; Wallace 2011, conclusion). Should we be allowed to say with Deleuze that the potentiality of interacting quantum structures manifests itself in the fact that they differen*t*iate and then differen*c*iate in concrete reciprocal determinations, in this case: emitting particles?

Modality with David Lewis

When David Kellogg Lewis, whose philosophy is primarily focused on the empirically real or on everything that is empirically conceivable, searches for an appropriate semantics for a modal logic, he formally considers a variability of concrete worlds.[5] His philosophical-historical reference here is not Spinoza, to whom he would presumably impute a primitive modality hypothesis included in the concept of God, but David Hume.

Hume is known for his view that the metaphysical idea of necessary connections is illusory, and thus so is any idea of necessity at all. This idea is nothing more than a copy of our impressions of cause and effect:

> The scenes of the universe are continually shifting, and one object follows another in an uninterrupted succession, but the power of force, which actuates the whole machine, is entirely concealed from us and never discovers itself in any of the sensible qualities of body. (Hume 2008, 46–7)

Lewis adheres paradigmatically to the dynamics assumed by Hume. It allows him to think of a principle of unrestricted recombination of physical point masses and to see in this combinatorial variability the condition that worlds can vary at all.[6]

> Humean supervenience is named in honour of the greater denier of necessary connections. It is the doctrine that all there is to the world is a vast mosaic of local matters of particular fact, just one little thing after another. (But it is no part of the thesis that these local matters are mental.) We have geometry: a system of external relations of spatio-temporal distances

> between points, maybe points of spacetime itself. Maybe point-sized bits of matter or aether or fields, maybe both. And at these points we have local qualities, perfectly natural intrinsic properties which need nothing bigger than a point to be instantiated. For short, we have an arrangement of qualities. And that is all. There is no difference without difference in the arrangement of qualities. All else supervenes on that. (Lewis 1986a, ix–x)

Lewis admits that this is a contingent and empirical claim that does not find undivided support in either classical physics or quantum physics. However, as he sees himself in a position to plausibly explain phenomena such as causation, change, persistence, and the laws of nature with the help of Hume's supervenience, he considers it justifiable as a working hypothesis. In particular, he considers it to be a practicable way of tracing proven findings of classical physics back to a single metaphysical basic principle. Causation would then be the form of dependence between counterfactual events and as such the transitivity between asymmetrical arrangements of physical qualities. Laws of nature and persistence would refer to regular arrangements, change and chance to irregular arrangements of physical qualities, whereby Lewis fervently hopes that chance as such might itself supervene on arrangements of qualities (Lewis 1986a, xiv).

Lewis rejects primitive modality and metaphysical necessity, but not the laws of nature and an essential basic configuration of the physical world: "There may be properties of mass, charge, quark colour and flavour, field strength, and the like, and maybe other besides, if physics as we know is inadequate to its descriptive task" (Lewis 1986b, 14). As previously mentioned, equipping the world in this way is a pragmatic decision. Perhaps it is because quantum physics was not yet sufficiently developed to serve as a secure metaphysical guide (Lewis 1986a, xi), or perhaps it is merely a reverence for the orthodoxy of classical physics, which clings to the Aristotelian perspective: "Really, what I uphold is not so much the truth of Humean supervenience as the *tenability* of it. If physics itself were to teach me that it is false, I wouldn't grieve" (Lewis 1986a, xi, emphasis in the original).

Regardless of the ongoing conceptual disputes in the metaphysics of physics and quantum physics, Lewis vigorously defends the purely philosophical claim that all worldly phenomena, without exception, supervene on arrangements of physical qualities (cf. Lewis 1986a, xi). He does not even consider the postulate of possible worlds separated from one another to be necessary to equip the semantics of modal logic with concepts; this could be done better by mathematical means. However, the assumption of possible worlds is helpful to assign a field of application to metalogical considerations in the sense of modal realism (cf. Lewis 1986b, 17). If the world we inhabit is regarded as actual, then all other worlds are unactualized possibilities, and vice versa. According to Lewis, a world is only actual if there are speakers in it who refer to it indexically, that is, referentially (Lewis 1986b, 92–3). In Lewis's modal realism, de dicto modality relates to the truth of propositions across possible worlds. In contrast, de re modality relates to the properties that an individual possesses in other worlds through its counterparts. Since there is no central position of God, quantification is only possible relative to a contingent position in a current world. The statement that this or that modality *de re* exists therefore always refers to counterparts in other possible worlds (Lewis 1986b, 22).

So, if, according to Lewis, concrete worlds supervene on arrangements of physical properties in a multiply superimposed way, then all concrete worlds based on them are thinkable. At the same time, their thinkability gives them a place in a logical space in which they exist unrestrictedly and in isolation from one another. A world in Lewis's sense is therefore a closed context of references and relationships. In this logical space, modality, as far as it concerns human individuals, is conceived of as a variable state, dependent on their belief and conviction systems and on their way of being in the world:

> The content of someone's knowledge of the world is given by his class of *epistemically accessible* worlds. [...] Likewise the content of someone's system of belief about the world (encompassing both belief that qualifies as knowledge and belief that fails to qualify) is given by his class of *doxastically accessible* worlds. (Lewis 1986b, 27, emphasis in the original)

> These individuals are the believer's doxastic possibilities. But they are not different possible ways for the world to be; rather they are different possible ways for an individual to be and many of them may coexist within a single world. (Lewis 1986b, 29)

While individuals inhabiting the same actual world are connected spatially, temporally, and contextually, possible worlds are distinct spatiotemporal entities with different ontological structures. Because David Lewis's concept of modality involves quantification across concrete worlds, modality is ontologically grounded and comparable between different worlds (Lewis 1986b, 3–5, 7–8, 14, 70–1, 86–8).

As constructed as the metaphysics of modal realism might appear to a non-logician, it proves to be practicable for analyzing local worldly entities such as semantics, languages, cultures, systems of knowledge, belief, and conviction. If we allow the knowledge systems to coincide with the sciences, the belief systems with the religions, the conviction systems with the doxa, the meaning systems with the aesthetics and ethics, the languages and cultures with the forms of civilization—then Lewisian metaphysics offers a rich field of application, and the structural neutrality of the logical space guarantees effective protection against hegemonic claims to interpretation. The concept of metaphysics itself acquires a logical connotation and frees itself entirely from philosophical and theological paradigms. What has become so is given a place and a voice in the plenum of world facts. Lewis only needs counterparts in different worlds to show that de re modality is a metaphysical category that provides a semantic framework for empirical individuals. It seems to be his concern to give the proliferating diversity of the empirical a framework that transcends the format of a single autonomous world. Although modal truths vary from world to world, the modal category itself is world-invariant. The same modal structure applies to all actualizations. And the assumption of multiple worlds in which it is realized is a defensible metaphysical hypothesis, a "modal realism" in Lewis's words: "As other worlds are alternative possibilities for an entire world, so the parts of other worlds are possibilities for lesser individuals. Modality *de re*, the potentiality and essence of things, is quantification over possible individuals" (1986b, 8).

Application of the Lewisian Modality Hypothesis to a Metaphysics of Quantum Physics

If Lewis's analysis of what has become so can actually reach its last ascertainable presuppositions, it could become problematic. For the definition of what is physically presupposed is itself the subject of metaphysical disputes, not only in classical physics, but also in the metaphysics of quantum physics. Lewis was aware of this, but he was not interested in the details. On the other hand, judging by the extent of prominent criticism of the dot matrix and supervenience theories, Lewis's views met with great interest in physics.[7] And the methodological isomorphisms in the formation of certain metaphysical theories are unmistakable. A particularly noteworthy example of this is the theory of an emergent multiverse following Hugh Everett's Many Worlds theory. Everett himself advocates the idea of psycho-physical parallelism, that is, the physical measurements are regarded by him as distinct branches of a single universal wave function (Bell 1989, 133). The modified version of Everettism presented by David Wallace, on the other hand, defines quantum worlds—and this is very close to Lewis's thesis of possible worlds—as weakly correlating regions of a quantum world, as particular structures that "supervene" on a basic structure in the broadest sense. Here, "supervene" denotes structural dependence within the quantum state, rather than Lewis's Humean supervenience:

> This, in short, is the Everett interpretation. It consists of two very different parts: a contingent physical postulate, that the state of the Universe is faithfully represented by a unitarily evolving quantum state; and an a priori claim about that quantum state, that if it is interpreted realistically it must be understood as describing a multiplicity of approximately classical, approximately noninteracting regions that look very much like the "classical world." (Wallace 2013, 465)

According to Wallace, the "worlds" of Everettian quantum mechanics are not spatially separate regions, but rather, they are emergent, effectively autonomous branches of the universal quantum state produced by decoherence.[8] Decoherence is a time-dependent, dynamical process resulting from the unitary evolution of entangled quantum systems, not an interaction between distinct empirical domains.[9]

Appeals to a fundamental, non-temporal "structural causality" of entanglement go beyond Wallace's own views and reflect broader metaphysical debates about the nature of the quantum state. How one interprets the ontology of quantum mechanics depends on which metaphysical framework one adopts (realist, structuralist, nomological, etc.). Our conception of how fundamental matter behaves depends on these frameworks. As metaphysical ideas they are speculative, as concrete choices they are intuitive, and, in both cases, habitual or at least guided by thought conventions that express the common habitus of a group. Wallace's theory of the emergent multiverse is part of a spectrum of theories in the current metaphysics of quantum physics. By analogy with Lewis, this spectrum can be described as a logical space to which a theory only gains access if it does not collide with the currently established findings (Maudlin 1995, 7).[10]

At present, three types of theory are regarded as realistic in this sense, that is, as logically and empirically coherent. The first is the De Broglie–Bohm theory, which extends Schrödinger's wave function by further dynamical equations and can allow for nonlocal causation through hidden, that is, non-observable, variables (Bell 1989, 31–3). The second type concerns collapse theories, in particular the GRW theory, named after its authors Ghirardi, Rimini, and Weber, which introduces spontaneous discontinuities into the wave function, each of which leads to discrete events—a piece of matter would be "a galaxy of such events"

(Bell 1989, 205). The third type are the Everett theories, within which Wallace's multiverse theory is considered a promising candidate. The entire spectrum of theories "supervenes" on a quantum state in question as the "initial condition." Depending on how it is defined, the respective theory is designed as a structure of physical space, as world space. Here, too, there is a methodological isomorphism with Lewis: the possibility of thinking generates the possibility of worlds, different thinking prejudices different possible worlds.

Continental or Analytical Metaphysics: Gilles Deleuze or David Lewis?

In the second half of the twentieth century both Deleuze and Lewis ventured a new beginning in metaphysics from a continental and an analytical perspective. Is it a coincidence that both thematize categorial forms of modality? Deleuze under the guiding idea of a *transcendental empiricism* with the assumption of a multiplicity of structural modes as ideas (in *Différence et répétition*, 1968) and Lewis under the label of a *modal realism* with the assumption of a multiplicity of real worlds (Lewis 1986b)? What metaphysical presuppositions does Lewis's assumption of the plurality of worlds imply? Does Deleuze's multiplicity of structures correspond to Lewis's multiplicity of worlds? Would such a metaphysical determination be conceivable as an analogy to the Spinozistic dualism of the virtual modes of God and the empirical modes caused by specific laws?

Although the equivalences are surprising, Lewis refers neither to possible worlds in Leibniz's departure nor to Spinoza's modes, be they virtual or empirically real. His motive is due to an analytic thinking that transgresses toward an epistemic fiction, that is, it is a convenient theory for truth conditions of phenomena in axiomatic Hilbert space. Just as the Hilbert space allows scientific axioms to be connected to each other to establish them as a system of knowledge, Lewis proposes to assume possible worlds formally so as to be able to assign truth conditions to real phenomena.[11] For Lewis, to say X is true means that a real world is conceivable in which X is a fact. For Lewis, to say X is necessary implies that X is a reality in all possible conceivable worlds. To say X is contingent means, accordingly, that X is true only in some worlds, but not in all. To say X is untrue means that there is no conceivable world in which X occurs (Lewis 1986b, 7). The modality statement thus becomes a quantity statement for phenomena in possible worlds (Lewis 1986b, 5, 20).

In contrast, Deleuze's transcendental empiricism is committed to a metaphysical difference. It is differences of intensity that actualize the virtual, the hidden, transcendental dimension of the empirical.[12] In this paradigm, the intensive is the deeper reality, since it problematizes communication between different structures, thereby allowing multiple structural varieties to be actualized.[13] Lewis also attempts to locate the facticity of spatiotemporal phenomena in relations when he thinks of them in a space of real possibilities as the result of changing chains of causation, or discusses them as logical possibilities, as with the assumption of their different situatedness in plural worlds isolated from one another. By proceeding in this way, he assumes a concept of identity, because the phenomenon is 1) identifiable and 2) repeatable.[14] For Lewis, making a difference in today's world means choosing between options:

> This is something like change. We make a difference. But it is not literally change, for the difference we make is between actuality and other possibilities, not between successive actualities. The literal truth is simply that the future depends counterfactually on the present. It depends in part on what we do now. (Lewis 1986a, 38)

It can be ruled out that the option to choose between different real possibilities also refers to the option to change the conditions of these possibilities themselves, that is, to override the structures that generate them. The former option would be a Lewisian solution, the latter a Deleuzian one. For Deleuze is concerned with the genesis of structures—in his paradigm, modal differences insist and cause the symbolically created structures to actualize themselves, that is, to differ from themselves.[15] The openness here does not lie in causal transitivity within the empirical possible, nor in the possibility of thinking plural worlds in which empirical phenomena occur, but in the actualization of structures which, as symbolic relations within a world, vary the empirical possible in that world.

The metaphysical horizons of Deleuze's transcendental empiricism and Lewis's modal realism are not convertible in a first approach: Deleuze reflects the immanence of the structural dimension of empirical phenomena in a given actual world, Lewis the truth conditions of empirical phenomena in the actual as well as in possible worlds. Notwithstanding their philosophical achievements, the two differ in the direction they take. The analytic perspective focuses on the actual and thinks of a structural variance extraterritorially as different worlds; the continental perspective focuses on the immanence of the virtual in the actual as its structure. While Deleuze's approach addresses modal diversity in the variety of actualized structures of a single world, Lewis focuses on modal comparison across different worlds. For Lewis, the causal histories and contexts of a phenomenon, or the worlds in which it occurs, differ, but not the phenomenon itself. To quantify the empirical counterparts in possible worlds, is therefore to think with Lewis, whereas to qualify the structural variation of a single empirical world as becoming is to think with Deleuze.[16]

A metaphysics of the genesis of the modal and a metaphysics of the analysis of the modal aspects of the empirical can certainly be combined if one equates the hinge between them with a change of perspective: here the view of the empirically real of possible worlds (Lewis), there the view of the virtually real of structures (Deleuze). On the one hand, a logical analysis of the empirically possible in the difference to the empirically given (Lewis), on the other hand, the virtuality of structures in the difference to their empirical actualizations (Deleuze). If we agree with both authors that modality is something so-or-so conditioned or so-or-so possible, then we must also grant each of them their own particular philosophical perspective. The genetic speculation of the one meets the speculative logic of the other. Both approaches are heuristically productive, and both thinkers are empiricist fatalists in their own way. Once the modal landscape is exposed, it is already determined and open to analysis. In Deleuze's approach, the old Europe breathes and searches for new structures; in Lewis's approach, a new world is formed and searches for modal liabilities. Both metaphysics are situated in their respective sociocultural contexts and are each specifically motivated on this basis: Becoming through structural change is difficult to convey in a world that is in the process of giving itself structures (Lewis), while merely logically analyzing a structurally ossified world would amount to a philosophical affirmation of powerlessness (Deleuze).

Looking at the modal character of virtual structures and possible worlds, however, implies the metaphysical question of whether—in both conceptions—the same modal preconditions are not merely situated differently, both regarding the virtual variety of structures and the empirical variability of worlds. Insofar as structural variety precedes the empirically possible worlds, it is itself based on a difference-hypothesis that also assumes the "eternal recurrence," that is the repetition of formless differential being, in the structural variations.

> The order of time has broken the circle of the Same and arranged time in a series only in order to re-form a circle of the Other at the end of the series. The "once and for all" of the order is there only for the "every time" in the final esoteric circle. The form of time is there only for the revelation of the formless in the eternal return. (Deleuze 1994, 91)

> Repetition in the eternal return excludes both the becoming-equal or the becoming-similar in the concept, and being conditioned by lack of such a becoming. It concerns instead excessive systems which link the different with the different, the multiple with the multiple, the fortuitous with the fortuitous, in a complex of affirmations always coextensive with the questions posed and the decisions taken. (Deleuze 1994, 115)

For the metaphysics of science, this means providing the civilizational process with instruments of knowledge that relate to the modal preconditions of evolution and civilization. The methodological connection of the philosophical metaphysics of Deleuze and Lewis with the spectrum of the current metaphysics of quantum mechanics, which makes a contemporary cosmology possible, can therefore certainly be regarded as an epistemological success.[17]

Conclusion

This chapter has examined the modal metaphysics of Gilles Deleuze and David Lewis and related their accounts to selected developments in contemporary metaphysics of quantum physics. I have shown that Deleuze understands modal structure as virtual, immanent, and internally differentiated, while Lewis understands modality as empirical diversity distributed across a multitude of concrete possible worlds. Deleuze's conception of becoming as the structural transformation of a single world thus stands in stark contrast to Lewis's modal quantification over empirical counterparts across different worlds. Examining the points of contact between these two frameworks shows that each model exhibits significant structural affinities with contemporary scientific metaphysics. These affinities suggest that both approaches offer valuable heuristic resources for cosmology, even though they operate at different levels of metaphysical abstraction and rely on different explanatory strategies.

Notes

1. "The virtual has an ideality that is proper to it, but which does not merge with any possible image, any abstract idea. We will say of structure: *real without being actual, ideal without being abstract.* [...] To discern the structure of a domain is to determine an entire virtuality of coexistence which pre-exists the beings, objects and works of this domain. Every structure is a multiplicity of virtual coexistence. [...] Of the structure as virtuality, we must say that it is still undifferentiated (c), even though it is totally and completely differential (t). Of structures which are embodied in a particular actual form (present or past), we must say that they are differentiated, and that for them

to be actualized is precisely to be differentiated. [...] Genesis, like time, goes from the virtual to the actual, from the structure to its actualization; the two notions of multiple internal time and static ordinal genesis are in this sense inseparable from the play of structures" (Deleuze 2003, 179–80; emphasis in the original).

2. "As we have seen, every structure is primarily a structurally stable and (self-) regulated system of connections between positional values" (Petitot 2004, 54).
3. The following Deleuze interpretation is based on my dissertation (Schweitzer 2022).
4. When Deleuze and Guattari polemicize in *What is Philosophy?* against scientific functions and perhaps also against the algebraic functions of scientific concepts, they mean that both—despite transcendental mathematics—have empirical levels of reference; even statements about uncertainty and probability still aim at the functionality of matter.
5. In my attempt, as a non-logician, to interpret David Lewis appropriately, I have found Phillip Bricker's (2006) text on Lewis very helpful.
6. Aristotle's theory of categories stands behind the concept of property-laden point masses. In the amalgamation of substance and accident, Aristotle ascribes to each indivisible *this* not only a nominal, but also an essential existence, which is defined by a *quality*, as well as localized by space and time determinations.
7. See Saunders (2003, 132); Sider (2006); Maudlin (2007, 182); Ladyman (2017, 144).
8. "Decoherence is a dynamical process by which two components of a complex entity (the quantum state) come to evolve independently of one another, and it occurs due to rather high-level, emergent consequences of the particular dynamics and initial state of our universe" (Wallace 2010, 62).
9. "So—on pain of rejecting the coherence of the special sciences as a whole—we should accept that unitary quantum mechanics is *already* a many-worlds theory: not a many-exact-worlds theory in which the worlds are *part* of the basic mathematical structure, but an emergent-worlds theory in which the worlds are instantiated as higher-level structures within the basic structure. In this sense, advocacy of the Everett interpretation has come full circle: the rise and fall of many-exact-worlds and many-minds theories has returned us to Everett's original insight that unitary quantum mechanics should be understood as, not modified to become, a many-worlds theory" (Wallace 2013, 474; emphasis in the original).
10. On the plurality of quantum theories, see Wallace (2020).
11. Lewis weighs the advantages and disadvantages of such a procedure and argues in favor of accepting the ontological assumption of isolated worlds to gain a logical space: "As the realm of sets is for mathematicians, so logical space is a paradise for philosophers" (Lewis 1986b, 4).
12. "L'être univoque postule la communication *immédiate* et *transversale* de toutes choses qui, selon leur modalité virtuelle, communiquent par leur différence" (Jacques 2014, 101, emphasis in the original).
13. "In this sense, difference in depth is composed of distances, 'distance' being not an extensive quantity but an indivisible asymmetrical relation, ordinal and intensive in character, which is established between series of heterogeneous terms and expresses each moment the nature of that which does not divide without changing its nature" (Deleuze 1994, 237–8).
14. The identity of a mental phenomenon, for example, is measured by the state of the brain. Lewis professes the hypothesis "that—not necessarily, but factually—every experience is identical with some physical state. Specifically, with a neurochemical state" (1966, 17). Just as a mental state supervenes on

a brain state, so physical properties supervene on a pattern of points defining them, while the laws of nature rise above these properties: "Are the laws, chances, and causal relationships nothing but patterns which supervene on this point by point distribution of properties? Could two worlds differ in their law without differing, somehow, somewhere, in local qualitative character?" (Lewis 1986b, 14).

15. The differences in kind and the gradual differences are expressions of one and the same heterogeneity of being. If all the degrees of difference lie between the two, then beneath them lies the difference of intensity that makes them possible. This is the difference between the virtual modalities/structures (Deleuze 1994, 238–9).
16. In contrasting the metaphysical concepts of Lewis and Deleuze, I have largely followed James Williams's chapter "Deleuze and Lewis: The Real Virtual or Real Possible Worlds?" (2005, 101–28). However, I found it equally productive to leave both schools of thought in their cultural contexts, which define modality differently, and apply them independently to a quantum physics-based metaphysical context.
17. A foundational work on the philosophy of cosmology, published in 2017, provides a definition of this emerging academic discipline in the preface: "Cosmology is the unifying discipline par excellence, combining theories of gravity, thermodynamics and quantum field theory with theories of structure formation, nuclear physics, and condensed matter physics. [...] It draws them together in ways that put pressure on each: whether because, as in quantum mechanics, cosmology is an application in which there is no 'external observer'; or because, as with the standard model of particle physics and general relativity, there is tension between their basic principles; or because as in statistical mechanics, it highlights the extraordinary importance of the initial conditions of the universe for local physics" (Chamcham et al. 2017, x).

References

Bell, John S. 1989. *Speakable and Unspeakable in Quantum Mechanics: Collected Papers in Quantum Philosophy*. Cambridge: Cambridge University Press.

Bricker, Phillip. 2006. "David Lewis: On the Plurality of Worlds and Concrete Possible Worlds." https://websites.umass.edu/bricker/research/. Accessed September 10, 2025.

Chamcham, Khalil, Joseph Silk, John D. Barrow, and Simon Saunders, eds. 2017. *The Philosophy of Cosmology*. Cambridge: Cambridge University Press.

Deleuze, Gilles. 1992. *Expressionism in Philosophy: Spinoza*. Translated by Martin Joughin. New York: Zone Books.

Deleuze, Gilles. 1994. *Difference and Repetition*. Translated by Paul Patton. New York: Columbia University Press.

Deleuze, Gilles. 2003. "How Do We Recognize Structuralism?" Translated by Michael Taormina. In *Desert Islands and Other Texts, 1963–1974*, edited by David Lapoujade, 170–92. Los Angeles: Semiotext(e).

Deleuze, Gilles, and Félix Guattari. 2010 [1987]. *A Thousand Plateaus: Capitalism and Schizophrenia*. Translated by Brian Massumi. London: Continuum.

Deleuze, Gilles, and Claire Parnet. 2007. *Dialogue II: Revised Edition*. Translated by Hugh Tomlinson and Barbara Habberjam. New York: Columbia University Press.

French, Steven. 2014. *The Structure of the World. Metaphysics and Representation*. Oxford: Oxford University Press.

Hume, David. 2008 [1748]. *An Enquiry Concerning Human Understanding.* Oxford: Oxford University Press.
Jacques, Vincent. 2014. *Deleuze pas à pas.* Paris: Éditions Ellipses.
Ladyman, James. 2017. "An Apology for Naturalized Metaphysics." In *Metaphysics and the Philosophy of Science: New Essays*, edited by Mathew Slater and Zanja Yudell, 141–61. Oxford: Oxford University Press.
Ladyman, James, and Don Ross. 2007. *Every Thing Must Go: Metaphysics Naturalized.* Oxford: Oxford University Press.
Lewis, David. 1966. "An Argument for the Identity Theory." *The Journal of Philosophy* 63, no. 1: 17–25.
Lewis, David. 1986a. *Philosophical Papers*, vol. II. Oxford: Oxford University Press.
Lewis, David. 1986b. *On the Plurality of Worlds.* Oxford: Blackwell.
Maudlin, Tim. 1995. "Three Measurement Problem." *Topoi* 14, no. 1: 7–15.
Maudlin, Tim. 2007. *The Metaphysics within Physics.* Oxford: Oxford University Press.
Petitot, Jean. 2004. *Morphogenesis of Meaning.* Translated by Franson Manjali. Bern: Peter Lang.
Saunders, Simon. 2003. "Structural Realism, Again." *Synthese* 36, no. 1: 127–33.
Schweitzer, Ilona. 2022. *Prozessdenken bei Alfred North Whitehead und Gilles Deleuze.* PhD thesis, Universität Hildesheim. https://hilpub.uni-hildesheim.de/entities/publication/cb75f6ac-cd4e-4341-b4e5-f034bbd73ec8/details. Accessed September 10, 2025.
Sider, Theodore. 2006. "Bare Particulars." *Philosophical Perspectives* 20: 387–97.
Spinoza, Baruch. 2003 [1677]. *The Ethics.* Translated by Robert H. M. Elwes. Project Gutenberg. https://www.gutenberg.org/cache/epub/3800/pg3800-images.html. Accessed September 10, 2025.
Wallace, David. 2010. "Decoherence and Ontology." In *Many Worlds? Everett, Quantum Theory, and Reality*, edited by Simon Saunders, Jonathan Barrett, Adrian Kent, and David Wallace, 53–72. Oxford: Oxford University Press.
Wallace, David. 2011. "Taking Particle Physics Seriously: A Critique of the Algebraic Approach to Quantum Field Theory." *Studies in History and Philosophy of Science Part B* 42, no. 2: 116–25.
Wallace, David. 2013. "The Everett Interpretation." In *The Oxford Handbook of Philosophy of Physics*, edited by Robert Batterman, 460–88. Oxford: Oxford University Press.
Wallace, David. 2020. "On the Plurality of Quantum Theories: Quantum Theory as a Framework, and its Implications for the Quantum Measurement Problem." In *Scientific Realism and the Quantum*, edited by Steven French and Juha Saatsi, 78–102. Oxford: Oxford University Press.
Williams, James. 2005. *The Transversal Thought of Gilles Deleuze: Encounters and Influences.* Manchester: Clinamen.

Chapter 16

Toward a Process Cosmology: Deleuze, Whitehead, and Spacetime

Keith Robinson

Every instant is at once a giver and a plunderer...

Bachelard (2013, 8)

In *Deleuze: A Philosophy of the Event*, Francois Zourabichvili claims that "there is no ontology of Deleuze," and argues that "if there is an orientation of the philosophy of Deleuze, this is it: *the extinction of the term 'being' and therefore of ontology*" (2012, 36, 37, emphasis original). Although we agree that Deleuze does aim to substitute becoming for being, this does not "therefore" eliminate ontology or metaphysics. In elaborating on the "no ontology" claim, Zourabichvili interprets this to mean that there is no metaphysical discourse in the "vulgar sense," that is, a discourse that would tell us "what there is of reality," and nor is there a primacy of being over knowledge (that Zourabichvilli attributes to Heidegger or Merleau-Ponty). We agree that there is no ontology in the "old sense" of being as substance with universal categories applicable to what Deleuze calls "sedentary proportionality" and "sedentary distribution" (Deleuze 1994, 284–5). Rather, being appears through the event of its happening with the help of new categories, more "nomadic" and "phantastical" as Deleuze says, closely bound up with their existential coordinates (1994, 285). Indeed, we think that it is precisely because of the conceptual transformations that Deleuze carries out that he is still able to make both metaphysical and ontological claims, claims that are best situated within a process metaphysics and an event ontology.

These conceptual transformations extend to cosmological speculations, and Deleuze's "chaosmos" is, in many respects, close to A. N. Whitehead's cosmology. Whitehead's *Process and Reality: An Essay in Cosmology* is an attempt to "elaborate an adequate cosmology in terms of which all particular topics find their interconnections" (1978, xii). An adequate cosmology must also be coherent, such that, as Whitehead defines it, no entity can be conceived in complete abstraction from the system. Deleuze's chaosmos operates in much the same manner: as a self-organizing system that creatively advances through the immanent construction of its own generative principles. One aspect of their shared cosmology is the claim that we cannot treat the cosmos as one object made up of parts. This is a projection of our macroscopic experience, a projection undermined by, for example, special relativity in its refutation of a cosmic "Now." If there is no cosmic now, then treating the universe as one very large mega-object made up of simultaneously coexisting parts makes little sense. Equally, the idea of many universes, or a "multiverse," composed of many simultaneously coexisting universes cannot be entertained for the same reason. This is a version of Lee Smolin's "cosmological fallacy."[1] Thus, their chaosmos is neither one nor many but what Deleuze calls a qualitative or "fusional multiplicity," the elements of which interpenetrate. This interpenetration is spatiotemporal and forms the basis of the conceptual transformations that ground a process cosmology.

This chapter provides an account of the merging of spacetime in Deleuze and Whitehead and briefly illustrates this with the example of life as one aspect of these conceptual transformations. The first part of this essay is a basic summary of Whitehead's and Deleuze's critical diagnosis of mechanistic and dogmatic notions of space and time, as well as their continuation in some contemporary philosophical and scientific interpretations, including certain interpretations of the spacetime of relativity. The second part outlines Deleuze's and Whitehead's fusional and rhythmic account of spacetime and life. The significance of this lies in its contribution to the loosening up of spatialized and substantialized modes of thinking that continue to operate in contemporary cosmology and cosmogony.

Mechanistic Cosmology: Simple Location and Substance-Quality

Deleuze's and Whitehead's emphasis on timespace is in tension with numerous cosmologists, philosophers, and physicists who reject the reality of time. The rejection of the reality of time has a long history in philosophy and science and, in the modern context, is shaped by Newtonian cosmology. At the base of Newton's cosmology are absolute space and absolute time. Newton's absolute space is a homogeneous medium or container for physical matter. Euclidean in metric and three-dimensional, absolute space is independent of what it contains and is fundamentally empty, qualitatively undifferentiated, and immutable. All points and positions are equivalent and distinguished only by their externality to each other. Any given straight line in absolute space can be extended to infinity and is infinitely divisible, or mathematically continuous. Absolute time is, apart from its one-dimensionality, strictly analogous to absolute space. Space and time here are independent of each other and set within a basically permanent and unchanging cosmological framework with fundamental physical constituents that cannot be created or destroyed, that are discontinuous, discrete, and inert, and governed by invariant physical laws. Gravity operates instantaneously as do inertial forces, and the motions of particles are, on this view, potentially reversible.

Whitehead critically summarizes this mechanistic cosmological picture in terms of two basic features: simple location and substance-quality metaphysics. These are both examples of what Whitehead called the "fallacy of misplaced concreteness." Simple location is, for Whitehead, an answer to the question: What is nature made of?

> The answer is couched in terms of stuff or matter or material—the particular name chosen is indifferent—which has the property of simple location in space and time [...] The characteristic common both to space and time is that material can be said to be *here* in space and *here* in time, or here in space-time, in a perfectly definite sense which does not require for its explanation any reference to other regions of space-time. (Whitehead 1967, 48–9, emphasis original)

Simple location is, then, the idea that something called matter is located in a self-identical present in which times are conceived as instants and spaces are conceived as points fully present to themselves in accordance with the logic of identity. One immediate problem, among others, that Whitehead and Deleuze have with this is that nothing like this exists in nature or our experience of nature. In Whitehead's words, "we deny this immediately given instantaneous present. There is no such thing to be found in nature. As an ultimate fact it is a non-entity. What is immediate for sense awareness is a duration" (1920, 65).

In addition, Whitehead explains that, if matter has existed in any period, then it has existed in any portion of the period. This means that dividing the time does not divide the material, whereas dividing the space does divide the material such that dividing the volume will result in less of the material. This difference leads to the idea that matter is fully itself in any sub-period and so the transition of time is regarded as superficial and inessential. Whitehead writes

> this fact that the material is indifferent to the division of time leads to the conclusion that the lapse of time is an accident, rather than of the essence, of the material [...] Thus the transition of time has nothing to do with the character of the material. The material is equally itself at an instant of time. Here an instant of time is conceived as in itself without transition, since the temporal transition is the succession of instants. (1967, 50)

This is a key point that Deleuze will take up and develop in great detail in his own way in several of his books. In the terms Deleuze uses, extensive quantities (area, length, volume, etc.) can be divided, but intensive quantities (temperature, pressure, etc.) are indivisible, or are what cannot be divided without changing nature. This is a distinction in philosophical modernity that can be traced back at least to Kant, where extensive magnitudes are mathematically continuous, that is, defined by the successive addition of homogeneous units, whereas intensive magnitudes cannot be divided without changing their nature as, for example, in an ordinal series (first, second, third, etc.). In any case, for Whitehead, mechanistic cosmology posits the idea of the most real as a set of extensive magnitudes, a succession of simply located instantaneous configurations of matter. Any given instantaneous configuration implies both the future configuration and the past configuration, which are determinately bound up with each other in a mathematical continuity representative of the universe as a whole. The conceptual persona that best captures all of this is Laplace's demon. If this figure knows the positions and velocities of everything at one instant, then they can determine, through the application of mechanistic laws, complete knowledge of the past and the future. As Bergson would say, in this scheme time is nothing because it does nothing.

The elimination, or rather spatialization, of time as a simply located key feature of modern cosmology is part of what Whitehead calls a "substance and quality metaphysics." Take any given object at some point, and observe its properties. Whitehead's example is something that is "hard and blue and round and noisy" (1967, 52). The object that has these properties is the substance or substratum, and the properties it has are hard and blue, etc. Some of these properties are deemed essential or "primary" for the object to be what it is. Other properties are regarded as accidental and changeable, or "secondary." In the mechanistic scheme, it is the quantitative mass and simple location which are considered primary or essential qualities. What then becomes of the qualities of blue and noisiness? They are considered psychic additions of the mind and not part of the nature of things. They are superficial byproducts produced in the mind by the instantaneous configurations of matter. As Whitehead pithily puts it, "nature gets credit which should in truth be reserved for ourselves: the rose for its scent, the nightingale for its song and the sun for its radiance" (1967, 54).

Although simple location and the substance-quality scheme are pragmatic and remain among the guiding principles of science, they are simplified logical abstractions that exhibit Whitehead's fallacy of misplaced concreteness. In other words, "we have mistaken our abstractions for concrete realities" (Whitehead 1967, 55). These features are not "false," as such, but they work, as Whitehead says, by suppressing what are taken to be "irrelevant details" (1967, 52, 58). Matter and motion as simply located in space and time, and mind, with its subject–predicate structure, are high abstractions producing a bifurcated universe that ignores the supposed irrelevant details. The differences slip through the net. For Deleuze, the differences are canceled in extensity, creating an illusion, what Deleuze (following Kant) calls a "transcendental illusion" (1994, 228). Difference remains implicated in itself just as it is explicated in extension. The illusion is that difference in itself is canceled at the moment that it is explicated in extensive qualities. As with Whitehead, for Deleuze, this is not just a sensory or perceptual illusion but a transcendental physical illusion that relates to entropy and energy in general.

For Deleuze, these abstractions exemplify what he calls a dogmatic image of thought, what thinking itself presupposes. The dogmatic image is an image of what thinking is, a set of assumptions or presuppositions that ground a metaphysics, an ontology, and a cosmology. The critique of this image is key for articulating a chaosmos and a process cosmology. Simple location and substance-quality enable what Deleuze calls recognition, or a "harmonious exercise of all the faculties upon a supposed same object" (1994, 133). An object is recognized when all the faculties (perception, memory, understanding) submit the given (blue, hard, round, noisy, etc.) to a form of identity in the object. This is common sense. And common sense as the form of identity in the object is supported by the unity in a thinking subject, where the thinking subject is one substance among others with qualities. As Deleuze puts it, "it is the identity of the self in the 'I think' which grounds the harmony of all the faculties and their agreement on the form of a supposed same object" (1995, 133). Recognition leads to what Deleuze calls "representation," where the supposed "irrelevant details," in Deleuze's terms difference or sense, are suppressed in favor of the same, the similar, the analogous, and the opposed, all measures of the extent to which an object is adequately represented in the unity of the "I think." Representation is a subjective structure of thinking (subject–predicate) which is presumed to be isomorphic with the objects recognized (the unity of substance-quality). Recognition, common sense, and representation are just some of the "postulates" of Deleuze's dogmatic image of thought. Newtonian cosmology, and mechanism generally, is based on the dogmatic image, a set of commonsense concepts, easily visualized, pictured, and represented in a model with the caveat that nearly all simple sense perceptions (color, sound, flavor, scent, etc.) are relegated to the private and subjective sphere of the perceiving mind. By the end of the nineteenth century Whitehead was arguing that the mechanistic scheme organizes the scientific picture of the world. It is remarkably efficient and without rival.

Relativity, Philosophy, and the Spatialization of Time

The first major blow to the mechanistic, Newtonian cosmological scheme, with its key conceptions of space and time, was Einstein's theory of relativity. Relativistic cosmology applies Riemannian geometry to give an account of the metrical structure of space. In Einstein's universe, space is said to have positive curvature that bends in on itself. With this, Newton's distinction between space and its contents is eliminated, as matter-energy is constitutive of the curvature of spacetime in Einstein's general theory. But even the cosmic spacetime curvature varies with time. The movement of matter becomes the disappearance of spacetime curvature in one region and the reappearance of curvature in another. The observation of "red shifts"—the stretching of wavelengths of light toward the red spectrum understood as a Doppler effect—seems to provide empirical support for this. On this view the cause of galactic red shifts is the famous recession of galaxies from each other and the expansion of the universe. Indeed, there is strong empirical evidence that the rate at which the universe is expanding accelerates at greater distances from us. Now known as Hubble's law, the velocity of the recession is proportional to the distance of the object from us. The expansion of spacetime understood as a continuous increase of the radius of spacetime curvature provides support for a thoroughly processual cosmology, particularly given the role of the expansion of the universe as a kind of universal "first principle" in cosmology.

However, many cosmologists, philosophers, and scientists continue to interpret the fusion of space and time in terms of simple location, resulting in the rejection and spatialization of time and a static cosmology. One important reason for the rejection of time across these groups is that they think the reality of time contradicts Einstein's theory of relativity.

Perhaps a statement from Julian Barbour, an acclaimed physicist and leading cosmologist, can be taken as representative of this group:

> Einstein's general theory of relativity [...] is telling us that time does not exist as an independent thing [...] When the fact that time has no independent existence is combined with the basic facts of quantum mechanics in the simplest possible way, the implications are startling: The quantum universe is static. Only timeless Nows exist. The quantum rules give them different probabilities. We experience the most probable Nows as individual instants of time. The appearance of motion and the flow of time are both illusions created by the very special structure of the instants that we experience. (Durie 2000, 19)

Barbour's view is a classical dualism of a timeless physical world and temporal consciousness whose experiences of motion and the flow of time are illusory psychic additions, and Barbour cites Einstein's general theory (and quantum theory) in support of this. Here Barbour grafts a specific metaphysics, a metaphysics that goes back to Parmenides, onto the science, as if it were somehow obviously derived or derivable from it. In Einstein's relativity, spacetime is understood as a four-dimensional structure (Minkowski spacetime) which is sometimes interpreted, as appears to be the case with Barbour, as simply located. In the spatializing versions of Minkowski spacetime, movement becomes timeless phenomena in four-dimensional space. The metaphysical meaning conferred upon Minkowski's and Einstein's four-dimensional spacetime, where spacetime is presented as static and time just is the fourth dimension of space, is conflated with its mathematical and scientific expression. In the derivation of equivalence between the mathematical expressions and the metaphysical reality they are said to convey, Barbour implies that the fourth dimension is given at least the same attributes as the other three. The main criticism here from a process point of view is that succession amounts to the points of a line all given at once.

Einstein himself is caught up in this problem and plays a key role in these conflicting metaphysical interpretations of relativity. On the one hand, Einstein was aware of the problem of spatializing interpretations, as his discussion with the great French philosopher and historian of science Emile Meyerson demonstrated at the legendary meeting on April 6, 1922. On the other hand, there is Einstein's infamous response to Bergson, at that same meeting: "the time of the philosophers does not exist" (Canales 2015, 5). In the decades that followed, Einstein seems to have moved closer to the spatializing interpretations. The now famous remark about his friend Michel Besso that the distinction between past, present, and future is illusory is supported by Karl Popper's claims in his book *The World of Parmenides*. Popper claims that he in fact discussed this point—that Einstein was a Parmenidean—with Einstein himself, who apparently agreed with Popper's characterization of him. According to Popper, during their conversation he even referred to Einstein as "Parmenides." Popper goes on to say that "Einstein's deterministic cosmology is that of a four-dimensional Parmenidean block universe" (1998, 165–6). Although Carnap's remarks on his conversations with Einstein and the problem of the "Now" seem to push the pendulum back the other way,[2] late Einstein did seem to lean toward the idea of time, as he famously said, as a "stubborn human illusion."

Some philosophers, especially in the tradition of analytic philosophy, also offer similar interpretations. In terms of analytic metaphysical positions on time, Einstein is placed squarely in the "eternalist" camp. Eternalism is usually set up in opposition to presentism, and both positions derive from McTaggart's A and B series, developed in his famous "The Unreality of Time" paper, published in 1908 in the journal *Mind*. The A series corresponds with a psychological experience of succession ordered around the present or the now, from which the past and the future receive their designation as earlier than or later than. Eternalism is the view, closely aligned with McTaggart's B series, that relations between times are equally real and exist permanently, irrespective of temporal locations. On this view the past, present, and future are compelling subjective illusions. Translated into Minkowski spacetime, this is the 4D Block Universe. Perhaps the central argument, or one of them, that links eternalism and relativity is the claim that temporal order is destroyed by relativity, that there is no objective now that is common to all observers, and so change is illusory. This is very close to Barbour's view. For a process view these examples attest to a wider and persistent problem within Western thought: the denial of genuine succession, including its irreversibility, and the refusal to acknowledge the reality of time.

No doubt the spatialization of time and, indeed, the rejection of time are part of our intellectual subconscious, or what Deleuze calls the dogmatic image. As soon as we have acquired the dogmatic habit of converting time into space we form the idea of an "instant" and an instantaneous configuration which is simply the classical or mechanistic view of time as a one-dimensional continuum of point-like instants. Although this is practically useful and has a negligible effect at the macroscopic level, no such instantaneous configuration can be part of the vast spacetimes dealt with in cosmology. What we find in Deleuze and Whitehead is a new image, consistent with relativity and an expanding universe, of a fusing of spacetime with its physical content (matter-energy) into a field of irreversible events. The increase of entropy is at one with the creation of the new, causing the expansion of the universe. This spatiotemporal fusion creates what Whitehead called the "extensive continuum" and Deleuze the "plane of immanence." The image of times and spaces extending over each other with variable degrees of memory and anticipation and varying degrees of patterning and coordination corresponds with the cosmological tendency toward entropy and contraction on one side, and the intensive creation and expansion of the new on the other. In this speculative cosmology, or chaosmos, the positive entropy gradient and the expansion of the universe are part of one irreversible process that began with the bursting open of what Deleuze and Guattari called the "cosmic egg" (1987, 164), and the cosmic microwave background rays are the fossilized remnants of the dynamics of the egg, the disintegration of the original quantum (the so-called Big Bang) into smaller and smaller quanta. On this view spacetime/matter-energy are created *with* the quantum explosion.

Whitehead, Spacetime, and Interstitial Life

The mechanistic and spatialized view leads to what Whitehead calls the "bifurcation of nature," which is the division of reality into the nonhuman causal system of electrons and molecules on the one hand, and the human awareness of warmth and redness on the other. As Whitehead puts it, we must "reject the distinction between nature as it really is and experiences of it [...] Our experiences of the apparent world are nature itself" (1920, 27). The bifurcation of nature has been "disastrous," for philosophy and science, Whitehead argues, not only because it has led to various fallacies, such as the fallacies of misplaced concreteness and the fallacy of simple location discussed earlier, but because "it has transformed the grand question of the relations between nature and mind into the petty form of the interaction between the human body and mind" (1920, 27). Whitehead's metaphysics opens the possibility of a new image of nature and the cosmos that does not rely on a bifurcation of the human and the nonhuman, in favor of the notions of passage, becoming, and process: in short timespace.

Whitehead responds to bifurcation in his later works with his "one genus" theory of dipolar actual occasions designed to circumvent or escape the difficulties associated with the various fallacies embedded in the dualisms and materialisms of the tradition. Whitehead takes his cue in part from Leibniz's famous thesis of the *mens momentanea.*

For Leibniz, matter is "instantaneous mind" without recollection of a preceding moment. The nature of the physical for Leibniz is that at each moment it is external to the past and to the future. Leibniz contrasted this with mind as a present that merges with immediate memory or a mind that repeats the past with a difference. However, by calling matter instantaneous or momentary mind, Leibniz points to, but without taking up, a possible new conception of the relation between the physical and the mental, the inorganic and the organic. Rather than panpsychism, which Whitehead never fully subscribed to, at least if we define panpsychism as a generalization of mind or consciousness, Whitehead's theory ascribes a physical pole to every occasion as well as a more or less recessive mental pole. Thus, experience or perception is a contrast, an integration and synthesis of physical inheritance and a conceptual reaction. Conceptual appetition here should not be identified with consciousness. For Whitehead, consciousness presupposes experience. Consciousness is contingent and derivative, an evolutionary later form of integration. This isn't just a naturalizing of consciousness, but a displacement and downgrading of the human intellect and the capacities usually associated with it (cognition, language, etc.). Moreover, as a fully fledged process theory, the Whiteheadian occasion does not just passively have or undergo experience, is not just a static perceptual experience of the real in the manner of a substance qualified by predicates, but *is* experience, an active experience of passage and becoming between interrelated processes that "influence each other, require each other and lead on to each other" (Whitehead 1966, 157). The world is in the occasion and the occasion is in the world. This is Whitehead's Leibnizian-inspired doctrine of "mutual immanence" (Whitehead 1966, 157), the creative "reciprocal insistence" (Whitehead 1982, 14), as he says, between the occasion and the rest of nature.

Here, Whitehead undermines any sharp separations and ruptures between the human and the nonhuman, or mind and matter, by developing the idea of a "passage of nature" made up of events, occasions, or rhythms that extend over each other. Time and space are the constitutive relations of this passage and this extending or passing over. Rather than the fallacy of simple location that we discussed earlier, Whitehead argues:

> The passage of nature which is only another name for the creative force of existence has no narrow ledge of instantaneous present within which to operate. Its operative presence which is now urging nature forward must be sought for throughout the whole, in the remotest past as well as in the narrowest breadth of any present duration. (1920, 66)

Durations are pasts and futures, times and spaces, that are inclusive and pass over into each other. They are memories and anticipations with varying degrees of what Whitehead calls "recessiveness and dominance" that correlate with the basic factors of experience (1947, 91). In Whitehead's speculative image of nature, developed in numerous texts, there is no need to posit essentially different types of actualities or activities (e.g, human and nonhuman, mind and matter, etc.). Rather, Whitehead posits one type of creative experiencing or vibratory existence in which we find, at one end, a tendency toward conformity with the present and repetition of the immediate past with the merest thread of memory approaching zero; and, at the other end, the inverse tendency: the confrontation between the widening and deepening of memory and the breadth of possibility as the past fills the present in anticipation of the future. Whitehead writes:

> When there is memory, however feeble and short-lived, the average influence of the immediate past, or future, ceases to dominate exclusively. There is then reaction against mere average material domination. Thus, the universe is material in proportion to the restriction of memory and anticipation. (1947, 91)

This image of times and spaces extending over each other with variable degrees of memory and anticipation and varying degrees of patterning and coordination is carried into Whitehead's analysis of groupings of events, or sets of occasions, that are said to form different types of nexus, societies, and "enduring objects." However, Whitehead cautions us not to confuse the real actualities that endure, the societies, with actual occasions or events: "it is this mistake that has thwarted European metaphysics from the time of the Greeks" (Whitehead 1961, 204).

One example of real actualities that endure is, of course, living societies or occasions in which lower levels of experience interpenetrate and are integrated into higher levels all the way up and down—life as a spatiotemporal actuality in the extensive continuum. As Whitehead puts it: "Life is a characteristic of 'empty space' [...] and not of space occupied by any corpuscular society [...] Life lurks in the interstices of each living cell, and in the interstices of the brain" (1978, 105–6). Life does not occupy the space of a corpuscular society as such but runs through its relations and interstices. Life lurks in the spacetimes between material bodies, an originative element that is "canalized" in material bodies and, in its higher forms, is productive of intensities that draw out novelties in the satisfactions of the occasions involved. Life is a "bid for freedom" (Whitehead 1978, 104), freeing itself from the shackles of mere repetition, but, without its canalization in material form, too much originality of conceptual response would spell disaster (Whitehead 1978, 107). Thus, life in its essence gains intensity through freedom of originality, but in its passage through material form it benefits from order. How much freedom or originality, and how much order, condition the grades of structured societies, but without what Whitehead calls "anarchic appetition" lurking in the interstices to counter the relapse into matter, there would be only a race to the grave. Life is a movement from physical order to mental originality, and mental originality is bound in physical form with each bud, drop, or quantum exhibiting a dipolar range of physical and mental tendencies. At one end, matter is a conformity to and repetition of what is with the originative element approaching zero, the same eternal objects repeated with negligible variation, yet retaining a capacity for mentality. "Mere repetition," Whitehead claims, is the "baffling of opportunity" (1958, 23), but, as he insists, it is a capacity for mentality rather than mentality itself, and in this capacity "it *is* mentality" (1958, 33). At the next step, as the past is carried into the present, life lures a potential or difference from matter providing a basis for a reaction. The potential or difference expands with greater activity at the mental pole, generating phases or stages of cyclic variation and novel contrast. In several books Whitehead calls this interstitial or abstract movement "rhythm," and "it pervades all life and indeed all physical existence" (1958, 21). Thus, there is a tendency toward matter where organic life shades off into the inorganic and slow decay; but there is also what Whitehead calls a "counter-agency" that ascends or pushes toward more complex forms of mentation and life. These two tendencies, rhythms, or spatiotemporalities can be followed up or down the life/matter scale, but the more complex forms are able to sustain broader and more intense rhythms, just as the less complex exemplify briefer rhythms approaching zero intensity.

Life is a tendency toward originality or novelty of reaction through a "*capture* of intensity" (Whitehead 1978, 105). We can describe the capture of intensity as a threefold process: first, the occasions (or societies) in becoming appropriate the processes of others that have achieved satisfaction; in a second step, they are fused with unrealized potentials that in a (third) moment produce what Whitehead calls variously "vivid immediacy" (1978, 105) or "immediate self-enjoyment" (1966, 151). The immediacy or enjoyment of the capture of intensity integrates the inheritance of the past with a novelty of definiteness not found in that past, adding depth to the satisfaction. But, in the account in *Modes of Thought*, Whitehead insists on yet another factor: "we must yet add another character to our description of life. This missing characteristic is aim" (1966, 152). The aim is the enjoyment of those data in that particular way, what Whitehead in *Process and Reality* calls "subjective form." This brings out another element: the aim includes that particular conceptual novelty with that data, but it also excludes a "boundless wealth of alternative potentiality" (Whitehead 1966, 151). Life as the capture of intensity requires the exclusion of possibilities and the destruction of other living societies. The achieved or actualized satisfactions are appropriated into the living reality or immediate self-enjoyments of the societies in becoming, which are then absorbed into the "consequent nature" of God. Whitehead sums this up by saying that "life is robbery" (1978, 105), a stealing away of the living present, capturing its intensities and separating it from its power to aim and to act. In its bid for freedom life is always subject to the possible capture of its immediate intensities, necessarily open to the robbery that might alter or destroy it. This is not an accidental feature of the living, but a necessary condition for its becoming.

Deleuze, Spacetime, and Abstract Life

Like Whitehead, Deleuze also cautions us not to confuse actualities with virtual events. Deleuze's own response to bifurcation or transcendence is to describe timespace or multiplicities in terms of a becoming-space of time and a becoming-time of space. Deleuze describes this in his *Bergsonism* book, and repeats it in various books including *Cinema 2* as the division of the present moment into two asymmetrical jets, "one oriented and dilated towards the past and the other contracted, contracting towards the future" (Deleuze 1989, 52). This splitting of time is an ongoing activity which cannot be recomposed from the two jets and added together; nor can the splitting be understood independently of space. The splitting of time is necessarily spatial, since succession tends toward extension for the relation between past and future to be expressed. This opens the possibility of a radical reconceptualization of matter, life, and the cosmos along a spatiotemporal continuum, or what Deleuze and Guattari call "the plane of immanence," the movement of the infinite. The basic principle at work here for Deleuze is that philosophical problems exclusively conceptualized in terms of either spatial or temporal abstractions must be rethought and rearticulated as spatiotemporal multiplicities to avoid all kinds of "illusions" (not unlike Whitehead's fallacies) that surround the plane.

Deleuze sometimes begins, like Whitehead, by citing Leibniz's thesis on matter as instantaneous mind to open a discussion on spacetime. Here Deleuze likes to describe how intensive quantities in their approach to zero are always separated from zero by an infinity of degrees, such that there can be no empty space or empty time and no instantaneous mind; or take the opening pages of chapter 2 in *Difference and Repetition*, where Deleuze explicitly references Leibniz's *mens momentanea*. Although the changes in matter remain the same (from the point of view of the perceiving mind), there is change in the mind that perceives, since it takes a certain time to register the diversity of mental content. Deleuze picks up this notion and considerably deepens it in terms of a repetition for itself that takes place in the mind prior to any active memory or reflection. This is what Deleuze calls a lived or living present, a synthesis of time that contracts the supposed successive independent instants into one another so that the past and the future are not separate instants distinct from the present, but dimensions of the present itself. Every organism is the sum of its contractions, retentions, and expectations and is defined by the problem of "habit," the habits that we contract and that we are; but also for Deleuze, habits are contractions in the contemplating mind: "it is simultaneously through contractions that we are habits, but through contemplation that we contract" (Deleuze 1994, 74). The spatiotemporal rhythms vary "according to the natural contractile range of its contemplative souls" (Deleuze 1994, 77). Deleuze describes rhythms of a present that vary from species, individuals, organisms, and the parts of organisms. Deleuze uses the example of wheat as a contraction of earth and humidity bound to rhythms of contemplations, just as the lily of the field sings the glory of the elements that it contracts in contemplating. Deleuze writes:

> All our rhythms, our reserves, our reaction times, the thousand intertwinings, the presents and the fatigues of which we are composed, are defined on the basis of our contemplations. The rule is that one cannot go faster than one's own present—or rather one's presents. (1994, 77)

As is well known, the living present, or the first passive synthesis of *Difference and Repetition*, is described by Deleuze in other books, especially *The Logic of Sense*, as Chronos. Here Deleuze develops the idea not just of a variable contractile range of the present according to individuals, but of an expanding and contracting cosmic present within which all presents are folded or coiled up. As Deleuze puts it, "whatever is future or past in relation to a certain present belongs to a more vast present which has greater extension or duration. There is always a more vast present that absorbs the past and the future" (1990, 162). God is the extreme end or cosmic circle of all the concentric circles, and what God experiences as present is for me already past and future. The divine present or eternal order gives a unity to the all, and as Deleuze says, it "measures the activity of the cosmic period in which everything is simultaneous" (1990, 163). This is the *totum simul* (as Augustine said), the all at once. But, as Deleuze says, the divine present is not unlimited. Indeed, we could say it is complete because it contains all the other presents. Yet it could also be limitless. The limitless past and future is part of the greatest present such that in its circular encompassing it begins again, measuring off a new cosmic period after the preceding one. Deleuze calls this movement "absolute," a movement that contracts and dilates the relative presents in the play of cosmic periods.

In contrast to the greatest or limitless present, there is also the infinitely small present, or rather, the smallest present is that which is infinitely divided. This is what Deleuze famously calls Aion. Rather than the eternal order of bodies and causes, the eternity of Aion is of the order of the event and the incorporeals that populate it. The infinitely divisible event is eternally what has just happened and what is about to happen—but never what is happening. There is an eternity here but of a different order from the eternity of chronos. Aion is the infinitely divisible present that stretches without limit toward the past and the future in what Deleuze and Guattari call a "non-pulsed time," a floating time of speeds and slownesses. Perhaps the characteristic mark of Aion is its pure indeterminacy and eternal newness or variability, a chaos into which determinations of the present perpetually perish at infinite speed.

Aion and Chronos have been described as two complete and ontologically distinct forms of time. Deleuze also talks in this way in *The Logic of Sense*, especially when he's thinking with the Stoics. Here we get language like "reciprocal exclusion," where each form of time is said to be complete in and of itself. Elsewhere, Deleuze places the temporal syntheses together, intratemporally as he says, with each grounding and ungrounding the other (1994). In *A Thousand Plateaus*, Deleuze and Guattari state that we must avoid what they call an "oversimplified conciliation," or a simple opposition between the forms of time or the planes of which they are an element, even though the illusion of opposition is what they call a "well-founded abstraction" (1987, 267). Rather, they construct a field of polarities (1987, 270) marked by degrees and thresholds. Here we could make distinctions of degrees between the intensive and extensive, the smooth and the striated, or between the numbering number and the numbered number, etc. Indeed, I want to suggest that one can think of the two forms of time as doubles of each other, with each as the inverted mirror image of the other.

With this in mind, it is perhaps in *Cinema 2* that Deleuze offers the clearest expression. Here we have the conception of the past being constituted not after the present that it was, but at the same time, which is essentially what Chronos is or, as Deleuze says, which amounts to the same thing, time has to split itself in two which is the moment of Aion. As I have already suggested, we need to be careful about how we understand this splitting. The splitting is not a division that separates as such into two externally related instants but unites the qualitatively different phases. There is a dynamic continuity and a heterogeneity that, as Deleuze says, "dilates" toward the past and "contracts" toward the future. Deleuze names the point of indiscernibility between these qualitatively distinct phases the "crystal image" (in fact the idea of the crystal of time belongs to Guattari) and, at one point, Deleuze even calls this "Cronos" (with the "h" removed).[3] Cronos and not Chronos. Thus, what we see in the crystal is Cronos (without the "h") where the actual present is doubled by its own contemporaneous virtual past; the past is preserved in itself, and there is a splitting between the present that passes and the past that is preserved. This dividing in two is what Deleuze calls a perpetual "self-distinguishing, a distinction in the process of being produced. [...] What we see in the crystal is time itself, a bit of time in the pure state" (1989, 82).

One of Deleuze's, and Deleuze and Guattari's, best examples of a bit of time in a pure state is life. For them we can live a bit of time in a pure state, what they sometimes call the abstract. As Deleuze says "the abstract is lived experience," a lived experience that represents nothing. What I live is a thousand spatiotemporal lines, plateaus, intensities, or dynamisms that are "completely abstract," and one way that these abstractions are expressed is through what Deleuze, and Deleuze and Guattari, call rhythms. As Deleuze says, "what could be more abstract than a rhythm" (1978, Lecture 3). Abstract rhythms are constitutive of organic and non-organic life. They are variable degrees of contraction and relaxation, territorialization and deterritorialization, that range across the strata of the physical, including the most elementary physical events—as Deleuze says, "one should therefore conceive of microintervals even in the primeval soup" (1986, 63)—as well as all the differential levels of consciousness and beyond. Deleuze and Guattari constantly return to this idea of life—both organic and non-organic—as constituted by intensifying and contracting spacetimes, movements, or rhythms throughout their work.

Much of this analysis is carried through, with little change in terminology, to the conclusion of Deleuze and Guattari's final book *What is Philosophy?* But what we also find in the later texts is that Deleuze, and Deleuze and Guattari, increasingly insist on speeds that exceed one's presents and, indeed, surpass the movements of organic life. Such speeds tend to expand the rhythms or spatiotemporal multiplicities that structure life away from the extensive pole, without ever leaving it behind completely. Rather than presents dividing into more extended or overlapping presents, the present intensifies and its volumes become more and more contracted, interpenetrating, and distended as the tendency toward heterogeneity and novelty becomes more pronounced and as substance and form become more and more destratified. From this perspective, the organism, Deleuze and Guattari claim, is a limitation of life, "that which life sets against itself in order to limit itself" (1987, 503). Not unlike Whitehead's interstitial life, in the deterritorializing movement between organisms an abstract line emerges—a "streaming, spiralling, zigzagging, snaking, feverish line of variation [that] liberates a power of life" (Deleuze and Guattari 1987, 499) that organisms had confined. This non-organic life that grips the world remains a spatiotemporal multiplicity that pushes the abstract rhythms to new expanded thresholds. In the cinema books, Deleuze connects non-organic life to the crystal image and the splitting of time that we described earlier as at the heart of abstract movement. In *A Thousand Plateaus*, Deleuze and Guattari tend to emphasize the spatial destratifications that characterize the smoothing of space, and the abstract lines, "ceaselessly bifurcating and diverging," that cut across assemblages. In all of these cases we see stratifying movements of the inorganic and the organic, and destratifying movements of the organic and the non-organic along a spatiotemporal continuum. In effect, Deleuze and Guattari redefine life as a new type of chronotopia, a constantly varying timespace.

In conclusion, both Deleuze and Whitehead present a critique of traditional conceptions of space and time and offer a new speculative image of spacetime, consistent with relativity, positive entropy, and an expanding universe that can contribute toward a process cosmology. A comprehensive process turn in cosmology could loosen up ingrained conceptual frameworks and provide traction for new questions and approaches. One aspect of this turn is an effort to free our thinking from simply located and spatialized conceptions of time in favor of a new image of spacetime as a rhythmic fusion or chronotopia. In contributing to what Deleuze has called a new metaphysical image of thought and being, we open cosmology to new directions and futures.

Notes

1. Smolin's cosmological fallacy is the idea of applying to the whole of the universe methods and ideas that only apply to part of the universe. This is a version of Whitehead's fallacy of misplaced concreteness and Deleuze's dogmatic image. Smolin is one of the contemporary cosmologists closest to process philosophy. See, for example, Smolin and Unger (2015).
2. Regarding his conversation with Einstein, see Carnap (1963, 37).
3. See, for example, chapters 4 and 5 in *Cinema 2* where Deleuze works out the implications of the perpetual self-splitting of time for modern cinema and distinguishes Chronos and Cronos as an important aspect of that self-splitting (1989, 81).

References

Bachelard, Gaston. 2013. *The Intuition of the Instant*. Translated by Eileen Rizo-Patron. Evanston, IL: Northwestern University Press.
Canales, Jimena. 2015. *The Physicist and the Philosopher: Einstein, Bergson and the Debate that Changed Our Understanding of Time*. Princeton, NJ; Princeton University Press.
Carnap, Rudolf. 1963. *The Philosophy of Rudolf Carnap*, edited by Paul A. Schilpp. La Salle, IL: Open Court.
Deleuze, Gilles. 1978. "Kant: Synthesis and Time." In *The Deleuze Seminars*. https://deleuze.cla.purdue.edu/lecture/lecture-03-6/. Accessed October 8, 2025.
Deleuze, Gilles. 1986. *Cinema 1: The Movement-Image*. Translated by Hugh Tomlinson and Barbara Habberjam. London: Athlone.
Deleuze, Gilles. 1988. *Bergsonism*. Translated by Hugh Tomlinson and Barbara Habberjam. New York: Zone Books.
Deleuze, Gilles. 1989. *Cinema 2: The Time-Image*. Translated by Hugh Tomlinson and Robert Galeta. London: Athlone.
Deleuze, Gilles. 1990. *The Logic of Sense*. Translated by Mark Lester and Charles Stivale. New York: Columbia University Press.
Deleuze, Gilles. 1994. *Difference and Repetition*. Translated by Paul Patton. London: Athlone.
Deleuze, Gilles, and Félix Guattari. 1987. *A Thousand Plateaus: Capitalism and Schizophrenia*. Translated by Brian Massumi. London: Athlone.
Deleuze, Gilles, and Félix Guattari. 1994. *What is Philosophy?*. Translated by Hugh Tomlinson and Graham Burchell. London: Verso.
Durie, Robin. 2000. *Time and the Instant: Essays in Physics and Philosophy of Time*. Manchester: Clinamen.
McTaggart, John. 1908. "The Unreality of Time." *Mind* 17: 456–73.
Popper, Karl. 1998. *The World of Parmenides*. London: Routledge.
Smolin, Lee. 1997. *Life of the Cosmos*. Oxford: Oxford University Press.
Smolin, Lee, and Roberto Unger. 2015. *The Singular Universe and the Reality of Time*. Cambridge: Cambridge University Press.
Whitehead, Alfred N. 1920. *The Concept of Nature*. Cambridge: Cambridge University Press.
Whitehead, Alfred N. 1947. *Essays in Science and Philosophy*. New York: Philosophical Library.
Whitehead, Alfred N. 1958 [1929]. *The Function of Reason*. Boston: Beacon Press.
Whitehead, Alfred N. 1961 [1933]. *Adventures of Ideas*. New York: The Free Press.
Whitehead, Alfred N. 1966 [1938]. *Modes of Thought*. New York: The Free Press.

Whitehead, Alfred N. 1967 [1925]. *Science and the Modern World*. New York: The Free Press.
Whitehead, Alfred N. 1978 [1929]. *Process and Reality*. New York: The Free Press.
Whitehead, Alfred N. 1982 [1919]. *An Enquiry Concerning the Principles of Natural Knowledge*. New York: Dover Publications.
Zourabichvili, François. 2012. *A Philosophy of the Event*. Translated by Kieran Aarons. Edinburgh: Edinburgh University Press.

Sequence 24.4.8 (5 minutes, color, silent, video projection) is part of an art project that was commissioned in anticipation of the total solar eclipse in North America on April 8, 2024. Typically articulated through uninterrupted events, my interest in process art is expanded upon here through editing. Moving through different "opsigns," there are indications of sun, moon, and Earth processes. Flowing material is guided by formal metamorphoses along such lines as color, shape, and number. So, change occurs apart from "sensory-motor schema," and the totality moment of the eclipse is considered as a difference in a multiplicity of interdependent differences. I'm currently reworking another version of this video to explore these concepts.

—Brian Fridge

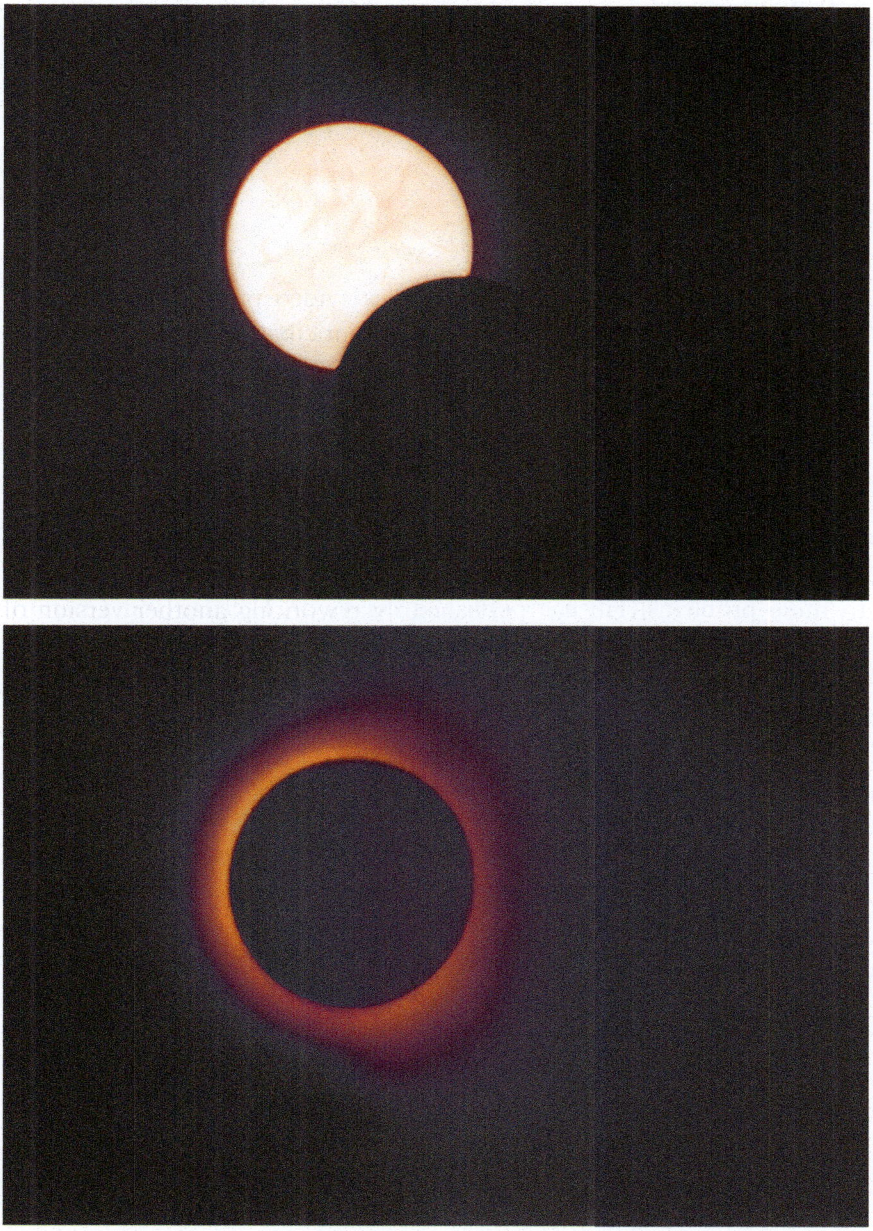

Brian Fridge, *Contact and Totality*, two production stills for "View Finder" project, 5 minutes, color video, 2023–24

Notes on Contributors

Michael J. Ardoline is Assistant Professor of Philosophy at Louisiana State University and former Post-Doctoral Fellow with the Gilles Deleuze and Cosmology Research Group. Michael's research focuses on problems at the intersection of metaphysics, the natural sciences, and mathematics, and how Deleuze and Guattari's philosophy provides novel solutions to these problems. He has published essays on Deleuze and Guattari, formal logic, the philosophy of physics, philosophy of mathematics, and metaphysics more generally.

Alain Beaulieu is Emeritus Professor of Philosophy at Laurentian University (Canada). He has a doctorat de 3e cycle in philosophy from the Université de Paris 8 with a thesis on *Gilles Deleuze et la phénoménologie* (Sils Maria/Vrin, 2nd ed. 2006), supervised by Alain Badiou. He has published *Cuerpo y acontecimiento. La estética de Gilles Deleuze* (Letra Viva, 2012) and *Gilles Deleuze et ses contemporains* (Harmattan, 2011), and has (co)edited the following works: *Gilles Deleuze and Metaphysics* (Lexington, 2014), *Abécédaire de Martin Heidegger* (Sils Maria/Vrin, 2008), *Michel Foucault and Power Today* (Lexington, 2006), *Gilles Deleuze. Héritage philosophique* (Presses Universitaires de France, 2005), as well as *Michel Foucault et le contrôle social* (Presses Universitaire Laval, 2nd ed. 2008).

Martin Calamari teaches modern and contemporary philosophy at a school for adult education in Bolzano (Italy). His research focuses on Deleuze's philosophy, process metaphysics, and the history, philosophy, and metaphysics of science, in particular fundamental physics. He has published in *Deleuze and Guattari Studies* and *Studies in History and Philosophy of Science*.

Russell Duvernoy is Associate Professor of Philosophy at King's University College in London, ON (Canada). He works at the intersection between process philosophies (Whitehead, Deleuze, James, Bergson), critical environmental philosophy, and contemporary continental philosophy, with additional interests in world philosophies (classical Daoist, Mahayana Buddhist, world Indigenous). His monograph *Affect and Attention after Deleuze and Whitehead: Ecological Attunement* (Edinburgh University Press, 2020) explores how a process metaphysical conception of subjectivity alters the experience of affect to argue that manner, style, and quality of attention have ontological consequences. He has published articles in a wide range of scholarly journals.

Brian Fridge is an American video artist, exploring light, matter, space, and time through his video installations. He earned a BA in Fine Arts from University of North Texas and a MFA from the University of Texas at Dallas. In 2005 and 2009 Fridge was a resident at CentralTrak, the University of Texas at Dallas artist residency. Exhibitions include the inaugural 2005 edition of the Turin Triennia at Museo d'Arte Contemporanea in Turin, Italy, and the 2000 Biennial Exhibition at the Whitney Museum of American Art in New York.

Mikey Georgeson is a multimedia artist and performative artist lecturer. Webpage: www.mikeygeorgeson.com

Chantelle Gray is a contemporary continental philosopher whose interests span critical algorithm studies, queer and critical feminist theories, experimental music studies, anarchism and continental philosophy, especially the work of Deleuze and Guattari. She is the Editor-in-Chief of the South African Journal of Philosophy, the co-convener of the South African Deleuze & Guattari Studies Conference, and an editorial board member of *Somatechnics*. Her books include *Deleuze and Anarchism*, co-edited with Aragorn Eloff (Edinburgh University Press, 2019) and *Anarchism after Deleuze and Guattari* (Bloomsbury, 2022). She currently works in the Department of Philosophy at the University of Johannesburg (South Africa).

jan jagodzinski is Emeritus Professor of Visual Art and Media Education, University of Alberta in Edmonton, Alberta (Canada), and series editor for *Educational Futures* (Palgrave-Springer). He is the author of twenty books, with a forthcoming book entitled *Pedagogical Encounters and the Post-Future Anthropocene* (Routledge). The recent decade has been spent on post-Anthropocene issues; he is the editor of *Interrogating the Anthropocene: Ecology, Art, Pedagogy, the Future in Question* (Springer-Palgrave, 2018), *Schizoanalytic Ventures at the End of the World: Film, Video, Art and Pedagogy* (Springer-Palgrave, 2019), and *Pedagogical Explorations in a Posthuman Age* (Springer-Palgrave, 2020); co-editor with Jessie Beier of *Ahuman Pedagogy: Multidisciplinary Perspectives for Education in the Anthropocene* (Springer, 2022); and co-editor with M. Paulsen and S. Hawke of *Pedagogy in the Anthropocene: Re-Wilding Education for a New Earth* (Palgrave-Springer, 2022).

Joseph Nechvatal's contemporary art and theory practice engages in the fragile wedding of image production and image resistance. His Open Humanities Press book *Immersion into Noise* (2011) was republished in a second edition in 2022. His book of essays *Towards an Immersive Intelligence* (2009) was published by Edgewise Press. He has also published three books with Punctum Press: *Minóy* (ed.) (2014), *Destroyer of Naivetés* (2015), and *Styling Sagaciousness* (2022). His audio works "Selected Sound Works" (2021) and "The Viral Tempest" (2022) have been published by Pentiments. Webpage: www.nechvatal.net

Bryan Noonan received his BA in Philosophy from Cal-State Fullerton before then completing his MA at Loyola Marymount University in Los Angeles, California. His primary research interests include French post-structuralism broadly, but most saliently the work of Gilles Deleuze. He also has deep interests in the history of metaphysics, medieval philosophy, and environmental philosophy. Outside of philosophy (*if there is such a thing*), Bryan enjoys playing board games, spending time with his family, and playing guitar.

Dorothea Olkowski is Distinguished Professor of Philosophy at the University of Colorado. She is the current Director of Humanities and Director of the Cognitive Studies Program. In addition, she has served as Chair of Philosophy and founding former Director of Women's Studies. She is the author of more than one hundred articles and fourteen books, including her most recent publication, *Deleuze, Bergson, and Merleau-Ponty: The Logic and Pragmatics of Affect, Perception, and Creation* (Indiana University Press, 2021). Her other publications include an edited collection with Julián Ferreyra, *Deleuze at the End of the World: An Argentinian Perspective on the Sources of his Thought* (2020); an edited collection with Eftechios Pirovolakis, *Deleuze and Guattari's Philosophy of Freedom: Freedom's Refrains* (2019); "Continental

Feminist Approaches to Philosophy of Science," in *The Oxford Handbook of Feminist Philosophy* (2021); "Time is Real, Continuity and the Arrow of Time," in *Gilles Deleuze and Time*, ed. Daniel Smith (2021); and "Deleuze's Oedipus," in *All About Father: Psychoanalysis, the Oedipus Complex, and the Modern Family* (2021).

Stefan Paulus is a multimedia artist. Webpage: www.2xb.org

Arkady Plotnitsky is the author ten books, has edited or co-edited eight volumes of essays and journal issues, and published over two hundred and fifty articles on continental philosophy, Romanticism, modernism and postmodernism, the philosophy of mathematics and science, and the relationships among philosophy, mathematics, and science. His most recent book is *Reality without Realism: Matter, Thought, and Technology in Quantum Physics* (Springer/Nature, Heidelberg 2021), and most recent coedited (with E. Haven) collection is *The Quantum-lie Revolution: A Festschrift for Andrei Khrennikov* (Cham, Switzerland: Springer, 2024).

Antoine Renzo is a doctoral student at Institut supérieur de philosophie (ISP)—under the supervision of Marc Maesschalck and Antoinette Rouvroy—as well as teaching assistant at the Université catholique de Louvain's Centre de philosophie du droit (CPDR) in Belgium. He pursues doctoral research on the notions of passage and transition at the intersection of Deleuze–Guattarian philosophy and the first generation of the Frankfurt School.

Macarena Rioseco (PhD, Lancaster University, UK) is an Associate Lecturer in Visual Arts at the Universidad Metropolitana de Ciencias de la Educación (UMCE), Chile. She is a scholar and art practitioner whose work engages practice-as-research methodologies to investigate material creative processes. Her research seeks to understand the role of artistic practices in the production and circulation of affective regimes of knowledge, and their potential contributions to contemporary art and education. Grounded principally in posthumanism and non-representational frameworks—including the work of Deleuze and Guattari and the new materialisms—her practice and scholarship explore the entanglements of matter, affect, and creative enquiry. She has presented her work internationally in exhibitions, academic conferences, and publications.

Keith Robinson is Professor of Philosophy at the University of Arkansas at Little Rock. He works on the interconnections between continental philosophy (Deleuze, Foucault, Nietzsche) and process philosophy (Bergson, Whitehead, James).

Debjyoti Sarkar is a PhD candidate at the School of Arts & Aesthetics, Jawaharlal Nehru University, New Delhi. He submitted his MPhil dissertation titled "Making of New Worlds: The Crisis of the Infinite and Late English Renaissance Theatre," on how the notion of an infinite universe has implications in theater. He is interested in the relation between science and the object of theater or art. For his PhD he is working on the notion of the "point" in art and philosophy, with specific focus on André Breton and Gilles Deleuze.

Ilona Schweitzer studied philosophy at the Humboldt University in Berlin and graduated with a thesis "On the Connection between Becoming a Subject and Appropriation," and completed her dissertation at the University of Hildesheim in 2022 on *Process Thought in Alfred North Whitehead and Gilles Deleuze*. She has worked as a lecturer at Berlin Universities of Applied Sciences (Philosophical Anthropology, Social Philosophy, Philosophy of Culture, Ethics, Philosophical Questions). Her interests include the metaphysics of science as well as metaphysics in both analytic and continental philosophy.

Janae Sholtz, Professor of Philosophy at Alvernia University, Director of General Education, Coordinator of Women's and Gender Studies and Research Associate at the University of Johannesburg, is an internationally recognized feminist and Deleuzian scholar, who has published on issues of social justice and gender politics, aesthetics, affect, and desire, and ontology, cosmology, and metaphysics. Sholtz is the author of *The Invention of a People: Heidegger and Deleuze on Art and the Political* (2015); co-editor of *Deleuze and the Schizoanalysis of Feminism* (2019); co-editor of *French and Italian Stoicisms: From Sartre to Agamben* (2020); and co-editor of *Infinite Eros: Deleuze, Guattari, and Feminist Couplings*, a special issue of *Deleuze and Guattari Studies* (2018).

Adrian Switzer is an Associate Professor in the Department of Political Science & Philosophy at the University of Missouri Kansas City (UMKC). He specializes in modern and contemporary continental philosophy and has written on such figures as Spinoza, Kant, Nietzsche, Merleau-Ponty, Foucault, Deleuze, and Derrida.

Ewa Szumilewicz is a PhD researcher in the Institute of Philosophy and

Sociology of the Polish Academy of Sciences. She is a laureate of scientific scholarship of the Polish Minister of Education and others. She is the author of *On the Paradox of Cognition* (Peter Lang, 2021) and a Polish poetical prose book, *In Neverwords* (Oficynka, 2020).

Samuel Talcott is Associate Professor of Philosophy in the Department of Philosophy at Saint Joseph's University, Philadelphia. Having done graduate studies at DePaul University, his work explores relations between continental European thought, medicine, and the sciences. In particular, his research focuses on introducing, interpreting, and extending the French epistemological tradition. His book, *Georges Canguilhem and the Problem of Error*, came out in 2019.

Arnaud Villani (Université de Nice-Sophia Antipolis, France) is a philosopher, poet, Hellenist, and translator of Parmenides and Whitehead, among others. Villani also had an enduring correspondence with Deleuze. He has taught philosophy for many years at the khâgne level (preparatory course for arts section of the École normale supérieure) at Lycée Masséna in Nice. His first book was on Kafka. He then became interested to Gilles Deleuze, to whom he dedicated two books, *La Guêpe et l'orchidée* (Belin, 1999) and *Le Vocabulaire de Gilles Deleuze* (co-ed., CHRI/Vrin, 2002), as well as many articles. He co-translated Alfred North Whitehead's *Process and Reality* (Gallimard, 1995). His recent publications include: *La Terre engloutie. Une philosophie de l'écologie* (Kimé, 2020), *Situations et atmosphères* (Unicité, 2022), *Être avec le sauvage* (La Salamandre, 2022), *L'Énigme de la philosophie grecque* (Les Belles Lettres, 2022), *Vers une pensée-mouvement* (Les Belles Lettres, 2023), *La recherche de l'absolu et le devenir des corps* (Les Belles Lettres, 2024) and *De Hölderlin à nos jours, "Le corps, une grande raison"?* (Les Belles Lettres, 2025).

Gabrielle Vitollo is an artist whose paintings and sculptures critically examine the effects of industrialization on natural environments. Vitollo recently expanded their practice by transforming geometric gaps from their satellite mosaic paintings into three-dimensional stainless steel sculptures. These sculptures incorporate digitally distorted giclée prints of their glacial paintings, challenging traditional notions of the flat rectangle and addressing themes of industrial destruction. Vitollo holds a BFA from the Maryland Institute College of Art and an MFA from NYU Steinhardt. Their awards include a German Fulbright Research Grant and a DAAD Scholarship. Vitollo currently teaches at Montclair State University and Pratt Institute.

Index

aberrant, 9, 56–7, 59
accident/s, 30, 33, 77, 108, 223, 258, 317
acoustic, 53–5, 62–5
actual, 28–9, 33, 43, 98–9, 108, 124–5, 136, 148, 175–6, 182, 183, 185–9, 192, 193, 198–202, 223, 226, 239–41, 245, 251, 263, 266, 274, 297–9, 304, 308, 322, 324, 329
actualization/s, 43, 78, 80, 130, 175, 181–3, 186–8, 201, 240–1, 245, 263, 272–3, 297, 298–9, 308–9; *see also* counter-actualization
aeon, 250–2
aesthetic/s, 4, 9, 53–4, 64, 66–7, 98, 246, 264, 271, 305
affect/s, 23, 33, 47, 75, 77, 79–80, 84, 90, 93, 94n, 132, 156, 216, 226, 240–1, 243, 245, 264, 271
  affected/ing/ive, 5–7, 9, 28, 55, 72–3, 75–6, 78, 80–1, 83–4, 88, 90, 92–3, 146, 157, 163–4, 172–3, 195, 216, 298
  affections, 93, 122, 132
agencement/s, 72, 75–8, 80–5, 88, 90, 94n
agency, 64, 72–7, 93
agential, 72, 75–6
  counter-agency, 325
agent/s, 29, 55, 57, 61, 77–8, 82, 84, 241, 248
Aion/aiôn, 24, 44, 115, 145–7, 158–67, 168n, 328, 329
Aion/Chronos, 9–10, 153–5, 157, 164–6; *see also* Chronos
aleatory, 115, 131, 163–4, 167n
alienation/s, 161, 273, 290
amplitude/s, 176, 211, 220
analogous/ly, 125, 194, 205, 220, 316, 318
Anaxagoras, 261
Anaximander, 37, 39, 49, 238, 261
Anaximenes, 37, 39, 261
animal/s, 38, 120, 137, 283–4, 290–2
annihilation, 139, 147, 219
anthropocene, 8, 236
anti-Cartesian/ism, 106, 129; *see also* Cartesianism; Descartes
anti-Platonists, 38; *see also* Plato
Antiphon/ian, 39–41, 48–9
apeiron, 39–41, 238, 261
apparatus/es, 57, 66, 140, 142–4, 176, 244–6
architecture, 102, 202–3, 210, 261
Ariadne, 56
Aristotle, 2, 23–4, 41, 122, 222–3, 311n
  Aristotelian, 223, 227, 230n, 303

arrangement/s, 19, 63, 102–4, 116, 127, 201, 301–4
Artaud, Antonin, 46, 236
arts, 7, 26, 32, 101, 300
  artist/ic, 5–8, 34, 64, 72, 75, 92–3, 105, 207, 236, 244, 264, 267, 270–1
assemblage/s, 5, 54, 58–60, 76, 206, 240, 242–3, 330
astronomers, 280, 294
astronomy, 172, 261, 281
atom/s, 33, 79, 120, 129, 133, 220, 246, 279
  atomic, 79, 246
  atomism, 128, 130, 205
automaton, 172
autonomous, 82, 283, 304–5
autopoiesis, 98, 99, 273–5
  autopoietic, 82, 84, 273–5
Axelos, Kostas 2

Bachelard, Gaston, 42, 262, 269, 280–2
background-dependent, 218–20, 222, 229
background-independent, 216, 219–20, 225
Bacon, Francis, 6, 23, 35n, 109
Badiou, Alain, 115n, 116n, 129
Barad, Karen, 72, 76, 77
Barbour, Julian, 221, 320
Baroque, 1, 45, 101–7, 109–10, 112–15, 119, 128–32
Bartleby, 276
Bayesian, 239
beauty, 105, 123, 142
Beckett, Samuel, 50n, 245–6
becoming/s, 1, 11, 32, 41–2, 45, 47, 57, 59–61, 65–6, 73–7, 82, 84, 90, 98, 103, 110, 114, 128, 130, 133, 146, 149, 163–4, 166–7, 179, 181, 196, 199, 215, 221–4, 228, 241–3, 248, 251–2, 272, 275, 282, 299, 301, 309–10, 314, 322–3, 325–6
Bell's and the Kochen–Specker theorems, 205
Benjamin, Walter, 109–10
Bergson, Henri, 1, 2, 27–31, 43, 47, 65, 223, 237, 239, 243, 245, 317, 320
bifurcation, 322, 326
  bifurcating, 330
Big Bang, 4, 11, 237–9, 250–2, 279, 322
biological, 25, 145, 236–7, 282–5, 287–9, 292, 300; *see also* cosmobiological; neurobiological
biology, 4, 102, 172, 282, 283–6, 289
biosphere, 244
biotechnology, 268
black holes, 4, 25, 84, 140, 214, 248–51, 279, 281, 293–4
Bohm, David, 248–9, 306
Bohmian mechanics, 178, 184
Bohr, Niels, 193, 239
Borges, Jorge Luis, 115
Born's rule, 210–11
boson, 143
bosonic theory, 143; *see also* Higgs boson
Boulez, Pierre, 54, 65
boundless, 60, 146, 261, 326
brain, 24–34, 45, 193, 198, 236, 243, 246, 248, 324
Bréhier, Émile, 159, 168n
Broglie-Bohm theory, 306
Bruno, Giordano, 2, 103
Butler, Samuel, 10, 118, 126–7

Cage, John, 7
Canguilhem, Georges, 12, 262, 269, 280–95
Carnap, Rudolf 153, 167n, 262
Carroll, Lewis, 38, 46
Carter, Brandon, 281
catastrophe, 23, 25, 166
causality, 123, 202–3, 209, 268, 306
causation, 303, 308
Cavaillès, Jean, 262
celestial, 3, 6, 20–1, 61
Cézanne's modulation, 40
chance, 7, 23, 25, 113–15, 185, 193, 198, 202–3, 239, 243, 303, 306; *see also* random
chaos, 12, 24, 28, 35n, 37, 39–49, 49n, 60, 72, 75, 80, 82–4, 88, 93, 106, 115, 130–3, 141, 180, 185, 192–4, 198–200, 202–4, 206, 212, 239–40, 244–5, 263–4, 267, 270–1, 276, 288–92, 328
  chaos and chaoids, 43
  chaos as randomness, 201, 203–4
  chaos theory, 4–5, 34n, 203–4, 212n
  chaoerrance, 290
  deterministic chaos, 22, 25, 27
  virtual chaos, 92, 185, 192–3, 198–200, 202–3, 240
  see also disorder
chaosmology 11, 204, 215, 227, 228
  chaosmological, 4, 8, 82–3
chaosmos, 1, 5, 12, 37, 47, 49, 51n, 82–3, 93, 114, 204, 212, 228, 238, 240, 243, 264, 290–3, 314–15, 318, 322
  chaosmic, 3, 5, 204, 212, 267
Chronos, 24, 115, 145–7, 158–69, 311n, 327, 331
  Cronos, 311n, 329
  chronotopia, 330
cinema, 27, 30–1, 33, 59, 237, 330
clinamen, 79, 239, 245
clone, 268–73
coexistence/s, 65, 147, 297, 299
cogito, 101, 106
cognition, 66, 152, 274, 323
  recognition, 28–30, 64, 105, 109, 247, 285, 318–19
coherence, 81, 82, 156, 187, 289
Collapse theories, 144, 176, 178–9, 184, 189, 243, 245–6, 248, 306
communication, 57, 98, 133, 176, 243, 247, 270, 308
compossible, 113–14
compossibility, 114
contemplation, 6, 126–7, 131–2, 270, 275, 327
continuum, 12, 32, 34, 59, 64, 72, 82, 93, 139, 147, 149, 216, 227, 264, 299, 321–2, 324, 326, 330
convergence/s, 10, 57, 110, 138, 146, 225, 236
convergent series, 108, 111, 114, 119, 129
Copenhagen interpretation, 177–8, 230
Copernicus/can, 63, 102–3, 261
corporeal/ity, 60, 81, 90, 92, 120, 158, 160–4
cosmic, 1–15, 43, 47, 60, 63, 65–6, 86, 90, 93, 98, 119, 121–2, 124, 129, 130–3, 154, 158–65, 172, 237–8, 240, 242, 249–52, 258, 261, 270,

279–80, 285–6, 289, 315, 319, 322, 327–8
cosmicity, 1, 5–6
cosmogenesis/tic, 53, 130
infra-cosmic, 5
cosmobiological, 159
counter-actualization, 270, 272–3
creation/s, 5, 9, 19, 21–8, 41, 45, 72–3, 76–7, 79, 90, 101–2, 108, 112, 114, 122, 194, 196, 202, 211–12, 219, 239, 247, 264, 272, 292, 322
co-creation/s, 83, 90, 93
creativity, 24, 26–7, 31, 72, 75–6, 82, 93, 125, 237–8, 251–2
self-creation, 73
crystal, 33, 329–30
cybernetic/s, 4, 237, 241, 273

dark energy, 4, 11, 237, 282
dark matter, 4, 11, 251, 281–2, 293–4
dark precursor, 43, 242–3, 246–7
dark universe, 238–9, 242–3, 251–2
Darwin, Charles, 261
Darwinian 69
death, 42, 45, 58, 63, 78, 123, 146, 159, 163, 166–7, 204, 238, 240, 282, 285, 289
decoherence, 175–7, 179–82, 184, 186–8, 190n, 244–5, 248, 306, 311n
density, 64, 78, 139, 176–7, 180–1, 183, 211
depotentialization, 266
Descartes, René, 101, 106–7, 129–30, 193
Cartesian/ism, 107, 129, 147, 223
see also Anti-Cartesianism
desire, 56, 58, 61, 70, 94n
destratification, 330; *see also* stratification
deterritorialization, 6, 55, 62, 83–4, 242, 263, 292, 329; *see also* re-territorialization; territorialization
diagram/s, 11, 236, 240, 243–7
dice-throw, 167, 239, 247
discontinuity, 25, 139, 245
discontinuous, 239, 246, 251, 316
discrete, 81, 132, 158, 161, 201, 204, 205, 220, 286, 306, 316
discreteness, 214, 219, 220
disjunction/s, 60, 66, 133, 138, 161, 163, 246–7
disjunctive-synthesis, 61
disorder/s, 27, 72, 75, 82–3, 88, 93, 192, 193, 198, 202–4, 238–9, 240, 243, 251; *see also* chaos
disordered, 28, 106
disparate/ly, 6, 84, 102, 118, 182
divergence/s, 10–11, 47, 110, 113–15, 133, 138, 163, 270, 294
divine, 113, 123, 158, 161, 298–9, 328; *see also* God
double/s, 33, 61, 66, 84, 102, 148, 161, 165, 224, 237, 258, *259*, 262, 265–6, 273, 329
double-slit experiment, 177, 180
dream/s, 28–30, 59, 102, 141

duration/s, 34, 65, 158, 223, 237–8, 243, 248, 316, 323–4, 327
durational/ess, 158
dynamic/s, 11–12, 22, 31, 39, 66–7, 72–7, 79, 81–4, 86, 90, 93, 119, 130–2, 148, 172, 215–16, 218–20, 224–7, 237, 242, 251, 273, 279, 302, 293, 322, 329
dynamical/ly, 9, 22–4, 27, 184, 186, 214, 216–17, 219–22, 224–8, 306
dynamism/s, 64, 74, 80, 92, 129, 156, 164, 186, 329

earth/ly, 2–3, 19, 21, 38, 42, 54, 62, 65, 67, 68n, 72, 74–8, 83–4, 90, 92, 102–3, 105–6, 119, 128 252, 261, 263, 279, 282, 284, 291–4, 294, 327, *334*
Einstein, Albert, 4, 10, 153, 154–7, 161, 165, 167n, 177, 205, 239, 261, 279, 319–21
electromagnetic/ism, 140, 141, 193, 195, 212n
electron/s, 139, 194, 196–7, 200, 208–9, 240, 282, 322
electronic, 7, 237
embryology, 126
emergence/s, 16, 53, 75–7, 84, 90, 98, 161, 196, 205, 216, 247, 273, 275
emergent, 77, 82, 149, 155–6, 160, 186–7, 216, 222, 250, 252, 273, 274, 291, 305–6
Empedocles, 39
encounter/s/ing, 5, 12, 41, 44–6, 48, 57, 59, 72, 84, 177, 188, 239–40, 264, 281–2, 284, 287–9, 293, 295
energy/ies, 11, 34, 47, 61, 82, 120, 136, 140–1, 143–4, 147–9, 193, 199–200, 209–10, 214–15, 217, 219–20, 237–8, 241–2, 247–8, 250–1, 273–4, 281, 284, 294, 302, 318–19, 322
energetic/izing, 40, 58
see also dark energy
entanglement/s, 61, 63, 73, 77, 82–3, 90, 92–3, 98, 153, 155, 162, 181, 220, 306
entropy, 45, 82–3, 238, 240, 242, 247, 249–51, 273, 285, 290, 318, 322, 330
entropic, 82, 249
negentropy/pic, 11, 45, 242, 249–50
epimaterialism, 16
epiphenomenalism, 16
epistemology, 98, 103, 156, 194, 265, 269, 286–7
epistemological/ly, 203, 222, *258*, 262, 265, 269, 285–6, 310
epistemic/ally, 154–5, 165, 177–8, 184, 223, 265, 269, 304, 307
equilibrium, 28, 82, 157, 238, 285
nonequilibrium, 82
far-from-equilibrium, 246
errant, 288, 290–2
essence/s, 43, 45, 48–9, 75, 79–80, 104, 110, 112, 122, 125, 129, 222–4, 243, 268, 270, 305, 317, 325
eternal return 159, 166, 237, 250, 309

eternal/ly/lism/list, 24, 44, 74–5, 103, 115, 120, 122, 145–6, 159–60, 223–4, 248, 261, 309, 321, 325, 328
ethics/cal, 53, 153, 155, 162, 164, 252, 305
Euclidean, 98, 129, 217–18, 263, 266, 269, 315
  non-Euclidean geometry, 262, 263, 266, 269
Euripides, 50
Everett, Hugh, 305–6
  Everettian, 184, 187
evolution/ary, 4, 12, 22, 25, 32, 75, 93, 176, 178, 186, 219, 220–2, 225, 227, 237, 243–4, 248–51, 285, 310, 323
exhaustion, 45, 245–7, 290
expansion, 249–50, 252, 319, 322
expression/s/ism/ist, 5, 10, 41, 53, 57, 60, 64, 74, 83, 92, 106, 109, 118, 120, 124–8, 167, 241–2, 275, 284, 299, 320, 329

fabulatory, 270
faculties, 64, 133, 318
Fardy, Jonathan, 272
Faucher, Kane, 242
Feynman, Richard, 220, 247
field theory (vector/gravitational/quantum/Higgs field theory), 22, 25, 29, 55, 139, 143, 175, 193, 214–20, 225, 240, 245, 247, 302–3
  transcendental field, 59–60, 112
film, 29, 30, 32–3, 59
finite, 65, 105, 107, 109–11, 123, 141, 143, 145, 160, 199, 210, 219, 237, 271, 280
  finite-dimensional, 210
  finitude, 63, 65, 294
flows, 9, 44–8, 55, 58, 73–4, 78–9, 81–2, 84, 86, 90, 93, 110, 119, 130, 155, 165, 218, 241, 244, 247
fluctuation/s, 40, 130, 219, 250
flux/es, 9, 82, 90, 125, 172, 241, 244, 274–5
fold/ing, 38, 44, 85, 87, 88, 90, 102, 106–8, 113–14, 118–33, 136–49; *see also* superfold; manifold
force/s, 1, 3, 5–7, 16, 19, 21, 31, 34, 43–4, 54–5, 57, 60, 62–7, 68n, 72–86, 90, 92–3, 118–21, 129–32, 140–3, 148, 166, 184, 186, 195, 214, 238–43, 247, 266, 284, 302, 316, 323
formalism/s, 177–80, 184–6, 194, 201, 208, 220
formless/ness, 43, 78, 247, 309
Foucault, Michel, 37, 48, 126, 237, 239, 262, 280
four-dimensional, 139, 216, 218, 220, 320–1
fractal/s/ality, 5, 35n, 131, 138, 269
frequency/ies, 31, 176, 178, 202, 208
Freud, Sigmund, 43, 146
friend/s/ship, 30, 33, 165, 212, 291, 320

galaxy/ies, 3, 54, 140, 237, 251, 293, 306, 319

genesis, 7, 53, 56, 78, 131–2, 239, 247, 262, 298, 308–9
  cosmogenesis, 130
  ontogenesis/tic, 238, 262
  see also morphogenetic
genetic/s, 12, 120, 127, 236, 239, 251, 309
  cosmogenetic, 53
geophilosophy, 3, 279, 291–2
geopolitical, 166
gesture/s, 44, 57, 67, 90, 101, 104, 109, 146, 162–3
God, 1, 103, 105, 107, 109–14, 115n, 120–1, 123, 128, 133, 158, 167n, 198, 238–9, 298–9, 302, 304, 307, 326, 328; *see also* divine
Gödel, Kurt, 20, 141, 150
Goldschmidt, Victor, 159, 168
gravitation, 279
gravitational, 58, 62, 216, 251, 281
  gravitational field, 139, 141, 219–20, 225
  gravitational forces, 142
  gravitational interactions, 219
graviton/s, 142–3
gravity, 21, 79, 140–3, 150n, 193, 195, 203, 206, 214–15, 217, 219, 227–8, 249, 251–2, 279, 316

Hamiltonian, 209, 229
haptic, 64–5, 244
harmony, 5, 40, 83, 90, 102, 113–14, 119, 121, 132–3, 163, 318
  harmonious, *98*, 115, 318
  harmonic/s, 114, 125
Hawking, Stephen, 2, 22, 141, 250–1
heaven/s, 2, 10, 64, 103, 105, 118, 128, 244
heavenly bodies, 19
Heidegger, Martin, 2, 3, 49n, 109, 211, 314
  Heideggerian, 38, 108
Heisenberg, Werner, 150n, 194–5, 201, 206–8, *209*, 210–12, 230n
Heisenberg's principle, 139, 209
heliocentric, 102
Heraclitus, 9, 40, 42, 159
  Heracliteanism, 221
heterogeneity/ties, 26, 66, 80, 74, 80, 85, 166, 238, 311n, 329–30
heteregeneous, 7, 47, 55, 58, 63, 77, 79–80, 82–4, 129–30, 158, 297
hierarchy/ies, 21, 102–3, 114, 227
hierarchical/ly, 23, 102–3, 105, 148, 156
Higgs boson, 197–8, 240; *see also* boson
Hilbert space, 176, 180–1, 207–211, 307
Hippias, 49
Hippocrates, 285
Hölderlin, Friedrich, 39, 49, 50n
hologram theory, 249
holographic principle, 248–9
homogeneous/nization, 74, 79, 82, 129, 218, 285, 315, 317
Hopkins, Gerard Manley, 40
house, 1, 107, 131–2
Hubble, Edwin, 237
  Hubble's law, 319
Hume, David, 126, 203, 223, 302–3
Husserl, Edmund, 2–3, 116n

hybrid, 53, 73, 75–7, 81, 166, 300
  hybridization, 53, 63
hypercube, 233

idealism, 162, 266
identity, 23, 32, 61, 103–4, 106, 113, 122, 163, 181–3, 200, *209*, 210, 224, 250–1, 267–8, 289, 298, 301, 308, 316, 318
imaginary, 33, 59, 145, 183, 209, 240, 245, 249
imagination, 1, 42, 106, 244, *258*
immanence, 72, 75, 127, 215, 218, 223, 225, 228, 240, 252, 263, 265, 267–8, 270–1, 275, 281, 289–90, 294–5, 308, 323
  plane/s of immanence, 3, 6, 7, 9, 45, 47, 73, 98–9, 127, 205, 218, 240, 252, 263–4, 270–1, 287, 292, 322, 326
  cosmic/cosmological immanence, *258*, 281
immanent/ly, 7, 38, 44, 46, 59–60, 72, 74, 76, 90, 93, 121, 140, 148, 160, 165, 201, 217, 220, 224, 228, 239, 241, 243, 245, 265, 267–8, 270, 281, 282, 284–6, 289, 290–1, 293, 315
immaterial, 3, 34, 120, 132
imperceptible, 7, 243, 252
implication/s, 44, 107, 118, 121, 128, 136, 154, 166, 215, 222, 225, 227, 263, 268, 270, 274–5, 279, 289–90, 320
in-between, 43, 57, 59, 252, *258*, 290
incompossible, 113, 184–5
  incompossibility, 133, 140, 184
incorporeal/s, 3, 7, 59, 67n, 132, 146, 158–60, 164–5, 328
indefinite, 41, 42, 45, 105, 247, 261
  indefinitely, 104, 285
indeterminacy, 7, 39, 40, 328
indetermination, 49, 245
  cosmological indetermination, 9, 21
indeterminism, 21, 179, 220, 245, 301
Indigenous/ness, 10, *98*, 156
indiscernibles/bility, 104, 111, 181, 329
individuation/s, 10, 60, 75, 175, 181–2, 186, 188, 241–3, 262–3, 266, 272, 275
infinite, 3, 6–7, 19, 29, 31–2, 34, 39, 45, 47, 60, 65, 83, 97, 101–15, 118–24, 129–33, 140, 142, 158, 160, 165, 192, 198, 199, 210, 219, 224, 237, 240, 246, 271, 279–80, 285, 289–90, 326, 328
  infinite-dimensional, 195, 209
  infinitesimal/ly, 48, 111, 127, 246–7, 263
infinity, 32, 102, 106, 108, 111, 119, 123, 125, 127, 128, 129, 131, 133, 141, 145, 165, 218, 299, 316, 327
informal, 43–6, 66
inhuman, 54, 65, 273
inorganic, 39, 57, 63, 84, 136, 159, 323, 325, 330
instant/s, 44–5, 154, 157–61, 163, 166, 239, 242, 263, 314, 317, 320–1, 327, 329
instantaneous/ly, 34, 45, 263, 316–17, 321, 323, 327
intelligence, 26, 31, 48, 246

intensities, 63, 83, 90, 98, 133, 181–2, 187, 241, 244–5, 292, 299, 324, 326, 329
intensity, 10, *16*, 32, 82, 125, 175–6, 181–3, 189n, 238, 241–2, 307, 325–6
intensive/ly, *10*, 44–5, 54, 63, 83, 118, 121, 127, 130–2, 161, 175, 182, 185, 239, 242, 299, 307, 317, 322, 327–8
interference/s, 8, 14, 177, 195, 200, 220
interstitial, 44, 62, 322, 325, 330
intratemporal/ly, 147, 328
intuition/s, 23, 37, 45, 47, 64, 137, 139, 146–7, 164, 244, 251–2, 285
invariant, 218, 265, 271, 273–4, 316
invention/s, 3, 19, 41, 44, 92, 194–6, 207, 211, 237
invisible, 5–7, 40, 42–4, 46, 67, 90, 245, 252
irrational, 56, 130–1
irreversible/bility, 41, 75, 83, 157, 249, 285, 321–2
isomorphism/s, 180–3, 187–8, 305, 307, 310
  isomorphic, 10, 182–3, 194, 205, 298, 319

Joyce, James, 1, 49, 82, 204
justice, 32, 165

Kafka, Franz, 48
Kandinsky, Wassily, 207
Kant, Immanuel, 2, 203, 266, 276, 317–18
  Kantian 64, 101, 240
Kepler, Johannes, 262
Klee, Paul, 5, 6, 42, 253
Kleist, Heinrich von, 51n
Klossowski, Pierre, 61
Koyré, Alexandre, 103, 280, 286

Lærke, Mogens, 110, 118
Lagrange, Joseph-Louis, 21
Lamarck, Jean-Baptiste, 285
Laplace, Pierre-Simon, 20, 317
Lapoujade, David, 56–7
Large Hadron Collider (LHC) 197
Laruelle, François, 11, 243, 262, 264–72, 275–6
Lautman, Albert, 217–18, 248, 262
Leibniz, Gottfried Wilhelm, 44, 49, 101–14, 115n, 116n, 118–33, 134n, 138, 223, 228, 238, 262, 307, 322–3, 327
  Leibzian/ism, 101, 107, 110, 113–14, 119, 122, 124, 128, 228, 323
Lewis, David, 297, 302–10, 311n
limit, 24, 35, 39, 56–7,130, 148, 152, 158, 184, 194, 227, 328, 330
limits, 14, 24, 33, 54, 57–8, 60, 146, 150, 158, 216, 242–4, 254, 289, 293–4
limitless, 160, 286, 289, 328; *see also* unlimited
linear/ity, 156–7, 160, 210–11, 237, 247
  rectilinear; curvilinear, 79
local/lly/azible/ization/ity, 26, 59, 63, 78, 220, 242, 247, 263, 294, 299, 301–4
logos, 101, 159
loop quantum gravity (LQG), 142, 214–16, 219, 220–2, 224, 227–8, 249

loop/s, 34, 78, 142–3, 214–15, 227–8, 249
love/lovers, 38, 46, 153
low-energy, 193, 200, 215
Lucretius, 79–81, 90, 245
Lyotard, Jean-François, 243

M-theory, 144
machine/s, 7, 41, 59, 80, 98, 120, 236, 242, 246, 302
  machinery, 209, 294
  machinic, 58
macroscopic, 140, 187, 315, 321
madness, 27, 30, 238
Maldacena paradox, 136, 139, 142–5, 149, 151n, 249; *see also* T-duality
Maldiney, Henry, 69
Mallarmé, Stéphane, 108, 113, 236
Mandelbrot sequence, 131
manifold/s, 22, 29, 127–9, 133, 134n, 143, 216–17, 219, 245, 252, 263
many-worlds theory, 190n, 311n
Marx, Karl, 268, 271
mass/es, *16*, 78, 82, 84, 119, 120, 125, 142, 144, 200, 221, 240, 302–3, 317
  massless, 250–1
material-discursive, 73, 77, 83, 90
material-discursive-affective, 77, 92–3
material-force, 73–4, 77, 79, 90
materialism/s, 73, 162, 322
mathematics, 3, 20, 27, 32, 35n, 58, 102, 108, 111, 113, 128, 130, 138, 194–5, 198–9, 205–7, 210–12, 217, 252, 262, 301
matter-energy, 82, 217, 319, 322
Maturana, Humberto, 12, 273–5
mechanosphere, 244
Meillassoux, Quentin, 266
Melville, Herman, 48, 123
memory, 1, 29–31, 38, 126, 146, 245, 247, 318, 322–4, 327
mental, 27, 29, 31, 33, 161, 193, 199, 291, 302, 323, 325, 327
Merleau-Ponty, Maurice, 246, 314
Messiaen, Olivier, 5
metalanguage, 246–7
metamorphosis/metamorphoses, 42, 48, 51, 58, 61, 63, 90, 334
metaphor/s, 38, 102–3, 105, 138, 148–9, 172
metastable, 84, 242
metastasis, 238, 241–3
microperceptions, 126–7, 132–3
microplastic/s, 73, 81, 82–4
microscopic, 119, 144
Milesian/s, 37, 39, 40, 42
milieu/s, 53–5, 84–5, 242, 243, 270, 274, 282–6, 288–92, 294
mind/s, 5, 27–8, 31, 33–4, 42, 53, 76, *98*, 120, 123, 126, 129, 146–7, 157, 211, 245–6, 280, 283, 317–19, 322–4, 327, 329
Minkowski spacetime, 218–19, 229n, 252, 320–1
Moby Dick, 48, 122–3
mode/s, 60, 65–6, 82–3, 154–5, 181, 298, 264, 292, 298–9, 307, 315
modernity, 106, 112, 317
modulation, 7, 40, 132, 137, 139, 149, 156
molar, 23, 33–4, 84, 90, 127

molecular, 7, 23, 33–4, 84, 127, 285
monad/s, 106–9, 112–14, 121, 124, 126–7, 132–3, 231
monadology, 114, 228
Mondrian, Piet, 207
monstrous, 73, 81–2, 92, 165
moon/honeymoon, 21, 61, 113, 279, 334
morphogenetic, 73, 77
morphological, 146, 251
movement/s, 7, *16*, 19, 20, 24, 28–9, 39–41, 44, 47, 53–7, 59, 61–2, 64, 79, 82, 85, 90, 121, 125, 129, 131, 140, 172, 176, 223, 240, 245–6, 248, 263–5, 270–1, 291–2, 298, 319–20, 325–6, 328–30
multidimensional, 72, 76, 93, 136, 145, 147
multiverse, 4, 12, 237, 305–6, 315
music, 2, 5–7, 9, 53–5, 60, 62–7, 68n, 101–2, 124, 244

n-dimensional, 25, 217, 252, 263
negentropy, 45, 242, 250
neo-Leibnizian, 228
neurological/neurobiological, 26, 27, 236
neuroscience, 9, 26–7
neutrinos, 150n, 196
Newton, Isaac, 19–21, 203, 218, 237, 262, 286, 315, 319
Nietzsche, 2, 37–8, 50n, 78, 105, 113, 115, 133, 137–8, 223–4, 230n, 243
  Nietzschean, 138, 159, 166, 169n, 237, 250
  Nietzsche's grand style, 47
nomad, 24, 55, 66, 78, 82, 134
  nomadic, 54–5, 67n, 83, 112, 114, 314
  nomad science, 74–5
  nomadology, 114
non-chaotic, 200, 203
non-commutativity, 267, 275
non-dynamical, 216, 219
non-Euclidean geometry, 262–3, 266, 269
non-localizable, 126, 183, 298, 306
non-philosophy, 12, 254, 265–76
nonsense, 46, 56–8, 182, 246–7
non-spatiotemporal, 215–16, 220–2, 227–8
non-standard philosophy, 264–5, 270–2, 275
non-terrestrial, 5–6, 62, 72, 92, 93
noncommutative, 195
nonhuman, 72, 93, 98, 156, 283, 322–4

objectile, 103, 106, 110, 127, 132, 137, 148–9
obscure precursor, 61; *see also*, dark precursor
one-dimensional, 136, 147, 263, 321
ontogenesis, 262
  ontogenetic, 238
optical, 29, 31, 65, 244

panpsychism, 323, 323
paradox, 136, 139, 140–1, 161–2, 164, 238, 244–8, 251–2
  paradoxical, 22, 39, 146, 247–8
  paradoxical instant, 163
  see also Maldacena paradox

parallelism, 266, 305
Parmenides/ian, 9, 11, 39, 41, 157, 221, 320–1
Pascal, Blaise, 47
pathological, 287–8
Peirce, Charles Sanders, 31–2, 34, 35n, 245
Penrose, Roger, 11, 241, 244, 248–51
  conformal cyclic cosmology (CCC), 237, 250
  Orch OR theory of consciousness, 246, 248
perception/s, 7, 23, 29–31, 33, 66, 106, 109, 120–1, 125, 127, 130–1, 133, 163, 245, 305, 318, 323
  pure perception, 246
  see also microperceptions
performative, 55, 67
persona/ae 105, 264, 270–3, 291–2, 317
perspectivism, 105, 107, 132–3, 136–9, 142, 144, 148–9, 158, 160, 228
phantasmata, 146
phenomenology, 2,156, 262
philo-fiction, 267, 270
Planck, Max, 193–4
  Planck length/scale 21, 143–5, 214–16
  Planck constant, 194, 209
plane of immanence, 3, 12, 45, 127, 263, 271, 326
planet/s, 24, 62, 73, 78, 195, 279, 282, 293–4
planetary, 3, 20, 22, 293–4
planomenon, 54–5, 62, 67n, 263
Plato/nic, 2, 32, 38, 41, 43, 48, 51n, 60, 157, 159, 221; *see also* Anti-Platonist
Pleasure principle, 146
Plotinus, 39, 108, 126
poet/s, 34, 40, 42, 157, 236, 244
  poetic, 10, 48, 126, 163
Poincaré, Henri, 20–2, 34n, 218
Popper, Karl, 20, 321
positron, 200
potential/ity/ities, 3, 5–6, 61, 66, 72–6, 78, 80, 82, 88, 93, 94n, 103, 157, 168n, 176, 180, 182–3, 185–6, 241–2, 244, 246, 251, 266, 293, 299, 302, 305, 310, 325–6
pragmatics, 10, 155, 157, 165–6
pre-Socratic/s, 38, 42–3, 45, 47–9, 49n, 50n, 205, 211, 261
prehension, 124–6, 128
Prigogine, Ilya, 75, 82–3, 167n
primitive, 39, 119–20, 130, 192, 302–3
probabilities, 177–8, 187, 195, 209–11, 247, 320
probability, 139–40, 176–7, 179–83, 185–7, 200, 202–5, 210–11, 219–20, 239, 241
pseudo-Aristotle, 40, 52
pseudo-Euclidean, 218
pseudo-Riemannian, 216
psychological/psychiatry, 25, 27, 46, 153, 321
Ptolemaic cosmology, 103
pulsation, 55, 65
Pythagoras, 244
  Pythagoreans, 168, 213

quantum foam, 199
quantum gravity/QG, 132–43, 214, 221, 225, 249

quantum physics, 192, 198, 203–6, 209, 211, 212n, 245, 264, 300–1, 303, 305–6, 310, 312n
quark/s, 240, 303
quasi-particles, 185
quasi-subject, 26

radiation, 208, 238, 250–3, 282
Rancière, Jacques, 272
random, 41, 48, 201–5, 247, 251
randomness, 193–4, 198, 201–5, 219–20
  randomly, 45
  see also chance
rationalism, 122, 262
refrain, 6, 38, 54, 62, 182
repetition, 12, 24, 32, 75, 250, 263, 289, 309, 324–5, 327
resistance, 26, 41, 49, 64–5, 77, 149, 240
resonance, 10, 26, 90, 94n, 118, 126, 165, 182, 264, 269, 290
reterritorialization, 242, 263, 292
revolution, 10, 231, 301
  Copernican revolution, 63,103, 200
rhizomatic, 59, 217, 228
rhizome, 47, 50, 252
rhythm, 6, 12, 40, 53, 55, 65, 79, 84, 90, 128, 284, 323, 325, 327, 329–30
  rhythm-chaos 290
rhythmic, 6–7, 54–5, 85, 90, 315, 330
Riemann, Bernhard, 10, 118, 129, 133, 217, 229n, 262–3
Riemannian geometry, 113, 217, 319
Riemannian surface, 252
ritornello, 263, 267
Rovelli, Carlo, 189, 220–2, 225–7, 249, 261
rupture/s, 32, 47, 129, 280–1, 289, 323
Ruyer, Raymond, 126, 133
RWR-type, 194, 204

Sarnel, Romain, 58
Sartre, Jean-Paul, 48
Schrödinger, Erwin, 189n, 242, 247
  Schrödinger's equation, 176, 180, 206, 209, 219, 230n, 247, 306
screen, 131–2, 239, 244, 246–7, 251
semiotics, 46–8, 53
sensation/s, 31, 47, 61, 63, 73, 76–7, 92–3, 121
sense, 33, 56, 64, 146, 153–4, 159,160, 163–4, 316, 318
  common sense, 318–19
  sense-event, 160, 164–7
  sense-making, 153, 156, 160–1, 164, 166–7
  senses, 236
  see also nonsense
sensibility, 2, 8, 34, 53, 67, 239
sensible, 40, 45, 60, 66, 81, 92, 281, 302
sensible intuition, 23, 64
sensitivity, 48–9, 136, 148–9, 289
  cosmological sensitivity, 2–3, 8
settler-colonial, 156
Simondon, Gilbert, 44, 46, 241, 262
singularities, 44–6, 48, 74–5, 77, 79–80, 83, 85–6, 88, 90, 107–8, 111–15, 121, 127, 131–2, 185, 239–41, 248, 252

singularity, 64, 113, 122–3, 166, 212, 247
Smolin, Lee, 10, 157, 167n, 221–2, 237, 315, 331n
smooth, 129, 216, 219, 216, 228
  smooth space, 9, 54–6, 73, 78–82, 84–5, 90, 249, 252
Socrates, 40
sonic, 65
  sonic experimentation, 7
  sonic framework, 55–6, 59
Sophist, 37, 39, 41
soul, 10, 102, 106–9, 120–5, 127–9, 130–3, 134n, 136–7, 148
contemplative soul, 118–19, 131, 327
sound, 6, 7, 29, 31, 38, 62–3, 67, 125, 130, 149, 319
spacetime, 54–5, 62, 139, 142–3, 151, 154, 161, 214–19, 248, 279, 303, 315–16, 319–22, 326–7, 330
  Minkowski spacetime, 320–1
spatiotemporal, 49, 58, 222, 227, 298–9, 302, 308, 315, 322, 324–30; *see also* non-spatiotemporal
  spatio-temporal rhythms, 12
speed/s, 9, 30, 45–6, 48–9, 81, 90, 150, 242, 330
  absolute speed, 9, 45, 48
  infinite speeds, 34, 45, 193, 198–9, 240, 290, 328
  speed of light, 240, 242
  speeds and slownesses, 6–7, 328
Spinoza, Baruch, 3, 39, 44, 112, 153, 298–9, 302, 307
spiritual, 3, 7, 29, 33, 107, 291
  spiritual exercises, 155
Stengers, Isabelle, 75, 82–3, 155–6
Stockhausen, Karlheinz, 5, 6, 14
Stoics, 67n, 115, 154–5, 159, 162, 165–6, 168n, 223, 328
  Stoicism, 2, 159
strange attractors, 24
strata, 54, 60, 73, 79, 83–4, 90, 119, 328–9
stratification, 90; *see also* destratification
striated, 54–5, 73, 79, 85–6, 129
string theory, 4, 10, 136–49, 150n, 214, 248–50
structuralism, 54, 58, 297, 300–1
stupidity, 30–1
stuttering, 24, 246–7
style, 38, 47
subjectivation, 47, 248
subjectivity, 29–30, 46, 59–60, 64–5, 68, 101, 111, 118, 202
superfold, 237
superposition, 10, 175–80, 184, 190, 219–20, 243–6, 252
supersymmetry, 143
synesthetic, 65, 68n

T-duality, 142, 144; *see also* Maldacena Paradox
Tarde, Gabriel, 243
temporal, 29, 55, 66, 98, 98, 139, 157, 162, 218, 306, 317, 320–1, 326, 328
temporal modulation, 132, 137, 139, 149; *see also* spatiotemporal; intratemporal
territorialization, 6, 263, 329; *see also* de-territorialization; re-territorialization
Thales, 37, 116, 261

thermodynamics, 4, 238, 240, 249, 250, 253, 275, 285, 312
three-body problem, 20–2, 24
thresholds, 328, 330
time-image, 28, 237
time-reversible, 21, 249
transcendence, 12, 41, 60, 64, 218, 252, 265–7, 281–2, 326
transcendental, 44, 154, 166, 252, 266–70, 275–6, 298, 307, 318
  transcendental conditions, 155–6
  transcendental empiricism, 47, 307–8
  transcendental field, 59, 112
  transcendental illusion, 238, 293, 318
  transcendental subject, 154
  transcendental sensibility, 239
transference, 54, 105
transistantial, 53, 58–60, 62
transistantiality, 9, 56–8, 66
transmutation, 63, 90, 166
transversal field/vector, 59, 65
transversality, 57–8

Übermensch, 237, 239
unlimited, 39, 60, 63, 160, 237, 328
  not unlimited, 158, 328
  see also limitless
unpredictability, 22, 45, 83, 141
unthought, 34, 57, 238

vacuum, 138, 199
Varela, Francisco, 12, 273–5; *see also* Maturana, Humberto
vector, 59, 160, 176, 209–11, 245
  vector field, 22
  transversal vector, 65
vitalism, 239
void, 106, 113, 240, 288–90
obstetrical void, 40

Wallace, David, 180, 184–8, 305–6
Wheeler, John A., 199
Wheeler-DeWitt equation, 229n, 230n
Whitehead, Alfred North, 2, 10, 12, 104, 118, 124–6, 133, 135, 223, 243, 246, 297, 314–19, 321–30, 331n
Wittgenstein, Ludwig, 162, 205

zig-zag, 239
Zourabichvili, François, 314